**The Dynamic
Environment of the
Ocean Floor**

The Dynamic Environment of the Ocean Floor

Edited by
Kent A. Fanning
University of South Florida

Frank T. Manheim
U.S. Geological Survey

LexingtonBooks
D.C. Heath and Company
Lexington, Massachusetts
Toronto

Library of Congress Cataloging in Publication Data

Main entry under title:

The Dynamic environment of the ocean floor.

    1. Ocean bottom. I. Fanning, Kent A. II. Manheim, Frank T.
GC87.D95   551.4'608   78-24651
ISBN 0-669-02809-6

Copyright © 1982 by University of Miami

This book was prepared with the support of NSF Grant OCE 76-10515. However, any opinions, findings, conclusions, or recommendations herein are those of the contributors and do not necessarily reflect the views of NSF.

Both the publisher and author acknowledge and hereby grant to the U.S. Government a royalty-free, irrevocable, worldwide, nonexclusive license to reproduce, perform, translate, and otherwise use and to authorize others to use the work for government purposes only and not for commercial purposes such as sale.

All rights reserved. No part of this publication may be reproduced or transmitted in any form or by any means, electronic or mechanical, including photocopy, recording, or any information or retrieval system, without permission in writing from the publisher.

Published simultaneously in Canada.

Printed in the United States of America.

International Standard Book Number: 0-669-02809-6

Library of Congress Catalog Card Number: 78-24651

Contents

	Preface	ix
Part I	*The Sampling of Interstitial Waters*	1
Chapter 1	**Extraction and Investigative Techniques for Study of Interstitial Waters of Unconsolidated Sediments: A Review** *P.A. Kriukov* and *F.T. Manheim*	3
Part II	*The Dynamics of the Interactions between Sediments and Seawater*	27
Chapter 2	**Water Movement through Porous Sediments** *R.J. Riedl* and *J.A. Ott*	29
Chapter 3	**Oscillation of Continental Shelf Sediments Caused by Waves** *J.N. Suhayda, J.M. Coleman, Thomas Whelan*, and *L.E. Garrison*	57
Chapter 4	**Current-Induced Sediment Movement in the Deep Florida Straits: Observations** *Mark Wimbush, Laszlo Nemeth,* and *Barton Birdsall*	77
Chapter 5	**The Benthic Interface of Deep-Sea Carbonates: A Three-Tiered Sequence Controlled by Depth of Deposition** *W.H. Berger*	95
Chapter 6	**The Influence of a Diffusive Sublayer on Accretion, Dissolution, and Diagenesis at the Sea Floor** *Bernard P. Boudreau* and *Norman L. Guinasso, Jr.*	115
	Appendix 6A: Solution to the Diffusion Equation	143
Part III	*Biological Interactions at the Sea Floor*	147
Chapter 7	**Microbiological Studies of Decompressed and Undecompressed Water Samples Collected with a Deep-Ocean Sampler** *Rita R. Colwell* and *Paul S. Tabor*	149

Chapter 8	Direct Calorimetry of Benthic Metabolism *Mario M. Pamatmat*	161
Part IV	*Gases in Sediments*	173
Chapter 9	Inert Gas Gradients and Concentration Anomalies in Pacific Ocean Sediments *Ross. O. Barnes*	175
Chapter 10	Methane Production, Consumption, and Transport in the Interstitial Waters of Coastal Marine Sediments *Christopher S. Martens*	187
Chapter 11	A Major Sink and Flux Control for Methane in Marine Sediments: Anaerobic Consumption *William S. Reeburgh*	203
Chapter 12	The Presence of Methane Bubbles in the Acoustically Turbid Sediments of Eckernförder Bay, Baltic Sea *Michael J. Whiticar*	219
Part V	*Transition Metals in Deep-Sea Sediments*	237
Chapter 13	Pacific Sediments from Japan to Mexico: Some Redox Characteristics *A.G. Rozanov*	239
Chapter 14	Migration of Manganese in the Deep-Sea Sediments *Shizuo Tsunogai* and *Masashi Kusakabe*	257
Chapter 15	On the Mechanisms of the Formation of Ferromanganese Concretions in Recent Sediments *I.I. Volkov*	275
Chapter 16	Trace Metals in Interstitial Waters from Central Pacific Ocean Sediments *M. Hartmann* and *P.J. Müller*	285
Part VI	*Transition Metals in Nearshore Sediments*	303
Chapter 17	Diagenetic and Environmental Effects on Heavy-Metal Distribution in Sediments: A Hypothesis with an Illustration from the Baltic Sea *Rolf O. Hallberg*	305

Contents vii

Chapter 18	Depth Distributions of Copper in the Water Column and Interstitial Waters of an Alaskan Fjord *David T. Heggie* and *David C. Burrell*	317
Part VII	*The Influence of Seawater on Freshwater Deposits*	337
Chapter 19	Authigenic Barite in Varved Clays: Result of Marine Transgression over Freshwater Deposits and Associated Changes in Interstitial Water Chemistry *Erwin Suess*	
Part VIII	*Hydrothermal Interactions*	357
Chapter 20	"Heated" Bottom Water and Associated Mn-Fe-Oxide Crusts from the Clarion Fracture Zone Southeast of Hawaii *H. Beiersdorf, H. Gundlach, D. Heye, V. Marchig, H. Meyer,* and *C. Schnier*	359
Chapter 21	The Origin of the Volcanic Component in Active Ridge Sediments *Christer Löfgren* and *Kurt Boström*	369
Chapter 22	The Nature of Hydrothermal Exchange between Oceanic Crust and Seawater at 26°N Latitude, Mid-Atlantic Ridge *Robert B. Scott, Darcy G. Temple,* and *Phillipe R. Peron*	381
Chapter 23	On Mantle Helium, Argon, and Methane Discharge in Thermal Spring Waters of Ocean Margin and Ridge Areas *L.K. Gutsalo*	417
Part IX	*Geophysics of Hydrothermal Processes*	439
Chapter 24	"Active" and "Passive" Hydrothermal Systems in the Oceanic Crust: Predicted Physical Conditions *C.R.B. Lister*	441
Chapter 25	Numerical Models of Hydrothermal Circulation for the Intrusion Zone at an Ocean Ridge Axis *Patricia L. Patterson* and *Robert P. Lowell*	471

Index 493

List of Contributors 499

About the Editors 502

Preface

The sea floor is a major dynamic oceanic boundary that has critical importance for many marine processes. Characteristics of the water column, the underlying sediments, and sea life can be strongly affected. As knowledge of its importance has grown, research on benthic processes and features has become increasingly intense. An example is the book *The Benthic Boundary Layer* (edited by I. N. McCave) in which approximately the last third was devoted to recommendations for future research.

Several of those recommendations have now been pursued, and a broadly based conference called Benthic Processes and the Geochemistry of Interstitial Waters of Marine Deposits was held at the Joint Oceanographic Assembly in Edinburgh, Scotland in September of 1976. Several of the chapters in this book were presented in initial form at that conference, and other important ones have been added in an attempt to capture the full flavor of the research in this exciting field.

It has many intriguing aspects. Devices to record, photograph, measure, and sample the benthic interface in much greater detail have been built and tested. There has been an encouraging improvement in the sophistication of mathematical models of benthic and upper–sedimentary processes, with both numerical and analytical solutions being applied. The types and ranges of the required coefficients in those models are being refined as well. Knowledge of the importance of hydrothermal emanations at mid-ocean ridges and elsewhere has rather suddenly appeared, and unique deep-sea ecosystems seem to be solely supported by such flows at some locations in the East Pacific. Sediment traps have been deployed in an effort to quantify the delivery rates and composition of the debris reaching the sea floor. Within practical constraints, as many relevant advances as possible are included in our book.

The protocol of the book has the following rationale. After a review of sampling methods, there is a section on the description and quantitative studies of the most important or dynamic interactions between solid particles and water at the sea floor. The section begins at the beach with the models of Riedl and Ott and then steps seaward to the abyssal studies of Berger, Boudreau, and Guinasso. Next, there is a section on biological measurements in the bottom water and sediment, and that is followed by four chapters on dissolved interstitial gases in sediments, both "inert" and reactive. Then there are six chapters on interstitial trace metals, which are included because of the value of metals as indicators of chemical reactions. Two aspects of research on hydrothermal processes conclude the book: (1) chemical and geological studies, and (2) mathematical models of hydrothermal processes. Most of the ideas and data in the book's chapters are new, and some have appeared before in one form or another. When placed in a single volume, the chapters uniquely outline the tremendous scope of benthic processes and address the need of those interested in one kind of study to understand what other types of studies have discovered.

Some acknowledgments are certainly in order. The Conference and preparation of the book were generously supported by Grant No. OCE 76-10515 from the National Science Foundation to the University of Miami, Rosenstiel School of Marine and Atmospheric Science. Ms. Sharon Barnard played a key role in organizing the conference. Finally, Ms. Linda Bell was a cheerful, effective, and tireless editorial assistant without whom the compilation of these diverse manuscripts into a coherent book would not have happened.

Part I
The Sampling of Interstitial Waters

1 Extraction and Investigative Techniques for Study of Interstitial Waters of Unconsolidated Sediments: A Review

P.A. Kriukov and
F.T. Manheim

Abstract

Fluid extraction from sediments dates back more than 110 years. Three main types of interstitial fluid extraction systems have widespread current use and proven effectiveness: centrifugation, gas displacement, and mechanical or hydraulic pressure filtration. Sophisticated in situ measurements of fluids and gases have shown dramatic development in the past 10 years, but those do not invalidate the sampling and analysis of cores. Rather, they can serve to check, refine, and extend the use of more standard methods.

Introduction

The H.M.S. Challenger expedition marked the beginning of many new scientific techniques in oceanography. One was applied to investigation of the chemical composition of pore fluids in "blue," or terrigenous, muds off the coast of Scotland (Murray and Irvine, 1895). After Murray and Irvine's paper, however, more than half a century passed before major new insights about the chemical composition of interstitial fluids of marine sediments were achieved.

Well before Murray and Irvine's work, however, soil scientists had already recognized the importance of interstitial solutions in soils as major pathways of nutrient supply to plants. As early as 1866, a Dutch agronomist, H. Schloesing, extracted solutions from cylinders of tobacco soil by displacement with distilled water and carefully analyzed the effluent. Following Schloesing, soil scientists used a great variety of techniques in removing pore fluids from soils, some of which will be discussed in the following section.

Contributions to the study of interstitial fluids in "sediments" have been made by the materials–processing industry (slurry monitoring), by geotechnology, and by petroleum engineering. Analysis of the electrolyte content of clayey muds used in drilling for groundwater, waste disposal, and oil and gas became important in the 1940s and 1950s when quantitative methods for evaluating geophysical logs in boreholes were developed (see Schlumberger, 1972).

Since the late 1950s, however, the marine sciences have probably led other fields in the sophistication and accuracy with which the best interstitial water measurements have been made. Progress in interstitial water studies of general sediments has been reviewed by Kriukov (1971), and studies of marine sediments have been reviewed by Shishkina (1972), Manheim and Sayles (1974), Gieskes (1975), and Manheim (1976). The purpose of this chapter is to review and comment on the applicability of and problems associated with techniques employed in interstitial water studies, with special emphasis on extraction of fluids for marine geochemical purposes.

Techniques of Interstitial Fluid Extraction

Displacement Techniques

The earliest quantitative studies on pore fluids appear to have been performed by agronomists using fluid–displacement techniques. Schloesing (1866) displaced pore fluids from a column of soil stoppered on the bottom with cotton by using distilled water. The breakthrough point was monitored by observing decreases in effluent concentration. In discussing "lisimetry," or measurement of water in soils, Stiles and Jørgensen (1914) reviewed the objections to the technique of fluid extraction in unsaturated soils by water displacement. Apparently, fluid may pass through central channels and bypass moisture films on labyrinthine surfaces so that displacement in partly aerated soils could be equivalent to incomplete leaching. Ishcherekov (1907) reported that up to 95 percent of saturated soil fluids could be displaced by methyl and ethyl alcohol. The lighter displacing fluid was said to minimize channeling, and alcohol breakthrough is signaled by turbidity. A careful study by Parker (1921) examined displacement by water, alcohols, and acetone and compared those displacement techniques to leaching. He found ethyl alcohol displacement to be the best, achieving up to 36 percent recovery of fluids in unsaturated soils. Acetone displacement was poorest. The soil particles seemed to be unaffected by alcohol.

Immiscible displacement, using petroleum oil, was first employed by Morgan (1917) for soils. Scholl (1963) favored silicone oil having a specific gravity close to that of water to displace fluids from shallow marine and coastal sediments. However, immiscible displacement at low pressure is not applicable to clayey sediments, and the technique has been rarely used because of its messiness.

Gas displacement of interstitial fluid has been extensively used since its introduction by Richards (1941). Half-liter commercial mud presses use compressed gas from cylinders or CO_2 cartridges to force filtrate through filter paper. Those presses have been available in the United States since the 1940s to remove mud filtrates from oil–field drilling muds. Lusczynski (1961) used such a device to study fresh/salt groundwater interfaces by extracting interstitial water from cores on Long Island, New York. A gas-mechanical variant that placed a rubber membrane between the gas and sediment was described by Hartmann (1965). The device was used for studies of piston–cored sediments from the Baltic Sea, Atlantic Ocean, and other areas.

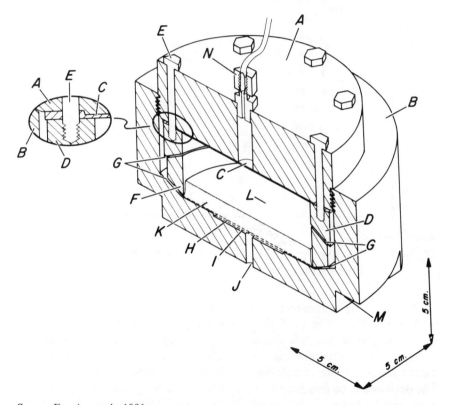

Source: Fanning et al., 1981.

Figure 1–1. Gas-displacement squeezer for use with multisqueezer manifold. A = threaded squeezer top; B = squeezer bottom with inner taper (G); C = rubber membrane; D = sealing ring with tapered edge (G); E = nylon screw; F = inner ring with two tapered edges (G); G = sealing tapers; H = paper filter; I = Nucleopore membrane; J = exit hole; K = sediment sample; L = plastic covering on sediment; M = locking notch for tightening top.

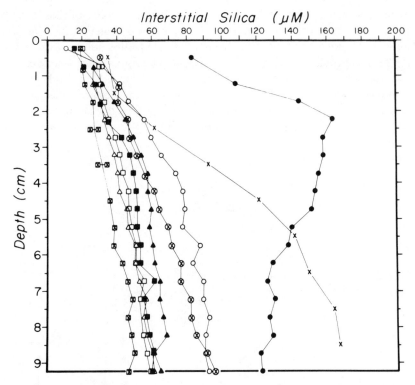

Source: Jones, 1977.

Figure 1–2. Distribution of interstitial silica in Mediterranean Sea cores from various sites.

An all-plastic extraction device utilizing gas displacement was developed by Reeburgh (1967), and a Teflon unit was reported by Presley et al. (1967). Reeburgh-type devices have received widespread use in marine, estuarine, and lake studies. Pressures up to about 200 psi (14 kg/cm^2) permit relatively rapid recovery of tens of milliliters of fluids from unconsolidated sediments. The devices can be free from metal surfaces, rendering them appropriate for work with trace metals and labile constituents. Fanning et al. (1981) have powered as many as twenty units simultaneously from a single gas cylinder by means of a manifold (figure 1–1). That approach permitted very closely spaced sampling of pore fluids from Mediterranean cores after core recovery on ship and thereby provided detailed knowledge of chemical gradients immediately beneath the sediment/water interface (figure 1–2; Jones, 1977; Jones and Fanning, 1981). Effects of temperature changes during retrieval were counteracted by keeping sediment samples in a walk-in refrigerator before and during extraction of pore water.

Centrifugation

Centrifugation was employed as early as 1907 by Briggs and McLane (1907) to extract fluids from soils with a steam turbine-driven machine. They also separated the fluid with a three-tube device that, along with other techniques, was later improved by Kriukov and Komarova (1956). A much-quoted investigation of nutrient distributions in southern California offshore sediments by Emery and Rittenberg (1952) used centrifugation techniques. Centrifugation also was used to extract pore fluids from cores during the first "Mohole" ocean drilling project near Guadalupe Island, Mexico (Rittenberg et al., 1963; see also Edmunds and Bath, 1976).

Earlier techniques often suffered from speeds that were too slow to promote efficient extraction but that tended to heat core samples. With the availability of powerful cooled ultracentrifuges achieving more than 50,000 r/m (largely used for medical and clinical purposes), rapid and highly efficient extraction at desired temperatures is possible. Capping of centrifuge tubes prevents evaporation of samples, and use of syringes permits accurate transfer of fluids to containers for storage or analysis. Sediment can be loaded (under inert gas, if necessary) in centrifuge tubes in the field and analyzed after return to the laboratory.

Vacuum Filtration

This technique, using Buchner funnels, filter candles, or other variants, has been used by soil scientists dating back to Briggs and McCall (1904) and Whitney and Cameron (1905), as well as by workers interested in surficial nearshore sediments (Johnson, 1967). Recognizing the problem of surface evaporation, Van Suchtelen (1912) poured light paraffin oil over soil so that the surface was covered. Interstitial water was then separated from the lighter oil/water mixture by means of a separatory funnel. Kuzmin (1922) and Zhorikov (1930) simultaneously applied pressure and vacuum to remove fluids. Considering the fundamental problem of evaporation of fluid, as well as the inefficiency of the vacuum-extraction process in comparison with the pressure–filtration systems discussed later, this method cannot be recommended.

Leaching

Leaching sediments to determine the composition of interstitial fluids appears at first sight a simple procedure, but following early studies by Gedroiz (1906), Cameron (1911), and others, soil scientists established clearly the fact that leaching involved potential error not only by solubilization of solids, but also by changing sediment/water equilibria in soils. Thus for example, Parker (1921) showed that calcium concentrations obtained by leaching techniques could be twice as high as those obtained by alcohol displacement, whereas Mattson et al. (1949) showed that, because of Donnan equilibria and other factors, phosphate could be immobilized and diminished in pore fluids by dilution of soil sediments. Sulfates may be increased by oxidation of sulfide to sulfate with consequent reduction of alkalinity and solubilization of Ca from $CaCO_3$. Later authors, including Tsyba and Kriukov (1959) and Arslanbekova et al. (1962), found leachate compositions of drill cores grossly different from those extracted by pressure filtration. The only constituents that can be safely determined by simple leaching techniques are relatively unreactive anions such as Cl, as has been done by Swarzenski (1959), Kullenberg (1952), Arrhenius (1952), and others. Gripenberg (1934) used leaching and chloride analysis and the assumption of uniform interstitial chlorinity to determine the original porosity of dried cores from the Baltic Sea. Without careful technique, leaching may sustain errors as high as 10 to 30 percent or higher.

Leaching works best with undesiccated sediment. When clayey sediment is first dried, chlorides may be tightly occluded. Walters (1967) found that five to eight boiling leaches and ultrasonic treatment were required to remove all soluble Cl^- and suggested that anion exchange bound Cl^-. That possibility was also raised by Bischoff et al. (1970). But temperature experiments on Caribbean clays by Sayles et al. (1973a) showed that anion-exchange did not detectably influence interstitial solutions, possibly due to the negative charge of most clay micelle surfaces at typical pH values in the range 6 to 8.

Reviewing the influence of cation-exchange capacity of marine sediments on leaching studies, Sayles and Mangelsdorf (1978) recently concluded that all methods based on preliminary leaching or washing of sediments gave invalid results for the cation-exchange capacity of marine or saline sediments and that only the method of Zaitseva (1958) and their own "difference chromatographic" methods yielded reliable values.

Pressure Filtration

Pressure-filtration has several favorable attributes as a fluid-extraction method. The major ones are lessened contamination and the potential efficiency and convenience in removing fluid, even from consolidated low-porosity sediments. Pioneering studies by Ramann et al. (1916) required 40 kg soil in eleven to twelve squeezings with a "hydrostatic press" at 300 kg/cm^2 to obtain 1 liter of pore fluid. Also using a hydraulic press, Lipman (1918) extracted fluid at 3,000 kg/cm^2. Those studies drew criticism from

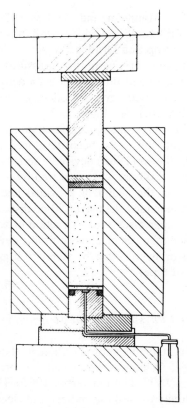

Figure 1-3. High-pressure squeezer.

Northrup (1918), who pointed out that such high pressure may affect physicochemical equilibrium in soils and influence specific gravity, viscosity, surface tension, osmotic pressure, specific conductivity, and chemical composition of fluids. Kriukov and co-workers and others have investigated those paramenters in considerable detail (see later) and found that, for marine sediments, the pressure of filtration introduces, in practice, insignificant error under most conditions.

Pressure-filtration devices are highly dependent for their practicality on design. Application of self-sealing free gaskets to steel squeezers by Kriukov (1947) greatly improved the technique of hydraulic pressure filtration. Those squeezers are simple and effective enough to allow widespread use in the Soviet Union, both for agronomic and geological purposes. From the 1950s onward, Soviet workers used Kriukov-type squeezers to achieve major advances in knowledge of the interstitial chemistry of oceanic sediments (Bruevich, 1966; Shishkina, 1972 and publications cited). Later modifications (Manheim, 1966; Manheim and Sayles, 1974) were successfully employed on oceanic drill-cores and, since 1968, have become standard on squeezers used by the Deep Sea Drilling Project.

Thick-walled steel squeezers, permitting pressures up to 10,000 kg/cm^2 (Kriukov, 1971), have obtained fluids from even dense sedimentary rocks. However, the most widely used devices produce pressures from 200 to 700 kg/cm^2 (3,000 to 10,000 psi). See figure 1–3.

In addition to the high-pressure squeezers, manually operated devices for unconsolidated sediments have also been developed by soil scientists (for example, Gola, 1910). Siever (1962) devised a screw-driven sediment press that could develop up to 300 psi when powered by a "muscular" scientist. Pencils of filter paper in the sediment provided better effluent recovery. Manheim (1972) recommended the use of 50-cc disposable plastic syringes with screen and filter paper inserts for extraction of small amounts of fluid from unconsolidated sediment. Such easily portable devices can be pressured by large screw clamps or, even better, by standard trigger-operated caulking "guns" (manual ratchet type).

Electrode Methods and Physicochemical Measurements

Electrode methods for analyzing natural sediments date from the nineteenth century. For example, Whitney and Means reported on the electrical conductivity of soils in 1897 (cited in Jerbo, 1965). However, neither total conductivity nor resistivity is a reliable indicator of electrolyte concentration. Difficulties result from the effects of variable porosity, tortuosity, and, at low porosities, the surface conductance of clay solids and other minerals possessing cation-exchange capacity (Waxman and Smits, 1968). Complex electrical logging devices for measuring sediment resistivity, porosity, permeability, oil saturation, formation fluid resistivity, and other parameters in deep boreholes have been standard in the oil industry since the 1940s (Schlumberger, 1972–1978). Investigation of formation factor (ratio of sediment resistivity to pore-fluid resistivity) and other electrical properties of marine sediments is much more recent.

Kermabon et al. (1968) and publications cited therein studied the porosity and density of surficial Mediterranean muds. They used the "formation factor" (f) and assumed the pore-fluid resistivity to be similar to that of Mediterranean bottom water and relatively constant.

Manheim and Waterman (1974) determined formation factors for a quite different purpose. They employed a linear four-electrode configuration on a probe with a Schlumberger resistivity bridge and also extracted pore fluid by pressure filtration, determining fluid resistivity directly or by interpolation with optical refractive index measurements. F has varied between 1.5 and 2.5 for surficial sediments and may exceed 10 at depths to 300 m.

Based on the analogy between diffusion of salts and conduction of electricity in permeable pathways, the effective diffusion coefficient in unconsolidated sediments is

$$D_{\text{eff}} = \frac{D_0}{F}$$

where D_0 is the rate of diffusion for a given ion in free solution at comparable temperature conditions, and F is the formation factor. Turk (1976) confirmed the validity of the formation factor–diffusivity relationship for many ions. McDuff and Gieskes (1976) and references cited have modeled the distribution of Cl, Mg, Ca, and Sr ions in several Deep Sea Drilling Project (DSDP) cores using effective diffusion coefficients derived from the formation factor.

Jones (1977) used data obtained by Kermabon et al. (1968) to interpret diffusive permeability in surficial Tyrrhenian Sea sediments (upper 30 cm) via the formation factor and achieved close agreement with values obtained by chemical-diffusion experiments for dissolved silica.

Duursma and Hoede (1967) studied diffusion models for ions in recent sediments. Li and Gregory (1974) presented a valuable summary of ion-diffusion parameters, both in free solution and surficial Pacific sediments. Fanning and Pilson (1974) and Stoessel and Hanor (1975) measured in situ diffusion coefficients in marine sediments and porous media, respectively.

Glass electrodes with a sodium function were successfully used directly on sediments by Komarova and Kriukov (1959) and Siever et al. (1961, 1965). A new possibility is determination of sodium chloride activity with a sodium-active glass electrode and a chloride membrane electrode. Such a measurement in a cell without a liquid junction should permit measurement in sediments or in extracted solutions more accurately than measurements of sodium and chloride ion activities separately.

The problem with interstitial pH measurement is not that glass electrodes are difficult to use in sediments, but that the values obtained may differ from in situ values because of changes in temperature, pressure, loss of gases, oxidation, and "suspension effects," and similar problems related to carbonate equilibria. Calculations of in situ pH through measurement of controlling parameters such as calcium, alkalinity, total CO_2, temperature, and pressure may yield better values than direct measurements (Gieskes, 1974). Thus Manheim and Schug (1978) calculated pH values of 5.8 for Black Sea strata, whereas direct measurements under surface conditions yielded values between 7 and 8. The difference was interpreted as being due to CO_2

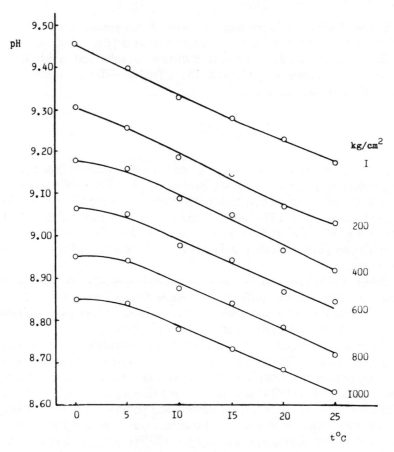

Figure 1–4. pH value of phosphate buffer solution at 0 to 25°C and 1 to 1,000 kg/cm².

pressure that had leaked away before shipboard measurement could take place.

pH equilibria are especially sensitive to pressure changes, as has been shown by Culberson and Pytkowicz (1968) and Disteche (1974). That proposition may be confirmed by figures 1–4 and 1–5, which represent new data for the dependence of the pH of buffer solutions on both temperature and pressure. The data are obtained in pressure cells equipped with palladium hydrogen and silver chloride electrodes (Kriukov and Zarubina, 1977). Apparatus for simulation of in situ parameters in seawater may be used for potentiometric (e.g., ion-selective electrode) measurements in sediments and for electrode calibration under deep-water conditions. Such devices need not be the main basis of investigation, because extraction of interstitial solution

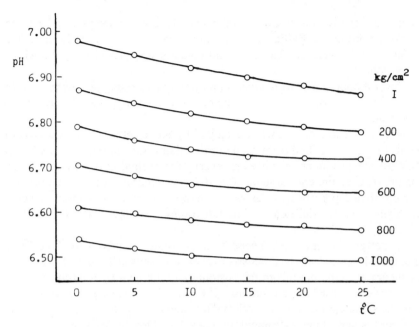

Figure 1-5. pH values of borax buffer solutions at 0 to 25°C and 1 to 1,000 kg/cm².

may be simpler and more convenient, but they are important as a check on the applicability and validity of the extraction techniques.

Measurements with Eh electrodes can yield useful operational measures of the relative redox condition of sediments (Bruevich, 1938; Whitfield, 1969; see also Rozanov, 1981, chapter 13 of this book), but the electrodes may not produce a valid thermodynamic (Nernst) response, especially in oxygenated systems (Stumm, 1966).

In Situ Measurements

As discussed in a later section of this chapter, systematic error or enhanced random error may occur when pore fluids are extracted or studied after the cores or sediment samples are removed from their natural environment. To avoid those errors, efforts have been made to perform measurements or extract fluids "in situ" at the sea floor. An early study was that of Pirogova (1950), who left a weighted and tubulated glass chamber on the bottom of the shallow aerated Black Sea floor for up to 100 days and monitored bottom water (influenced by diffusion from the sediment) for oxygen, nutrients, and other salts. For Eh measurement, Manheim (1961) lowered a glass pH electrode fitted with a platinum collar into the bottom sediment of a 16-m brackish fjord of the Baltic Sea. The reference electrode was placed at the surface of the water, so that the liquid junction was provided by the whole water column.

Several approaches to recovering pore fluid by driving porous pipes into soils (Brown, 1965) or stream sediments (Husmann, 1971) have been suggested.

A sophisticated device was developed by Sayles et al. (1973b; 1976) to recover in situ pore fluid from deep North Atlantic Ocean sediment. A 1.5-m stainless steel sampler permitted recovery of pore fluid in five filter-covered chambers. Calcium enrichments of up to 6 percent and potassium depletions of more than 1 percent with respect to seawater were subsequently detected by highly precise difference chromatography (Mangelsdorf and Wilson, 1971).

A newer concept, the "peeper," incorporates a tube with fluid-filled chambers inserted into deep ocean or other sediments (Hesslein, 1976). The chambers are separated from the surrounding sediment by micropore filters that serve as diffusion membranes. The chamber is allowed to equilibrate with the sediment and then is retrieved by a delayed release mechanism. The "peeper" is part of the instrumentation developed for the Manganese Nodule Program (MANOP) sponsored by the International Decade of Ocean Exploration (IDOE) of the U.S. National Science Foundation.

Hallberg et al. (1973) described chambers made to be emplaced on the sea floor and sediment chambers that simulate sea-floor conditions and can be continuously monitored in the laboratory for a variety of biochemical and inorganic changes.

Barnes (1973) devised a probe that achieves the in situ recovery of interstitial gases from sediment. A variant of that instrument is being used in DSDP drill holes both to capture in situ fluids and to take heat-flow measurements (personal communication).

In the Pacific Ocean off Chile, Pamatmat (1971) succeeded in making the first in situ oxygen-uptake measurements in deep sediments. In chapter 8 of this book, he reports some of the first straightforward in situ measurements of the heat budget for a sediment/organism community (in situ calorimetry).

General in situ physical and geotechnical measurements are beyond the scope of this chapter (see, for example, Suhayda et al., in chapter 3 of this book), but brief mention should be made of the electric-resistivity sonde perfected by Kermabon et al. (1968 and references cited). They were able to record resistivity for a 7.4-m penetration in Mediterranean sediments under nearly 3,000 m. The remarkable rapidity of the sonde's operation allowed thirty measurements at intervals of 100-m in only 90 minutes of recording time. By calibrating cores at the beginning and end of the traverse and correlating the runs, porosity and density could be interpreted from the data with a resolution of about 5 cm. Bouma et al. (1971) and Chmelik and Bouma (1970) discussed similar studies, and Richards et al. (1972) described other in situ techniques.

Operational Problems

Although analytical methods per se are not treated in this chapter, some fundamental problems common to most pore-fluid studies should be mentioned. Special consideration will be given to effects caused by extraction pressure.

One of the most common sources of erratic data is insufficient care in fluid transfer, including evaporation and pipetting error. The first can be reduced by rapid handling, e.g., by syringe and by quick sealing of the storage container. The second can be avoided by weighing all aliquots with a semimicrobalance or microbalance. Results are thereby obtained on a weight basis, avoiding density corrections that vary with temperature. For example, a 100-μl sample may be weighed to a precision of 0.1 percent on a semimicrobalance having a reproducibility of 0.1 mg. Use of check weights is essential to detect secular balance drift. Filtration by syringe-attached "Swinney" filter (Lockey, 1962) is quick and effective.

Choice of storage container is often critical. Polystyrene plastic permits evaporation, whereas polyethylene, linear polyethylene, and polypropylene are increasingly impervious to water vapor but still permit significant gaseous diffusion (Stannett, 1962). Thus, although many plastics are non-contaminating (if precleaned), their gaseous permeability renders them unsuitable for isotopic sampling. Since pore fluids often have an excess of ammonia over phosphate, algal and bacterial growth generally consume all available phosphate unless samples are poisoned (see also Bray et al., 1973, regarding phosphate behavior). Fused-glass ampules are effective barriers to gaseous permeation (except helium) but tend to react with solutions, liberating silica and cations and increasing total alkalinity. Rubber stoppers are often highly contaminated with elements like calcium and zinc. Thus a complete or extensive set of analyses may require several types of storage containers and careful determination of blanks.

Interstitial constituents that are sensitive to reaction require special preparation. Emery and Hoggan (1958) recovered gases from shallow marine sediments by disassembling a special coring tube in a large plastic bag filled with helium, whereupon the gases were extracted under vacuum and determined directly in a mass spectrometer. Horowitz et al. (1973) describe elaborate manifolds and a variety of precautions used to extract interstitial gases and pore fluids under inert gas and otherwise controlled conditions. In this book, Reeburgh (1981) and Martens (1981) discuss the technical problems of studying the distribution and origin of methane gas in sediments. Barnes (1981) describes a new in situ sampler for gas recovery from sea-floor sediments. Volkov (1961, and references cited therein) took special precautions in working with labile sulfur components in anoxic Black Sea sediments and environments. An indication of potential hazards is provided by Östlund and Alexander (1963), who found that H_2S was oxidized with a half-life of 20 minutes in intertidal sediments containing active bacteria. One attempt to avoid excess exposure to atmospheric gases was a pore-fluid press that operated on an entire sediment core in its liner (Kalil and Goldhaber, 1973).

The significance of thermal effects caused by change from in situ temperature was pointed out by Mangelsdorf et al. (1969). Their findings, which were confirmed by other workers, showed that increasing temperature caused largely reversible shifts in exchange coefficients in clayey materials, resulting in preferential loss of soluble monovalent cations and uptake of divalent cations. Thus sediments taken from low ocean temperatures, say 2 to 4°C, and raised to 20°C may upon extraction yield fluids with relative changes as large as -7 percent (Ca, Mg), -19 percent (Sr), $+1.5$ percent (Na), and $+25$ percent (K). The major anions chloride and sulfate did not show significant temperature effects (< 0.5 percent) in experiments by Sayles et al. (1973a). However, Fanning and Pilson (1971) found up to 50 percent increase in dissolved silicate concentration, and Sayles et al. (1973a) found up to a 60 percent increase for borate. Fanning and Pilson (1971) found no effect for phosphate. Trace cations (e.g., Mn, Ba, Li) follow the basic patterns of their respective valences and lyotropic parameters with respect to exchange properties. Since many of the changes are largely reversible, they may be corrected by storage at in situ temperature prior to extraction or by applying an appropriate correction factor.

Effects During Pressure Filtration

Some workers have suggested that electrolyte content may be increased when solutions are forced through clay membranes. Experiments have shown such effects in the laboratory (see Kharaka and Berry, 1973, and references cited therein). In contrast, Kriukov and Komarova (1956) and Kriukov and Zhuchkova (1962) have demonstrated that electrolyte concentrations sharply decreased in pore fluids extracted from clays near the end of a squeezing sequence. The effect occurred when the squeezing pressure was high or the interstitial ionic strength was low (figure 1-6). Membrane filtration was not responsible because the residual pore water had even lower electrolyte concentrations than the extracted pore water (table 1-1; Kriukov, 1971).

The "squeezing effect" arose from two major factors: (1) distribution of

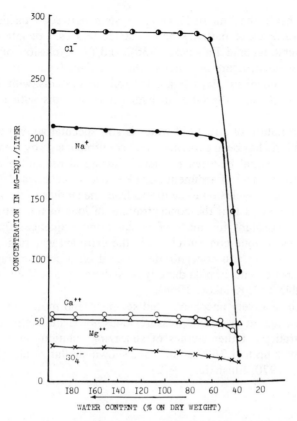

Figure 1-6. Change in concentration of solutions squeezed out from standard Askangel Clay.

Table 1-1.
Electrolyte Concentration in Original and Extracted Sediment

Type of Material	Water Weight (Dry Basis) (Percent Dry Weight)		Pressure (kg/cm^2)	Concentration of Solution (mole/L)	
	Initial	Residual		Initial	Residual
"Askangel"	150.3	19.7	4350	1.07	0.192
Bentonite	97.1	16.4	8160	1.04	0.307
	101.0	23.3	8160	0.0118	0.0019
Kaolinite	33.2	3.2	8160	0.106	0.070

electrolytes inside the "micro-Donnan" system between "inside phases" nearest the surfaces of particles and "outside phases," of intermicellular solution (Overstreet and Babcock, 1936), and (2) exclusion of dissolved substances in the structure of water in the boundary layer (Horne, 1969). The former mechanism plays a greater role in systems with large ion-exchange capacity and the latter in hydrophilic systems with low charge density.

The exact nature of structural water is still speculative, but the role of Donnan equilibria has been experimentally confirmed, e.g., by measuring the difference in potential between extracted fluids and remaining electrolyte during the "squeezing" of sediment (Kriukov and Zhuchkova, 1962).

The important conclusion to be drawn from the studies of the artifacts of squeezing pressure is that the concentrations of ions in the extruded pore fluids remain constant for most of a squeezing sequence (figure 1-6). Further, the electrolyte concentrations in the extruded waters appear to be quite close to the in situ concentrations based on a comparison of glass-electrode measurements of pNa directly on sediment "gels" and or extruded solutions (table 1-2; Kriukov, 1964).

Thus with relatively unconsolidated sediment and moderate electrolyte content, squeezing at moderate pressures does not significantly affect solution concentrations. Other studies of squeezing solutions at successively higher pressures up to 1,260 kg/cm^2 have shown similar results (Manheim, 1966; Sayles, 1970; Shishkina, 1972).

Modeling

Space does not permit an extensive discussion of modeling methods or results, but brief mention should be made of the increases in sophistication with which migration of dissolved contituents in sediments is being treated. Recent mathematical treatments incorporate diffusion, compaction, advec-

Table 1-2
Comparison of pNa Values in Sediments and Extracted Fluids

Material Added	Solution	Sediment	Difference
Na-kaolinite	.37	.37	.00
Na-kaolinite	1.91	1.91	.00
Na-"askangel"[a]	.49	.49	.00
Na-"askangel"	1.87	1.65	.22
Bentonite	.16	.16	.00
Pacific Ocean mud	.31	.31	.00

[a]A naturally occurring Russian clay with high absorbent qualities and exchange capacity.

tion, adsorption, and reaction rates, often with ingenious simplifications (for example, see, Berner, 1975; McDuff and Gieskes, 1976; Imboden, 1976; Lasaga and Holland, 1976, Boudreau and Guinasso, 1981; and especially the review by Lerman, 1977). Those and other models have great practical importance for predicting the interactions between sediments and overlying water bodies in freshwater and estuarine environments. A recent bibliography covering sediment/water interaction and its effect on water quality listed no fewer than 140 reports and in-house publications in the U.S. National Technical Information Service computer files (Lehmann, 1978).

References

Arrhenius, G.O.S. 1952. Sediment cores from the east Pacific. *Rept. Swedish Deep Sea Exped.* 5, 1947-1948.

Arslanbekova, Z.A., and Kriukov, P.A. 1962. Some data on the composition of rock solutions of oil fields Selly and Gasha (Russ.). *Trudy inst. geologii Dagestanskogo filiala Akad. Nauk SSSR* 3:3-12.

Barnes, R.O. 1981. Inert gas gradients and concentration anomalies in idated sediments. *Deep Sea Res.* 20:1125-1128.

Barnes, R.O. 1980. Inert gas gradients and concentration anomalies in Pacific Ocean sediments (chapter 9 of this volume).

Berner, R.A. 1975. Diagenetic models of dissolved species in the interstitial water of compacting sediments. *Am. J. Sci.* 275:88-96.

Bischoff, J.L., Greer, R.E., and Luistro, A.O. 1970. Composition of interstitial waters of marine sediments: Temperature of squeezing effect. *Science* 167:1245-1246.

Boudreau, B.P., and Guinasso, N.L., Jr. 1981. The influence of a diffusive sublayer on accretion, dissolution, and diagenesis at the sea floor. (chapter 6 of this volume).

Bouma, A.H., Sweet, W.E., Jr., Chmelik, F.B., and Huebner, G.L., Jr. 1971. Shipboard and in situ electrical resistivity logging of unconsolidated marine sediments. Offshore Tech. Conf. Paper OTC 1351, pp. 253-259.

Bray, J., Bricker, J.T., and Troup, O.P. 1973. Phosphate in interstitial waters of anoxic sediments. *Earth Planet. Sci. Lett.* 18:1362-1364.

Briggs, L.J., and McCall, A.G. 1904. An artificial root for inducing capillary movement of soil moisture. *Science* 20:566-569.

Briggs, L.J., and McLane, J.W. 1907. The moisture equivalent of soils. *U.S. Dept. Agriculture, Bureau of Soils Bull.* 45:1-23.

Brown, J. 1965. In situ interstitial water sampler for radioactive waste disposal into the ground: *International Atomic Energy Agency, Vienna: Safety Series* 15(3):47-52.

Bruevich, S.V. 1938. Oxidation-reduction potential of sediments from the Barents and Karsa Seas. *Compte Rendus Acad. Sci. URSS* 19(8):637–640.

Bruevich, S.V. (Ed.). 1966. Chemistry of interstitial waters in sediments of the Pacific Ocean (Russ.). In *Khimiya Tikhogo Okeana* Moscow: Izdat. Akad. Nauk, pp. 253–358.

Cameron, F.K. 1911. The soil solution: The nutrient medium for plant growth. Easton, Pa., cited in Kriukov, 1971.

Chmelik, F.B., and Bouma, A.H. 1970. Electrical logging in recent sediments. Offshore Technology Conf. Paper 1147I, pp. 49–54.

Culberson, C., and Pytkowicz, R.M. 1968. Effect of pressure on carbonic acid, boric acid, and the pH in sea water. *Limnol. Oceanogr.* 13:403–417.

Disteche, A. 1974. The effect of pressure on dissociation constants and its temperature dependency. In E.D. Goldberg (Ed.), *The Sea,* Vol. 5. New York: Wiley-Interscience, pp. 81–121.

Duursma, E., and Hoede, C. 1967. Theoretical, experimental and field studies concerning molecular diffusion of radioisotopes in sediments and suspended solid particles of the sea. A. Theories and mathematical calculation. *Neth. J. Sea Res.* 3:423–457.

Edmunds, W.M., and Bath, A.H. 1976. Centrifuge extraction and chemical analysis of interstitial waters. *Envir. Sci. Technology* 10:467–472.

Emery, K.O., and Hoggan, D. 1958. Gases in marine sediments. *Bull. Am. Assoc. Petroleum Geologists* 42:2174–2188.

Emery, K.O., and Rittenberg, S.C. 1952. California basin sediments and origin of oil. *Bull. Am. Assoc. Petroleum Geologists* 36:735–806.

Fanning, K.A., Barnard, L., and Jones, S.L. 1981. On the chemistry of interstitial waters of Antarctic sediments. In preparation.

Fanning, K.A., and Pilson, M.E. 1971. Interstitial silica and pH in marine sediments: Some effects of sampling procedures. *Science* 173:1228–1231.

Fanning, K.A. and Pilson, M.E.Q. 1974. The diffusion of dissolved silica out of sediments. *J. Geophys. Res* 79:1293–1297.

Gedroiz, K.K. 1906. On the changeability of composition of soil solution (Russ.). *Zhurnal Opyt. Agron.* 7(5):521–545.

Gieskes, J.M. 1974. The alkalinity-total carbon dioxide system in seawater. In E.D. Goldberg (Ed.), *The Sea*, Vol. 5. New York: Wiley-Interscience, pp. 123–151.

Gieskes, J.M. 1975. Chemistry of interstitial waters of marine sediments. *Ann. Rev. Earth Planet. Sci.* 3:443–453.

Gola, G. 1910. Saggio d'una terra osmotica dell'edafismo. *Annali di Botanica* 8:275–548.

Gripenberg, S. 1934. A study of the sediments of the North Baltic and

adjoining seas. *Havsforskningsinstitutets Skrift* (Helsinki) 96:1–231.
Hallberg, R.O., Bågander, L.E., Engvall, A.G., Lindström, M., Oden, S., and Schippel, F.A. 1973. The chemical-microbiological dynamics of the sediment-water interface. *Contr. Askö Laboratory, Univ. Stockholm, Sweden* 2:1–117.
Hartmann, M. 1965. An apparatus for the recovery of interstitial water from recent sediments. *Deep Sea Res.* 12:225–226.
Hesslein, R.H. 1976. An in situ sampler for close interval pore water studies. *Limnol. Oceanogr.* 6:912–914.
Horne, R.A. 1969. *The Structure of Water and the Hydrosphere.* New York: Wiley-Interscience.
Horowitz, R.M., Waterman, L.S., and Broecker, W.S. 1973. Interstitial water studies, Leg 15. New procedures and equipment. In B.C. Heezen, et al. (Eds.), *Initial Reports Deep Sea Drilling Project.* Washington: U.S. Govt. Printing Office, pp. 757–763.
Husmann, S. 1971. Eine neue Methode zur Entnahme von Interstitial Wasser aus subaquatischen Lockergesteinen. *Arch. Hydrobiol.* 68: 519–527.
Imboden, D.M. 1976. Interstitial transport of solutes in non-steady state accumulation and compacting sediments. *Earth Planet. Sci. Lett.* 27: 221–228.
Ishcherekov, V. 1907. Recovery of uncontaminated soil solutions (Russ.) *Zhurnal Opyt. Agron.* 8:147–166.
Jerbo, A. 1965. Bottniska lersediment, en geologisk-geoteknisk översikt. *Statens Järnvägars Centralförvaltning, Geotekniska Kontoret, Medd. (Sweden)* 11:159.
Johnson, R.G. 1967. Salinity of interstitial water in a sandy beach. *Limnol. Oceanogr.* 12:1–7.
Jones, S.L. 1977. Contribution from deep sediments to the dissolved silica in the deep water of the Mediterranean Sea. Masters thesis, Dept. of Marine Science, Univ. of South Florida, St. Petersburg, Florida.
Jones, S.L., and Fanning, K.A. 1981. Contribution from deep sediments to the dissolved silica in the deep water of the Mediterranean Sea. Manuscript in preparation.
Kalil, E.K., and Goldhaber, M. 1973. A sediment squeezer for removal of pore waters without air contact. *J. Sed. Petrol.* 43:553–557.
Kermabon, A., Gehin, C., and Blavier, P. 1968. A deep sea electrical resistivity probe for measuring porosity and density of unconsolidated sediments. *Geophysics* 34:554–571.
Kharaka, Y., and Berry, F.A.F. 1973. Simultaneous flow of water and solutes through geological membranes. I. Experimental investigation. *Geochim. Cosmochim. Acta* 37:2577–2603.
Komarova, N.A., and Kriukov, P.A. 1959. Determination of the activity of

sodium ions in dispersed systems (Russ.). *Kolloidnyi Zhurnal* 21:189–194.

Kriukov, P.A. 1947. Recent methods for physicochemical analysis of soils; Methods for separating soil solutions (Russ.). In *Rukovodstvo dlya polevykh i laboratornykh issledovanii pochv.* Moscow Izdat. Akad. Nauk, SSSR, pp. 3–15.

Kriukov, P.A. 1964. Some questions of the examination of rock solutions (Russ.). In *Chemistry of the Earth's Crust (Khimia zemnoi kory).* Moskva:Izdat. Nauka, pp. 456–468.

Kriukov, P.A. 1971. *Gornye pochvennye i ilovye rastvory (Interstitial waters of soils, rocks and sediments).* Novosibirsk: Izdat. Nauka, p. 219.

Kriukov, P.A., and Komarova, N.A. 1954. On squeezing fluids from clays at very high pressures (Russ.). *Doklady Akad. Nauk SSSR* 99(4):617–619.

Kriukov, P.A., and Komarova, N.A. 1956. Investigation of soil, mud, and rock solutions (Russ). *Mezhdunarodnomu kongressu pochvovedov, Moscow, 2nd Kommiss., Doklady* 6:151–184.

Kriukov, P.A., and Zarubina, S.A. 1977. pH of phosphate and borax buffer solutions at temperatures 0–25°C and pressures 1–1000 kg/sm. VINITI, Pub. N. 3665 Vsesoyuz. Inst. Nauchnoi i technich. informatsii.

Kriukov, P.A. and Zhuchkova, A.A. 1962. Phase potentials arising on pressing out solutions from gels (Russ.). *Izvestiya Sibirskogo Otdel. Akad. Nauk, SSSR, Ser. Khim.* 4:121–122.

Kullenberg, B. 1952. On the salinity of the water contained in marine sediments. *Oceanogr. Inst. Görborg Medd.* 21:1–37.

Kuzmin, M.S. 1922. On the extraction of soil solution (Russ.). *Zhurnal Opyt. Agron. of South-East* 1:2.

Lasaga, A.C., and Holland, H.D. 1976. Mathematical aspects of non-steady state diagenesis. *Geochim. Cosmochim. Acta* 40:257–266.

Lehmann, E. 1978. Sediment water interaction and its effect upon water quality (a bibliography with abstracts). *Nat. Tech. Inform. Svc. (NTIS)* PS-78/0015:1–140.

Lerman, A. 1977. Migrational processes and chemical reactions in interstitial waters. In E.D. Goldberg (Ed.), *The Sea*, Vol. 6. New York: Wiley-Interscience, pp. 695–738.

Li, Y.H., and Gregory, S. 1974. Diffusion of ions in sea water and in deep sea sediments. *Geochim. Cosmochim. Acta* 38:703–714.

Lipman, C.B. 1918. A new method of extracting the soil solution. *Univ. Calif. Pub. Agr. Sci.* 3:131–134.

Lockey, S.D. 1962. Preparation of sterile, small volume aqueous extracts by cold filtration. *Ann. Allergy* 20:189–192. (Cited in Millipore Filter Corp. bibliography of filtration, Bedford, Mass.)

Lusczynski, N.J. 1961. Filter-press method of extracting water samples for chloride analysis. U.S. Geol. Survey, Water Supply Paper 1544-A, pp. 1–8.
Mangelsdorf, P.C., Wilson, T.R.S., and Daniell, E. 1969. Potassium enrichments in interstitial waters of recent marine sediments. *Science* 165:171–174.
Mangelsdorf, P.C., Jr., and Wilson, T.R.S. 1971. Difference chromatography of seawater. *J. Phys. Chem.* 75:1418–1425.
Manheim, F.T. 1961. In situ measurements of pH and Eh in natural waters and sediments. Prelimary note on applications to a stagnant environment. *Stockholm Contr. Geol.* 8:27–36.
Manheim, F.T. 1966. A hydraulic squeezer for obtaining interstitial water from consolidated and unconsolidated sediments. U.S. Geol. Survey Prof. Paper 550-C, pp. 256–261.
Manheim, F.T. 1972. Disposable syringe techniques for obtaining small quantities of pore water from unconsolidated sediments. *J. Sed. Petrol.* 38:666–668.
Manheim, F.T. 1976. Interstitial waters of marine sediments. In J.P. Riley, and R. Chester (Eds.), *Chemical Oceanography*, 2d ed. London: Academic Press, pp. 115–185.
Manheim, F.T., and Sayles, F.L. 1974. Composition and origin of interstitial waters of marine sediments, based on deep sea drill cores. In E.D. Goldberg (Ed.), *The Sea*, Vol. 5. New York: Wiley-Interscience, pp. 527–568.
Manheim, F.T., and Schug, D.M. 1978. Interstitial waters of Black Sea Cores. In J.L. Usher and P. Supko (Eds.), *Initial Reports of the Deep Sea Drilling Project*. Washington: U.S. Govt. Printing Office, pp. 637–651.
Manheim, F.T., and Waterman, L.S. 1974. Diffusimetry (diffusion coefficient estimation) on sediment cores by resistivity probe. In C.C. von der Borch, J.G. Sclater, et al. (Eds.), *Initial Reports of the Deep Sea Drilling Project*. Washington: U.S. Govt. Printing Office, pp. 663–670.
Martens, C.,S. 1981. Methane production, consumption, and transport in the interstitial waters of coastal marine sediments. (chapter 10 of this volume).
Mattson, S., Eriksson, E., Vahtras, K., and Williams, E.G. 1949. Phosphate relationships of soil and plants. I. Membrane equilibria and phosphate uptake. *Ann. Royal Agric. College, Sweden* 16:457–484.
McDuff, R.E., and Gieskes, J.M. 1976. Calcium and magnesium profiles in DSDP interstitial waters. Diffusion or reaction? *Earth Plan. Sci. Lett.* 33:1–10.
Morgan, J.F. 1917. The soil solution obtained by the oil pressure method. *Soil Science* 3:531–545.

Murray, J., and Irvine,R. 1895. On the chemical changes which take place in the composition of the sea water associated with blue muds on the floor of the ocean. *Trans. Royal Soc. Edinburg* 37:481–507.

Northrup, Z. 1918. The true soil solution. *Science* 47:638–639.

Östlund, H.G., and Alexander, J. 1963. The oxidation rate of sulfide in sea water. *J. Geophys. Res.* 68:3995–3997.

Overstreet, R., and Babcock, K.L. 1936. Commentary on activities and Donnan effects. *Intl. Congress Soil Sci., Paris,* cited in Kriukov, 1971.

Pamatmat, M. 1971. Oxygen consumption by the sea bed. IV. Shipboard and laboratory experiments. *Limnol. Oceanogr.* 16: 536–550.

Pamatmat, M. 1981. Direct calorimetry in benthic metabolism (chapter 8 of this volume).

Parker, F.W. 1921. Methods of studying the concentration and composition of soil solution. *Soil Science.* 12:209.

Pirogova, M.V. 1950. On the chemical exchange between bottom sediment and water of the Black Sea (Russ.). *Gidrokhim. Materialy* 21:10–18.

Presley, B.J., Brooks, R.R., and Kappel, H.M. 1967. A simple squeezer for removal of interstitial water from ocean sediments. *J. Marine Res.* 25: 355–357.

Ramann, E. März, S., and Bauer, H. 1916. Uber Boden-Press-Säfte. *Int. Mitt. für Bodenkunde* 6(1):27–34.

Reeburgh, W.S. 1967. An improved interstitial water sampler. *Limnol. Oceanogr.* 12:163–165.

Reeburgh, W.S. 1981. A major sink flux control for methane in marine sediments: Anaerobic consumption (chapter 11 of this volume).

Rhoads, D.C. 1974. Organism-sediment relations on the muddy sea floor. *Oceanogr. Mar. Biol. Ann. Rev.* 12:263–300.

Richards, A.F., McDonald, V.J., Olson, R.E., and Keller, G.H. 1972. In-place measurement of deep sea soil shear strength. *Am. Soc. Testing and Materials Spec. Tech. Publ.* 501:55–68.

Richards, L.A. 1941. A pressure-membrane extraction apparatus for soil solutions. *Soil Science.* 51:377–386.

Riedl, R.J., and Ott, J.A. 1981. Water movement through porous sediments (chapter 2 of this volume).

Rittenberg, S.C., Emery, K.O., Hulsemann, J., Degens, E.T., Fan, R.C., Reuter, J.H., Grady, J.R., Richardson, S.H., and Bray, E.E. 1963. Biogeochemistry of sediments in experimental Mohole. *J. Sed. Petrol.* 33:140–172.

Rozanov, A.G. 1981. Pacific sediments from Japan to Mexico: some redox characteristics (chapter 13 of this volume).

Sayles, F.L. 1970. Preliminary geochemistry. In R.G. Bader, et al. (Eds.), *Initial Reports Deep Sea Drilling Project.* Washington: U.S. Govt. Printing Office, pp. 645–655.

Sayles, F.L. and Mangelsdorf, P.C., Jr. 1978. The equilibration of clay minerals with sea water. Exchange reactions. *Geochim. et Cosmochim. Acta* 41:951–960.

Sayles, F.L., Mangelsdorf, P.C., Jr., Wilson, T.R.S., and Hume, D.N. 1976. A sampler for the in situ collection of marine sedimentary pore waters. *Deep Sea Res.* 23:259–264.

Sayles, F.L., Manheim, F.T., and Waterman, L.S. 1973a. Interstitial water studies on small core samples, Leg. 15. In B.C. Heezen, et al. (Eds.), *Initial Reports Deep Sea Drilling Project.* Washington: U.S. Govt. Printing Office, pp. 783–804.

Sayles, F.L., Wilson, T.R.S., Hume, D.N., and Mangelsdorf, P.C., Jr. 1973b. In situ sampler for marine sedimentary pore waters: Evidence for potassium depletion and calcium enrichment. *Science* 181:154–156.

Schloesing, M.T. 1866. Sur l'analyse des principes solubles de la terre vegetale: *Compt. Rend. Acad. Sci.* 63:1007–1012.

Schlumberger Well Surveying Corp. 1972. *Log interpretation* 1; *Principles* 112; 1974, *Applications, 116;* 1978, *Log Interpretation Charts* 84. Houston, Texas.

Scholl, D.W. 1963. Techniques for removing interstitial water from coarse-grained sediments for chemical analysis. *Sedimentol.* 2:156–163.

Shishkina, O.V. 1972. *Geokhimiya morskikh i okeanicheskikh ilovykh vod. (Geochemistry of marine and ocean pore fluids).* Moscow: Izdat. Nauka, p. 227.

Siever, R. 1962. A squeezer for extracting interstitial water from modern sediments. *J. Sed. Petrol.* 32:329–331.

Siever, R., Beck, K.C., and Berner, R.A. 1965. Composition of interstitial waters of modern sediments. *J. Geol.* 73:39–73.

Siever, R., Garrels, R.M., Kanwisher, J., and Berner, R.A. 1961. Interstitial waters of recent marine muds off Cape Cod. *Science* 134:1071–1072.

Stannett, V. 1962. Permeability of plastic films and coated paper to gases and vapor. *Tappi Monograph Series* 23, Technical Assoc. Pulp Paper Industry, New York.

Stiles, W., and Jørgenson, I. 1914. The nature and methods of extraction of the soil solution. *J. Ecol.* 2:245–250.

Stoessel, R.K., and Hanor, J.S. 1975. A non-steady state method for determining diffusion coefficients in porous media. *J. Geophys. Res.* 80:4979–4982.

Stumm, W. 1966. Redox potential as an environmental parameter. Conceptual significance and operational limitation. *Third Int. Conf. Water Poll. Res.*, 1(13):1–16 *Munich.*

Suhayda, J.N., Coleman, J.M., Whelan, T., and Garrison, L.E. 1981. Oscillation of continental shelf sediments caused by waves (chapter 3 of this volume).

Swarzenski, W. 1959. Determination of chloride in water from core samples. *Am. Assoc. Petrol. Geol. Bull* 43:1995–1998.

Tsyba, P.N., and Kriukov, P.A. 1959. Comparison of methods of study of rock solutions (Russ.). *Gidrokhim. Materialy* 29:273–281.

Turk, J.T. 1976. A study of diffusion in clay-water systems by chemical and electrical methods. Doctoral thesis, Univ. of California, San Diego.

Van Suchtelen, F.H.H. 1912. Methode zur Gewinnung der naturlichen Bodenlosung. fur Landwirtschaft. *J. Landw.* 60:369–370.

Volkov, I.I. 1961. Formation and transformation of sulfur compounds in Black Sea sediments (Russ.). In N.M. Strakhov (Ed.), *Sovremennye osadki morei i okeanov*. Moscow: Izdat. Akad. Nauk, SSSR, pp. 557–596.

Walters, L.J., Jr. 1967. Bound halogens in sediments. Doctoral thesis, Dept. of Geol. Geophys., Massachusetts Institute of Technology.

Waxman, M.H., and Smits, L.J.M. 1968. Electrical conductivities in oil-bearing shaly sands. *Trans. A.I.M.E.* 243:107–122.

Whitfield, M. 1969. Eh as an operational parameter in estuarine studies. *Limno. Oceanogr.* 14:547–558.

Whitney, M., and Cameron, F.K. 1905. The chemistry of soils as related to crop production. *U.S. Dept. Agric. Bureau Soils Bull.* 22.

Zaitseva, E.D. 1958. Cation exchange capacity of marine sediments and methods of determination (Russ.). *Trudy Inst. Okeanol.* 46:181–204.

Zhorikov, E.A. 1930. K metodike izucheniya pochvennogo rastvora (On methods of study of soil solutions). *Trudy Nauchn. issl. instituta po khlopkovoi promyshlennosti* 1.

**Part II
The Dynamics of the
Interactions between
Sediment and Seawater**

2 Water Movement through Porous Sediments

R.J. Riedl and J.A. Ott

Abstract

Experimental studies and theoretical calculations show that water flow takes place in microscopic channels in surficial marine sediments to depths ranging from a few millimeters to a half-meter or more. These flows derive from tidal phenomena, pressure differences between wave crest and wave trough, and swash runs up the beach ahead of a tide. In each case, mean flow velocities are perturbed by characteristic fluctuations having the frequency of waves and swash. Such fluid movements have a fundamental effect on the depth of the redox potential discontinuity (RPD)*, on associated microbial and metabolic systems in the sediments, on particle movements, and on interstitial meiofauna.

For the entire east North American shelf (western Atlantic) between 27°N and 35°N from the coast to the 200-m isobath, Q, the total volume of water exchange due to intrasedimentary water flow from the subtidal pump, is 1,306 km^3 per year, or 33.2 L/m^2 of shelf area per day. The water overlying the inner shelf down to the 30-m isobath has $Q = 0.017$ km^3/km^2/yr, corresponding to a complete filtration of overlying water in 10 months under average weather conditions. Worldwide estimates of the combined effects of the subtidal pump and of surf and swash suggest that the flow of seawater through the nearshore sand filter system by those mechanisms equals one-third of the water evaporated annually from all oceans.

Introduction

Water movement is one of the important factors in aquatic ecosystems, either directly or indirectly affecting most other parameters and their distribution. The kinetic energy of moving water can do direct mechanical work on organisms and their substrate by creating mechanical stress and finally ending up as heat. However, its main effect is to transport dissolved and suspended material (e.g., nutrients, gases, and organic material) and thereby

---

This abstract was prepared by the editors.
*This is referred to as *redoxcline* in chapter 17.

provide a controlling mechanism that enhances and regulates the power flow through aquatic food chains (Odum, 1971). Without the advection and eddy diffusion of water, transportation of substances required by aquatic organisms would depend only on molecular diffusion, which is at least 10,000 times too slow to sustain life in this medium (Riedl, 1971a).

At the bottom/water interface, the complicated biologically relevant flow patterns are generally not well enough understood. One of us (Riedl, 1969, 1971a, 1971b; Riedl and Forstner, 1968) has recently summarized the knowledge of the biological aspect of water movement, emphasizing phenomena occurring in the hard-bottom littoral zone. The increase in knowledge of the biology of interstitial meiofauna, i.e., the animals living in the pore space between sand grains, stimulated a series of investigations on water movement in sediments and the associated distribution and dynamics of other parameters (Riedl, 1971c; Riedl and Machan, 1972; Riedl et al., 1972). This present contribution is based mainly on those papers.

Methods of Measuring Interstitial Water Movement

The probes used to measure flow velocities employ the constant-temperature hot-thermistor anemometer technique (Lumley, 1962; Eagleson and Van de Watering, 1964). Details of electronic circuitry and probe construction can be found in Riedl and Machan (1972).

Water flow in a porous medium actually takes place in microscopic channels. There are two possibilities for defining and measuring the flow. One is to build a device small enough not to disturb the local flow pattern. It could measure the flow in a channel between the sand grains and would yield results of great interest to biologists. However, that kind of device brings about technological problems too difficult to overcome. The other way is to assume the porous medium to be homogeneous and measure a mean water velocity with a probe large in comparison to the grain or pore size. A natural sand/water mixture, however, is not strictly homogeneous; therefore, to compensate for any natural heterogeneity, a housing was built around the probe (figure 2–1). It consisted of a piece of Plexiglass tubing big enough to contain both the sensing and compensation thermistors. The probe was inserted into the tubing at right angles from the side sealed with silicone rubber or an O-ring.

A given cross-sectional area of saturated beach is, of course, partially filled with sand grains. Also, the flow rate of water through a sand-free water junction must change inversely with the volume occupied by the water. If the probe just described is filled with water, closed at both ends with fine nylon grid to keep the sand out, and buried in the sand, the water velocity in the probe must of necessity be smaller than the average velocity in the sand.

Porous Sediments

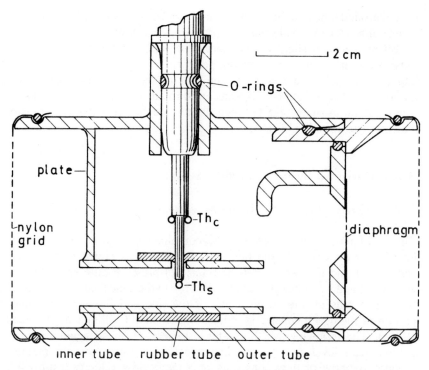

Source: Riedl and Machan, *Marine Biology* 13(3):179–209, 1972. Reprinted with permission).

Figure 2–1. Thermistor probe in probe housing. The thermistors are supported on an aluminum rod; their leads run through thin glass tubes into a handle where they are connected to the cable leading to the recording device. For the probe housing, a small tube is sealed into a bigger tube to increase flow velocity past sensing thermistor Th_s. The ends of the bigger tube are closed with fine nylon grid to keep out sand. A punched diaphragm can be inserted into probe housing to adapt the probe to the local permeability of sand. O-Ring fittings provide for easier handling. Th_c is compensation thermistor.

Therefore, the sensing thermistor was made to fit into a small tube through which the entire flow was forced to pass because of a plate between the two tubes. The resultant drastic reduction in cross-sectional area of flow within the chamber of the probe increases the velocity past the thermistor to an easily measurable value more in keeping with the velocity in the surrounding sediment. The internal configuration of the probe also helps to ensure that the flow measured by the thermistor is larger than the artificial convection that occurs around a heated thermistor bead. It therefore provides a way to adapt a current sensor to a wide range of problems and flow speeds.

By means of a slight modification, the probe also can be adapted to sensing the direction of flow within sediments. Three thermistors are built into one housing in a row along the axis of the inner tube. The central thermistor is heated and heat is transferred to one of the outside thermistors. The resultant error voltage between the two connected outside thermistors indicates the direction of flow.

Water Movement through Intertidal Sediments

Driving Forces of Intertidal Water Movement

Three different forces determine the complicated flow pattern in porous intertidal sediments: (1) the tide creates differences between the ground-water level and the sand/water interface; (2) pressure differences between wave crest and wave trough drive water through the sand (see below); and (3) the swash runs up the beach ahead of the tide and fills a wedge-shaped portion of the beach consisting of unsaturated sand ("filling wedge," Riedl, 1971c). The inflowing swash creates a hydrostatic head against both the ground-water level and the receding swash. The three forces combine to produce patterns of flow consisting of a mean flow velocity disturbed by characteristic fluctuations with the frequency of waves and swash.

Two basic current types are distinguished, differing in amplitude and in the frequency of changes in velocity and direction. They are *gravity currents* (Q) and *pulsing currents* (q). Table 2–1 gives a complete list of abbreviations for both types of currents. Gravity currents are the result of the tidal filling and draining of the beach; hence the changes in their speed and direction are correlated with tidal conditions. Pulsing currents are driven by surf and swash; thus their changes in speed and direction are correlated to rhythms of wave and swash (1 sec to 1 min, generally) and are 2 to 4 orders of magnitude faster than those of gravity currents. Their intensity is described by their amplitude (a), and the symbols for increasing (i), constant (c), and decreasing (d) refers to changes in amplitude alone.

Current Types

Q and q combine to form the main types of flow patterns observed. Figure 2–2 gives examples.

Swash pulsing is caused by those swashes which reach landward beyond the outcrop of the water table, which is the boundary between the shiny

Table 2–1
Abbreviations used in text and figures.

A	Operational amplifier
J	Constant
K	Constant
M	Meter
P	Electric power
Q_1, Q_1	Transistors
R_1, R_c, R_s	Resistors
T_a	Ambient temperature
T_e	*Element temperature*
Th_c	Compensation thermistor
Th_s	Sensing thermistor
U	Bridge driving voltage
U_c	Compensating voltage
V_s	Supply voltage
max	Maxima
mea	Mean
min	Minima
a	Amplitude of q phenomena (u)
a'	Amplitude of q phenomena (v)
u	Speed of water displacement (volumetric flow rate)
v	Actual particle speed
$q\downarrow, Q\downarrow$	Abbreviation for downward current
t	Time
x	Horizontal axis perpendicular to shore line
$y\uparrow, y\downarrow$	Upward or downward oriented
Q	Gravity current, in particular
C	Constant
D	Decreasing
I	Increasing
L	Landward
S	Seaward
q	Pulsing current, in particular
b	Boosting
c	Constant
d	Decreasing
f	Frequent
i	Increasing
n	Non-frequent
o	Oscillating
p	Pumping
r	Retarding
s	Swinging
ei	Early incoming-tide
eo	Early outgoing-tide
h	High tide
l	Low tide
li	Late incoming-tide
lo	Late outgoing-tide
mi	Mid-incoming-tide
mo	Mid-outgoing-tide
B	Filling bag
g	Saturation gap
N	Pore space
W	Filling wedge
w	Water table

Source: Riedl and Machan, *Marine Biology 13*(3): 179–209, 1972. Reprinted with permission.

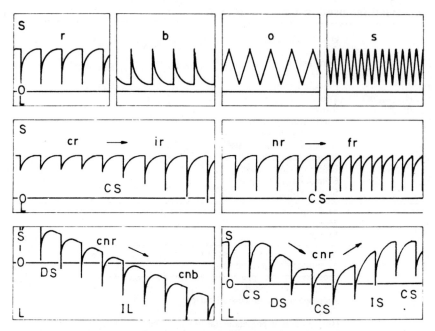

Source: Riedl and Machan, *Marine Biology* 13(3):179–209, 1972. Reprinted with permission).

Figure 2–2. Principal sketch of the most common temporal variations in amplitude of pulse currents, including some interactions with gravity currents. Top section shows retarding (*r*), boosting (*b*), oscillating (*o*), and swinging (*s*). Middle section: changes as pulsing currents pass fully or partially through zero (*O*), from seaward (*S*) to landward (*L*) due to interaction with gravity currents. Abbreviations of pulsing phenomena are given in lower case, and abbreviations of the gravity phenomena are given in upper case. For abbreviations, see table 2–1.

reflecting and dull non-reflecting sand surfaces. The swashes rhythmically inject water into the *filling wedge*, where it spreads downward and sideward and forms a baglike area of saturated sand, the *filling bag* (figure 2–3). Swash pulses are superimposed on the local gravity current. The kinds of swash pulsing are (see table 2–1):

> *r*: If the swash input pushes against *Q*, a *retarding* phenomenon results and is characterized by a sudden steep drop in flow velocity (within seconds) and then a slow buildup of local average speed (within minutes).
>
> *b*: If the swash input pushes in the direction of *Q*, a *boosting*

phenomenon is the result. There is a sudden increase of flow velocity and a slow return to original Q.

o: When the swash input changes its position relative to a point in the beach, the direction of q changes relative to Q, r becomes b, and vice versa. The transition is characterized by an *oscillating* pattern where spikes occur in both directions, and the flatter slopes change gradually from one side to the other.

Surf pulsing is caused by the waves above a submerged sand body because pressure and motion fields of the waves penetrate into the body due to its porosity. Depending on the relative magnitude of the amplitude a of q in relation to Q, we distinguish between two types:

s: *Swinging*, when $S > a/2$. In other words, the pulsing does not pass through zero, and the speed of a stable seaward flow only swings around a mean value.

p: *Pumping*, when $S < a/2$. The pumping now passes through zero, and seawater is pumped in and out of the interstices. There is not only a draining back toward the sea but also a permanent water exchange.

Combined Currents and Current Sequences

The basic current types described individually occur in a variety of combinations to form the complex and recurring current patterns within the sediment of the intertidal zone. In the following paragraphs we will describe the combinations (spatial correlations of current types) and sequences (temporal correlations).

The Q sequence. In a tidal beach without wave action, any given point within the intertidal sand body would experience a flow sequence of $IL \rightarrow CL \rightarrow DL \rightarrow IS \rightarrow CS \rightarrow DS \rightarrow IL$ during a tidal cycle, starting at the beginning of the rising tide (table 2–1).

The *r-o-b* or filling mechanism. Considering the swash zone at a fixed tidal position, several combinations of current are possible based on our sensor measurements in the filling bag (figures 2–1 and 2–3). At incoming tide with a predominantly landward Q, boosting b occurs above (or at the landward edge of) the swash zone, oscillation about zero flow occurs immediately underneath, and a retardation phenomenon r occurs at greater depth. At outgoing tide with predominantly seaward Q, the locations of b and r change place. With increasing wave action and consequently increased input through the wedge at incoming tide, net gravity flow in the lower end of

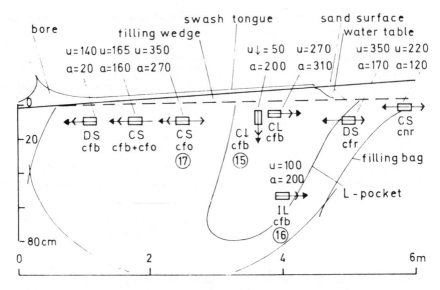

Source: Riedl and Machan, *Marine Biology* 13(3):179–209, 1972. Reprinted with permission.

Figure 2–3. Filling wedge and filling bag. Average conditions, particularly for incoming tide. Rectangles: sensor positions and direction of measurement. Arrows: recorded current direction (triangular arrowheads stand for Q, angular for q). Velocities in μm/s. Quantitative data of recordings are indicated above each sensor, qualitative data below. For abbreviations, see table 2–1.

the swash zone might be seaward S and, at the upper end, may have only a short period of landward flow L during the whole tidal cycle (L pocket).

The time sequence for gravity currents at a point near the mid line of the intertidal zone at incoming tide would now be $S \rightarrow DS$. When the first pushes of the swash reach the site, the pulsing current is nr. The gravity current now reverses from $DS \rightarrow IL$, and nr changes to nb and then to ifb. When the swash moves over the site, $fb \rightarrow fo \rightarrow fr$. In the case of strong input through the wedge, IL goes to DL and then to IS (see earlier discussion), so that $fr \rightarrow fb$. The amplitudes of the pulsing currents decrease through the sequences $ir \rightarrow co \rightarrow db$ at incoming tide and $ib \rightarrow co \rightarrow dr$ at outgoing tide, while the frequency goes from nonfrequent to frequent and back (e.g., $inr \rightarrow cfr \rightarrow cfo \rightarrow cfb \rightarrow dnb$ at incoming tide and $inb \rightarrow cfb \rightarrow cfo \rightarrow cfr \rightarrow dnr$ at outgoing tide). Deeper in the sediment, the higher frequency part is missing.

The b-s-p sequence. As soon as the zone of the filling wedges migrates with incoming tide beyond our point of observation, the boosting effect decreases, the frequency of pulsing phenomenon increases (from swash rhythm

Porous Sediments

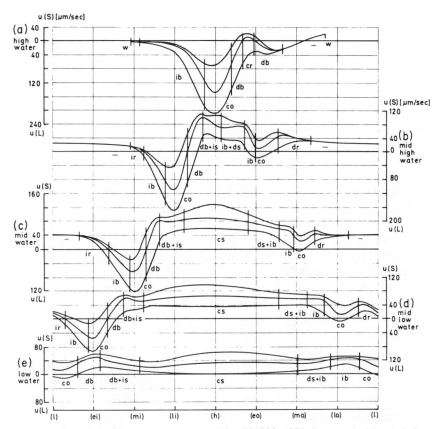

Source: Riedl and Machan, *Marine Biology* 13(3):179–209, 1972. Reprinted with permission.

Figure 2–4. Speed patterns within a tidal cycle, 60 cm beneath sand surface and in five positions relative to tide levels: (a) at high tide level; (b) between high and mid-water levels; (c) mid-water; (d) between mid and low water levels; (e) at low water level. All patterns are derived from recordings; time is from left to right in tidal stages (i = incoming; o = outgoing). The middle pattern is $u(Q)$; the two others indicated $a(q)$. The qualities of the a patterns are indicated below the patterns. Points w mark the time when water table passes through probe location, i.e., no q is present; u refers to fluid advection; Q refers to gravity current; q is pulsating current; other abbreviations are shown in table 2–1.

to wave rhythm), and directed effects (b or r) disappear. We find a harmonic swinging around Q, which is *IS* or *DS*. At incoming tide, a short period of *cfb* changes to *dnb*, which becomes superimposed by *is* to a pattern *dnb + is*.

As b gradually disappears, *is* becomes symmetrical. The complete sequence thus reads $cfb \to dnb + is \to is \to cs \to ds \to inb + ds \to cfb$. Near the subtidal zone during the middle part of the sequence, $S \to\, < a/2$ of s, and s changes therefore to p. The sequence there is $cfb \to dnb + is \to is \to cs \to ds + ip \to cp \to is + dp \to cs \to ds \to inb + ds \to cfb$.

In these sequences, there is a deviation from symmetry caused by two main differences. At incoming tide, the hydraulic head is weak, and the input forces are strong. At outgoing tide, the opposite is true. Therefore, the change in Q is much more pronounced at incoming tide, whereas at outgoing tide S prevails. In addition, $a(q)$ is usually much higher at incoming tide, and the swash pulsing sequence takes longer.

Comparing the current patterns in 60-cm depth under the sediment at five different positions in the intertidal zone, several trends are apparent (figure 2–4). In the high intertidal (figure 2–4a), the incoming bag tends to fuse with the outgoing bag, and the s part of the sequence disappears. Here a of q reaches high values, but q phenomena are restricted in time and interrupted by periods of constant flow. Toward the lower intertidal (figure 2–4e), the outgoing bag fuses with the incoming, and the s part of the sequence increases. Here a of q is generally smaller, but there are no periods of constant flow. In the high water position, Q is predominantly L; at the low water position, only S is recorded.

At different depths beneath the sediment surface at a midwater position, the r-o-b phenomena primarily change in a and time span, both decreasing with increasing depth. The amplitudes of s and p phenomena also decrease. Pumping occurs only near the sediment surface. Landward Q during the r-o-b sequence can be found down to 80 cm below the sediment surface, but time span and speed are reduced. The closer to the sand surface, the higher are $u(S)$ and $u(L)$.

Dynamics of the Total Pattern

To develop the total flow pattern resulting from the combination of r-o-b, s-p and Q phenomena as the filling bag migrates up and down the beach with the tide, the interstitial velocity fields of the beach have to be described in more detail.

The filling bag is limited to landward by the furthest penetration of the pulsing currents, to seaward by the point where $a(b) < a(s)$, and into the bottom by $a = 10$ μm/sec. It surpasses the filling wedge, both landward and seaward, and reaches about 1 m deep into the sediment. There are characteristic differences between incoming and outgoing filling bags at a mid-tide position (figures 2–5 and 2–6). The incoming filling bag has a zone of landward Q (L pocket) near its landward end, whereas an L pocket is usually

Porous Sediments

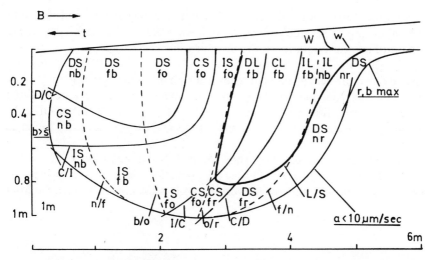

Source: Riedl and Machan, *Marine Biology* 13(3):179–209, 1972. Reprinted with permission.

Figure 2-5. Qualitative patterns in an incoming filling bag at mid-tide. Thin continuous lines: boundaries. Thicker lines: changes in Q characters. Thin broken lines: $r - o - b$ phenomena (heights exaggerated 4 times). Vertical scale is depth below sediment surface; W is water table. For other abbreviations, see table 2-1.

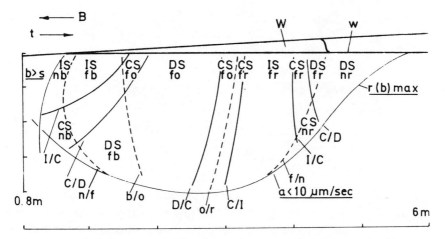

Source: Riedl and Machan, *Marine Biology* 13(3):179–209, 1972. Reprinted with permission.

Figure 2-6. Qualitative patterns in an outgoing filling bag at mid-tide. For further explanation and for comparison, see figure 2-5. For abbreviations, see table 2-1.

40   Dynamic Environment of the Ocean Floor

absent from the outgoing filling bag. The outgoing bag also extends less into the sand body; its velocity field is simpler, weaker, and lacks the strong landward-seaward polarity of the incoming bag.

Considering the $a(q)$ pattern (figure 2–7), we see the filling bag migrate up and down the beach, expanding toward high tide and shrinking toward low tide. The surf pulsing zone lags the bag and covers a great part of the intertidal at high tide, with the boundary between p and s lying very flat. As the tide recedes, the surf pulsing zone shrinks and the p-s boundary becomes almost vertical.

The Q pattern (figure 2–8) shows only S backflow at low tide, slightly interrupted by the filling bag. During incoming tide, the L pocket appears. At high tide, the upper backflow is overrun, the L pocket opens toward the front

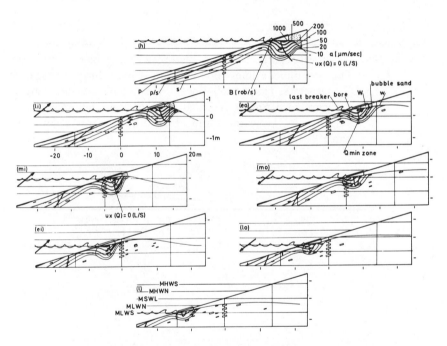

Source: Riedl and Machan, *Marine Biology* 13(3):179–209, 1972. Reprinted with permission.

Figure 2–7. Overall velocity field of pulsing phenomena $r - o - b, s - p$ and its change with tidal position (eight tidal positions in clockwise circle). Isotachs of q from $u = 10$ to $1,000$ μm/s. Seaward boundary of filling bag (B) is thin line; p/s boundary, thick line; L/S boundary, very thick line; and minimum speed of Q, broken line. Vertical scale is depth in meters. Abbreviations as in table 2–1.

Porous Sediments

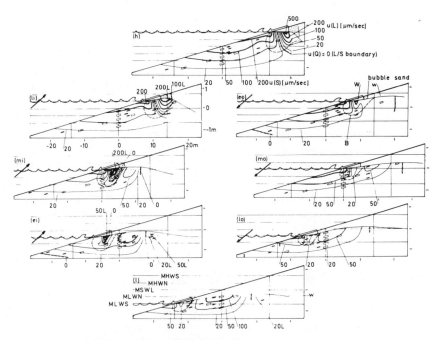

Source: Riedl and Machan, *Marine Biology* 13(3):179–209, 1972. Reprinted with permission.

Figure 2–8. Overall velocity field of gravity currents $ux(Q)$ (L,S) and its change with tidal position. Isotach of Q from $u = 0.200(S)$ and $0.500(L)$ in μm/s. Boundary line of the filling bag (B, $a = 10$ μm/s) is thin broken line; L/S boundary ($Q = o$) is thick broken line. The letter L after velocity value denotes landward current; all other values are seaward currents. For abbreviations, see table 2–1.

of the bag, and the outflow field widens. During outgoing tide, the L pocket disappears, and the outflow gradients grow steeper.

Pathways and Magnitudes of Flow

Input into the beach occurs via the filling wedge and the filling bag. Although those features cover only a certain portion of the intertidal zone at any given moment in the cycle, they migrate up and down between the high-tide and low-tide line, and thus the whole beach receives direct input. The output field extends from the seaward edge of the filling bag to where the gravity currents vanish. Its length increases and decreases with the tide as fields of higher output velocities are added landward at incoming tide and then disappear again.

Percolation pathways in our beaches ranged from 0 to about 80 m; the mean pathway length was from 14 to 29 m during the first 6 tidal hours. The overall mean percolation pathway was about 24 m. Average speed of water displacement u was calculated as 75 μm/s. Assuming a porosity of 25 percent, average speed of water particles v is therefore 300μm/s. Mean residence time of the water particles is 22.2 hours.

Percolation volume also varies strongly with tidal conditions (table 2–2). In our example, from 0.095 (at low tide) to 0.72 m^3/h (around high tide) drains back from a longshore meter of beach. Total percolation per day amounted to 4 m^3 during neap tides and to more than 8 m^3 during spring tides. Extreme values, with higher tides, flatter beaches, and strong wave action, might reach 15 to 20 m^3 per day per longshore meter.

Migration Parameters and Semistable Zones

The migration parameters of L, S, and Q, the a of r, o, b, s, and p, and the movement of the water table define the fields of input, output, pulsing, and aeration. Maximum input occurs in the higher intertidal, and the output field reaches from high water beyond the low-water line and also reaches much deeper than the input field. The pulsing field is even longer, showing stronger swash effect in the higher intertidal, whereas the surf effect is about equal over the whole area of its occurrence. The field of aeration is defined by the migration of the water table. It increases in thickness from low water landward to a maximum in the berm area.

Combinations of those fields, using maxima against maxima, maxima against minima, or more factor gradients now let us define certain semistable zones (figure 2–9) with biological significance for intertidal animals (see below).

Water Movement through Subtidal Sediments

Driving Forces of Subtidal Water Movement

In contrast to the complex pattern of intertidal water flow, water movement through subtidal porous sediments shows only one type of current, p (pumping), driven by one force, the pressure difference created under a moving wave. As defined earlier (see discussion on table 2–1), *pumping* is an oscillating motion of water particles resulting in water exchange in the upper layer of porous sediments.

Table 2-2
Volume of Seawater Percolation and Mean Flow Speeds for Spring Tide at Onslow Bay, North Carolina, U.S.A. Discharge values calculated for different speed groups in cubic meters per longshore meter of the coast per hour [$m^3/(m \cdot h)$], as computed from flow through a vertical transect on seaward side of the input zone. Mean flow rate u is averaged through the same cross-section as the output.

Groups Center Speed u ($\mu m/s$)	Height y (cm) of Output Zone per Speed Group and Tidal Hour												m^3 per tide
	1 h	2 h	3 h	4 h	5 h	6 h	7 h	8 h	9 h	10 h	11 h	12 h	
230	—	—	—	2.5	20	40	43	30	20	5	—	—	1.34
140	—	—	10	35	45	45	40	38	30	23	5	—	1.37
70	25	25	33	35	38	35	33	30	25	25	20	23	0.87
35	18	20	35	43	45	40	35	30	27	30	30	25	0.48
10	25	25	23	33	55	55	40	27	25	30	38	33	0.15
Output [$m^3/(m \cdot h)$]	0.0947	0.0972	0.186	0.352	0.565	0.716	0.716	0.563	0.423	0.269	0.127	0.102	4.2109
Mean u ($\mu m/s$)	38.7	38.6	51.2	65.8	77.3	92.5	104	101	91.8	66	38	34.8	

Source: Riedl and Machan, *Marine Biology* 13 (3): 179–209, 1972. Reprinted with permission.
Mean u per tide: 74.9 $\mu m/s$.

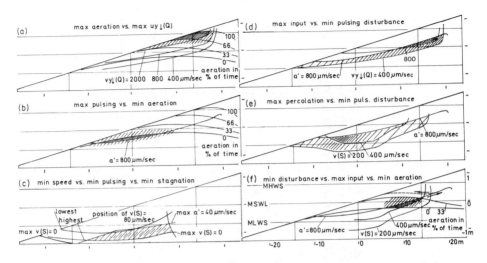

Source: Riedl and Machan, *Marine Biology* 13(3):179–209, 1972. Reprinted with permission.

Figure 2–9. Selected types of semistable zones. Hatching is more dense toward more stable or pronounced conditions: (a) field of maximum aeration and maximum pulsing disturbance; (b) field of maximum input and minimum pulsing disturbance; (c) field of maximum pulsing and minimum aeration; (d) field of maximum percolation and minimum pulsing disturbance; (e) field of minimum speed, pulsing, and stagnation; (f) field of minimum pulsing disturbance, maximum aeration. For abbreviations, see table 2–1.

The fluid-dynamic problem of wave-induced motion in porous beds has been studied by Putnam (1949), Reid and Kajiura (1957), Hunt (1959), and Murray (1965). Their emphasis has been on the effects of wave damping. Furthermore, all four analyses treated the fluid region as a whole, but with low viscosity the influence of viscous force should be confined to a thin boundary layer (e.g., see Schlichting, 1968). Therefore, boundary-layer technique should be used. Although the final result in flow pattern does not differ much, the rate of energy dissipation is different. A brief description of the calculation of the flow pattern is necessary.

The dynamic problem treated is essentially that of surface waves propagating over a porous bed. Viscosity in the fluid is small; therefore, boundary-layer analysis is that of Phillips (1966) and Huang (1970). The bed material is assumed to be uniform, so that Darcy's law is applicable. An arbitrary function $s(x)$ for the bottom slope is allowed, provided that $|s(x)| \ll 1$. Under these conditions, velocity q_s in the sediment is found to be:

$$q_s = \frac{\kappa\alpha\sigma^2 e^{k(z+d)}}{v \sinh kd} \{[-\sin\chi - s(x)\cos\chi]e_1 \qquad (2.1)$$
$$+ [\cos\chi - s(x)\coth kd \sin\chi]e_2\}$$

where κ is the permeability in Darcys; α, σ, k, and $\chi = kx - \sigma t$ are the amplitude, frequency, wave number, and phase function of the surface wave, respectively; v is the kinematic viscosity; d is the local water depth; z is the vertical location measured positive upward from the mean water surface; and e_1 and e_2 are the unit vectors in the horizontal and vertical directions, respectively.

The normal component of velocity at the bottom can be easily calculated as

$$u_{sn} = \frac{\kappa\alpha\sigma^2}{v \sinh kd} \{[-\sin\chi - s(x)\cos\chi]\sin\vartheta \qquad (2.2)$$
$$+ [\cos\chi - s(x)\coth kd \sin\chi]\cos\vartheta\}$$

where $\tan\vartheta = s(x)$.

Since $|s(x)| \ll 1$, u_{sn} can be approximated by

$$u_{sn} \approx \frac{\kappa\alpha\sigma^2}{v \sinh kd} [1 + \tfrac{1}{2}s^2(x)(\sin^2\vartheta$$
$$+ \cos^2\vartheta \coth^2 kd)] \cos(\chi + \vartheta + \varepsilon) \qquad (2.3)$$

with $\varepsilon = \tan^{-1}[\tan\vartheta(\sin^2\vartheta + \cos^2\vartheta \coth^2 kd)^{1/2}]$. This formula agrees with the results of Reid and Kajiura (1957) for the case $s(x) = 0$.

Equation 2.3 still represents an oscillatory motion; therefore net flow in and out of the bottom sediment must be zero. However, the amount of fluid Q_{sn} filtered by the porous media in the bed per unit area per unit time is not zero. It can be calculated as

$$Q_{sn} = \tfrac{1}{2} \frac{k\sigma}{(2\pi)^2} \int_0^{2\pi/k} \int_0^{2\pi/\sigma} u_{sn} \cdot \text{sgn}(u_{sn}) dx dt$$

$$= \frac{\kappa\alpha\sigma^2}{2\pi^2 v \sinh kd} [1 + s^2(x)(\sin^2\vartheta + \cos^2\vartheta \coth^2 kd)] \qquad (2.4)$$

A graphical solution of Q_{sn} for different wave conditions with $s(x) = 0$ was given in Riedl et al. (1972). That diagram was prepared for $\kappa = 1$ Darcy. However, since Q_{sn} increases linearly with κ for any given sediment, the

Table 2-3
Wave Length L (m) as a Function of Period T (s) and Depth d (m) according to Linear-Wave Theory.

T (s)	d (m)							
	5	10	20	30	40	50	100	200
<6	29	39.6	39.6	39.6	39.6	39.6	39.6	39.6
6–7	39.6	52	67	67	67	67	67	67
8–9	53.4	75	91.5	104	107	110	116	116
10–11	65.5	94.5	119	145	150	155	170	170
12–13	82.4	120	151	183	204	212	244	244
>13	104	146	187	234	262	274	338	356

Source: Riedl et al, *Marine Biology* 13 (3): 210–221, 1972. Reprinted with permission.

proper value can be obtained readily by multiplying the Q'_{sn} found in the diagram by a factor equal to the permeability in Darcys. Furthermore, since Q_{sn} stands for an average value over a wave length and a whole period, its value must be substantially lower than the maximum of u_{sn}. That difference is a factor associated with porosity. In addition to the adjustment discussed earlier, the flow speed decreases exponentially with depth in the sediment.

Computation of Subtidal Water Exchange

In order to compute Q_{sn}, the amount of fluid filtered per unit area, the wave parameters, the water depth, and the permeability of the bottom sediment must be known.

Wave parameters can be calculated from either wave period T or length L for a given depth (table 2–3) using linear wave theory. Knowing the probability of different classes of wave period or height for a given area, Q_{sn} for an interval of time can be calculated for a given depth or read from the graphical solution for Q_{sn}.

Local permeability must either be measured or may be computed with the empirical formula

$$\kappa[\text{Darcy}] \approx 760(GM)_\xi e^{2-1.31\sigma\phi}$$

where $(GM)_\xi$ is the geometric mean diameter of the sediment grains, and $\sigma\phi$ is the sorting in ϕ units.

As an example, water exchange on the North American shelf of the Atlantic Ocean between 27°N and 35°N (southern Florida to Cape Hatteras) was computed. The topography and sediments of that shelf and the wave spectrum above it are well known. To account for the differences in wave

Table 2-4
Probability (%) of Wave Condition Defined by Period T (s) and Height a (m) in Geographic Areas I to VIII (see Figure 2-10).

T (s)	Areas							
	a (m)	I	II	III	IV	V	VI	VII
>6	0.93	53.4	53.6	53.8	54.88	55.96	61.4	66.8
6–7	1.7	28.4	28.35	28.3	27.93	27.55	24.77	22.0
8–9	2.2	11.75	11.65	11.55	11.13	10.72	9.06	7.4
10–11	2.5	4.45	4.41	4.37	4.145	3.92	3.12	2.32
12–13	2.75	1.26	1.26	1.25	1.19	1.13	0.99	0.85
>13	3	0.74	0.735	0.73	0.725	0.72	0.675	0.63

Source: Riedl et al., *Marine Biology* 13 (3): 210–221, 1972. Reprinted with permission.

conditions, the shelf was subdivided into seven geographical areas (figure 2–10). The probabilities of waves of period groups and their dominant wave heights are given in table 2–4.

From the available data on mean grain size and sorting (figure 2–11), a permeability map could be constructed (figure 2–12). Most of the shelf plain shows values between 10 and 30 Darcys. Toward the continental slope, low-permeability isopleths tend to turn parallel to the isobaths.

Using the probabilities in table 2–4, water exchange or amount of water filtered per unit time and unit area can now be computed for different wave conditions. Also the average amount of water filtered over a given time span can be computed. Mean velocity perpendicular to the bottom/water interface is $2Q_{sn}$ because the actual exchange can be only 50 percent of the water passing up and down through the interface. Also, it is assumed that water leaving the interstices becomes mixed with the free water to such an extent that refiltration can be neglected.

In figure 2–13a, the Q_{sn} pattern for very small waves is shown ($T = 5$ s, $a = 93$ cm). The Q_{sn} isotachs follow the isobaths very closely, reflecting the rapid dissipation of smaller wave energies with depth. If we limit the effects of the subtidal pump to a Q_{sn} of 0.1 nm/s, the phenomenon is confined within the 200-m isobath (figure 2–13b).

The exchange pattern corresponding to maximum wave action (figure 2–13b; $T = 15$ s, $a = 3$ m) reflects the bottom permeability more closely, especially in the shallower part. Here Q_{sn} decreases only very little with increasing depth. On the outer shelf, however, Q_{sn} decreases sharply, and the isotachs again become parallel to the isobaths. More than 99 percent of the total water exchange is again confined within the 200-m isobath.

The pattern of $Q_{sn, tot}$ for an average year is depicted in figure 2–13c. Along the coast, $Q_{sn, tot}$ reaches 2,000 to 5,000 cm per year. The line for $Q_{sn, tot} = 10$ cm per year generally follows the 100-m isobath. If we limit the

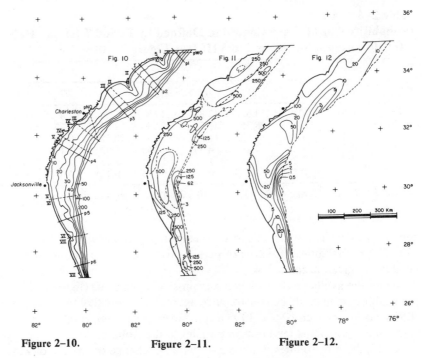

Figure 2–10. Figure 2–11. Figure 2–12.

Source: Riedl et al., *Marine Biology* 13(3):210–221, 1972. Reprinted with permission.

Figure 2–10. Bathymetry (m); geographic areas of differing wave actions.

Source: Riedl et al., *Marine Biology* 13(3):210–221, 1972. Reprinted with permission.

Figure 2–11. Grain size composition (solid line) in median diameter (μm), and sorting (dashed line) in $\sigma\phi\tilde{}$.

Source: Riedl et al., *Marine Biology* 13(3):210–221, 1972. Reprinted with permission.

Figure 2–12. Permeability in Darcys.

shelf to 200 m and the action of the subtidal pump to $Q_{sn,\ tot} = 0.1$ cm per year, 97.5 percent of the shelf experiences the action of the subtidal pump.

Magnitudes of Flow

For the portion of the North American shelf between 27°N and 35°N and from the coast down to the 200-m isobath, Q, or the annual total volume of water exchange, is

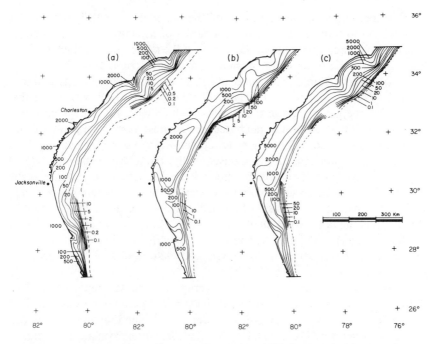

Source: Riedl et al., *Marine Biology* 13(3):210–221, 1092. Reprinted with permission.

Figure 2–13. Water exchange through surface of shelf sediments. Part (a) is computed on assumption of discrete, small, uniform wave conditions; (b) the same for long wave conditions; and (c) is the yearly mean computed from total wave spectrum and table 2-2.

$Q = 1306.4$ km^3 for the whole shelf, or

$Q = 1.117$ km^3/km of coastline

In terms of area, the volume of exchange is

$Q = 12.2$ m^3/m^2 of shelf per year, or

$Q = 33.2$ L/m^2 of shelf per day

Thus an amount equivalent to the complete volume of neritic water would be filtered in 3 years and 4 months. The water overlying the inner shelf down to the 30-m isobath, where the average value of Q is 0.017 km^3/km^2/yr could be completely filtered within 10 months under typical weather conditions.

Extrapolations for an estimate of the worldwide contribution of the subtidal pump have to be based on a series of assumptions since there is considerable uncertainty about the values of the necessary parameters: shelf area, shelf profile, sediment distribution, and wave conditions.

According to Menard and Smith (1966), the area between 0 and 200 m is 27.5×10^6 km^2; thus our studies have investigated 0.39 percent of that zone. The average profile of the world's shelf (Shepard, 1948) is narrower (average width, 78 km) and much deeper (greatest slope change at 132 m) than the profiles considered here (average width, 90 km; mean depth of greatest slope change at 55 m). Consequently, the average quantity filtered through the world's shelf will be much less. Average permeability of the world's shelf will be lower than on the eastern U.S. shelf, where gravel, shell, and sand constitute 92 percent of the sediment. The average contribution of those components to world shelf sediments is 52.48 percent (Hayes, 1967; Stoddart, 1969). Mean wave energy is probably on the low side in our example. However, that is the least well determined of the necessary parameters.

Assuming a minimum Q for an average square kilometer of shelf of $3,480 \times 10^3$ m^3 per year, we find that the subtidal pump moves a volume of 95,700 km^3 per year through bottom sediments. When that volume is added to the volume flowing through the intertidal filter system (see earlier discussion; Riedl, 1971c; Riedl and Machan, 1972), we calculate that 96,900 km^3 of seawater are filtered annually by porous marine sediments. The total water mass of the oceans would require 14,000 years to be so filtered. However, the sand filter system compares with other global water pathways (figure 2–14). For every 3.5 liter of seawater evaporated, about 1 liter passes through shelf or coast sediments. Sediment percolation volume is roughly equivalent to precipitation on land, but it occurs on only one-fifth of the area. Consequently, the yearly percolation on a unit area is more than 5 times higher than on land.

Bioclimatological Implications

Effects of Actual Water Speed

Water velocity in most sediments is generally on the order of magnitude of nanometers per second. Maximum current velocities v up to 2 mm/s were recorded in fully saturated sand in the intertidal zone. However, maximum speed of water displacement u_{sn} in the subtidal zone reached only 40 nm/s. Although those values appear to be low compared with the velocities found in the free-water column, they represent fairly high velocities for interstitial organisms living in the pore space of sediments.

Porous Sediments 51

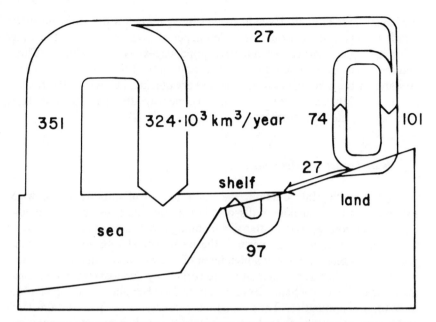

Source: Riedl and Machan, *Marine Biology* 13(3):179–209, 1972. Reprinted with permission.

Figure 2-14. Contribution of the subtidal pump to amounts involved in global water exchange. Values expressed 10^3 km^3/yr. Shelf data are from this chapter; other data are from Dietrich (1957).

The reason concerns velocity distribution in narrow channels (Scheidegger, 1960; Prandtl, 1966; Schlichting, 1968; De Wiest, 1969) and the relative size of many of the interstitial organisms compared with the width of the channels in which they live. Unlike small organisms at the sediment/water interface on the sea floor, they cannot hide in a well-developed boundary layer with velocities several orders of magnitude smaller than the free-water speed. In the interstitial channels they are fully exposed to the currents measured. Furthermore, owing to their small size, most of the mesopsammic organisms are exposed to currents that pass across their body length in less than 1 second.

The pulsing phenomena bring rapid changes in water velocity. At high tide at the center of the entrance to the filling bag, amplitudes up to $a = 2$ mm/s are common. When $v(Q) < a/2$, the pulsing current starts to reverse its direction, and the stress to which meiofauna may be exposed becomes especially obvious. In addition, the irregularities of the surf pulsing make adaptations of behavioral patterns to changes of speed and direction of water movement very difficult. Even in species living far below the sediment/water interface, where there would be little danger of being swept into the water column, a great variety of adhesive appendages have developed. Those features seem to be adaptations to "stormy" interstitial water movements. Apparently the adhesive appendages act very quickly, i.e., within fractions of seconds, and thus allow the organisms to cope with strong and unpredictable changes of water movement.

Effects of Transport Mechanisms

As stated in the Introduction, the main biological importance of water movement is the transport of dissolved or particulate substances. Although transport of both organic matter and metabolic wastes by water is important to interstitial communities, perhaps the most critical substance thus transported is oxygen. Because of a high input of organic matter into the sediment by such processes as sedimentation, reworking by endopsammic organisms, and descending food chains, oxygen can, in fact, become a limiting factor in interstitial communities. Thus oxidative processes stop at a certain depth under the sediment surface, below which reducing compounds accumulate. The boundary between the oxidized surface, with predominantly aerobic metabolism, and the deeper anoxic layers, with anaerobic metabolism, can be determined as the depth of the greatest rate of change of Eh and is called RPD (redox potential discontinuity) layer (Fenchel and Riedl, 1970).

The RPD layer has a characteristic mean depth, which can now be predicted in a given situation using water depth, sediment composition, wave energy, and, in the case of intertidal sediments, beach profile (Fenchel and Riedl, 1970; Riedl and Ott, 1971). Whereas in protected locations in the intertidal zone the RPD may even reach the sediment surface, it is absent from exposed beaches where the strong water movement associated with the migrating filling bag and the tidal filling and draining of the beach supply excess oxygen. Here the residence time of interstitial water is so short that only about 25 percent of the imported organic matter is oxidized during passage (J.R. Hall, personal communication). That narrow strip of exposed beach, extending with the output field toward the subtidal, has been named the *high-energy window* (Riedl and MacMahan, 1969).

As a consequence of the variations in intertidal water flow, the RPD shows a considerably dynamic pattern. Long-term Eh recordings and field experiments with special equipment (Machan and Ott, 1972) demonstrated that surface Eh dynamics are largely controlled by percolation and drainage even in very fine sand. In coarser sand, the thickness of the oxidized layer may double during a tidal cycle (Ott and Machan, 1971). In the subtidal zone, variability of Q_{sn} will account for the dynamics of the RPD, although on a different time scale. Those differences increase with depth. Assuming equal permeability, the variation of the mean wave condition from $T < 6$ to $T > 13$ results in an increase in Q_{sn} by a factor of 1.3 in 5 m water depth and by a factor of 38 in 50 m water depth. Sediment at greater water depths on the shelf will experience increasingly long periods of stagnation, with a corresponding rise of the RPD toward the sediment/water interface.

Seasonal differences in mean RPD depth in intertidal sand flats, as described in Riedl and Ott (1971), may also be attributed to the seasonally changing wave spectra. Those dynamics must have a profound effect on the interstitial meiofauna, which show surprisingly narrow tolerance limits (Wieser et al., 1974). Therefore, migrations are a common reaction in interstitial populations (for a classification of types see Rieger and Ott, 1971).

References

De Wiest, R. 1969. *Flow through Porous Media*. New York: Academic Press.

Dietrich, G. 1957. *General Oceanography, an Introduction*. New York: Wiley.

Eagleson, P.S., and Van De Watering, W. 1964. A thermistor probe for measuring particle orbital speed in water waves. U.S. Army Cstl. Engng. Res. Center, Washington, D.C., *Tech. Memo.* 3:1–44.

Fenchel, T.M., and Riedl, R.J. 1970. The sulfide system: A new biotic community underneath the oxidized layer of marine sand bottoms. *Mar. Biol.* 7:255–268.

Hayes, M.O. 1967. Relationship between coastal climate and bottom sediment type on the inner continental shelf. *Mar. Geol.* 5:111–132.

Huang, N.E. 1970. Mass transport induced by wave motion. *J. Mar. Res.* 28:35–50.

Hunt, J.N. 1959. On the damping of gravity waves propagated over a permeable surface. *J. Geophys. Res.* 64:437–442.

Lumley, J.L. 1962. The constant temperature hot thermistor anemometer. In *Symposium on Measurement in Unsteady Flow, Worcester, Mass.*, pp. 75–82.

Machan, R., and Ott, J.A. 1972. Problems and methods of continuous in situ measurements of redox potentials in marine sediments. *Limnol. Oceanogr.* 17:622–626.

Menard, H.W., and Smith, S.M. 1966. Hypsometry of ocean basin provinces. *J. Geophys. Res.* 71:4305–4325.

Murray, J.D. 1965. Viscous damping of gravity waves over a permeable bed. *J. Geophys. Res.* 70:2325–2331.

Odum, H.T. 1971. *Environment, Power and Society*. New York: Wiley-Interscience.
Ott, J.A., and Machan, R. 1971. Dynamics of climatic parameters in intertidal sediments. *Thalassia, Jugoslavia* 7(1):219–229.
Phillips, O.M. 1966. *The Dynamics of Upper Ocean*. London: Cambridge Univ. Press.
Prandtl, L. 1966. *Essentials of Fluid Dynamics*. London: Blackie.
Putnam, J.A. 1949. Loss of wave energy due to percolation in a permeable sea bottom. *Trans. Am. Geophys. Un.* 30:349–356.
Reid, R.O., and Kajiura, K. 1957. On the damping of gravity waves over a permeable sea bed. *Trans. Am. Geophys. Un.* 38:662–666.
Riedl, R.J. 1969. Marinebiologische Aspekte der Wasserbewegung. *Mar. Biol.* 4:62–78.
Riedl, R.J. 1971a. Water movement: General introduction. In O. Kinne (Ed.), *Marine Ecology*, Vol. 1. London: Wiley, pp. 1085–1088.
Riedl, R.J. 1971b. Water movement: Animals. In O. Kinne (Ed.), *Marine Ecology*, Vol. 1. London: Wiley, pp. 1123–1149.
Riedl, R.J. 1971c. How much seawater passes through intertidal interstices? *Int. Rev. Ges. Hydrobiol.* 56:923–946.
Riedl, R.J. 1971d. Energy exchange at the bottom water interface. *Thalassia, Jugoslavia* 7(1):329–339.
Riedl, R.J., and Forstner, H. 1968. Wasserbewegung im Mikrobereich des Benthos. *Sarsia* 34:163–188.
Riedl, R.J., Huang, N., and Machan, R. 1972. The subtidal pump, a mechanism of interstitial water exchange by wave action. *Mar. Biol.* 13:210–221.
Riedl, R.J., and Machan, R. 1972. Hydrodynamic patterns in lotic intertidal sands and their bioclimatological implications. *Mar. Biol.* 13:179–209.
Riedl, R.J., and MacMahan, E. 1969. High energy beaches. In H.T. Odum et al. (Eds.), *Coastal Ecological Systems of the United States*. Washington, D.C.: Federal Water Pollution Control Administration, pp. 197–269.
Riedl, R.J., and Ott, J.A. 1971. The suction corer, a device to yield electric potentials in coastal sediment layers. *Senckenberg. Marit.* 2:67–84.
Rieger, R., and Ott, J.A. 1971. Gezeitenbedingte Wanderung der Turbellarien und Nematoden eines nordadriatischen Sandstrandes: 3rd Europ. Symp. Mar. Biol. (Arcachon), 1968. *Vie Milieu (Suppl.)* 22:425–447.
Scheidegger, A. 1960. *The Physics of Flow through Porous Media*. New York: MacMillan.
Schlichting, H. 1968. *Boundary Layer Theory*. New York: McGraw-Hill.

Shepard, F.P. 1948. *Submarine Geology*. New York: Harper and Row.
Stoddart, D.R. 1969. World erosion and sedimentation. In R.J. Chorley *(Ed.), Water, Earth, and Man*. London: Methuen, pp. 43–64.
Wieser, W., Ott, J.A., Schiemer, F., and Gnaiger, E. 1974. An ecophysiological study of some meiofauna species inhabiting a sandy beach at Bermuda. *Mar. Biol.* 26:235–248.

3
Oscillation of Continental Shelf Sediments Caused by Waves

J.N. Suhayda,
J.M. Coleman,
Thomas Whelan,
and *L.E. Garrison*

Abstract

Measurements have been made of the oscillations of bottom sediments on the continental shelf induced by the passage of surface waves. A wave staff and pressure sensor were placed 45 m from a bottom-emplaced accelerometer in East Bay, Louisiana. Measurements were made in 20 m of water in an area where bottom sediments were composed of clays and silts. A sediment core was taken to a depth of 40 m. The results of the experiments indicate that these fine-grained bottom sediments move in a wave-like fashion under surface-wave action. Bottom oscillations on the order of 2 to 3 cm occurred under waves having a height of 1 m and a period of 5 seconds. The bottom motion appears to be an elastic-like response to wave pressures. Estimates of the amount of wave energy lost in forcing the mud wave indicate that the interaction can significantly affect surface-wave characteristics and the stability of bottom sediments.

Introduction

Bottom sediments on the continental shelves of the world are subject to the continual action of surface waves. Surface waves cause a variation in pressure on the bottom that affects the stability of bottom sediment, particularly in clays and muds. Previous work has suggested waves may cause or trigger submarine landslides on bottom slopes that are considerably smaller than the maximum allowed by gravitational equilibrium. While the interaction of waves and muds has been studied in the laboratory (e.g., Mitchell et al., 1972) and in theoretical studies (Gade, 1958, 1959; Wright and Dunham, 1972; Wright, 1976; Mallard and Dalrymple, 1977), few field studies have been conducted. This chapter presents the results of field studies involving direct observations of sediment properties, wave-induced pressure forces, and subsequent bottom movements.

Several important basic concepts that explain the role of waves in causing submarine mass movements were presented by Henkel (1970), who noted that submarine landslides have occurred on the Mississippi Delta between water depths of 9 and 63 m, on bottom slopes as low as 0.008, or 1/125. Measured values of sediment shear strength used in an infinite slope-stability analysis indicated that the soils should be stable under gravitational forces alone up to slopes of 0.032 to 0.048. Henkel shows through an arc-failure analysis that wave-induced bottom pressures could cause shear failure of the bottom sediments to a depth of tens of meters below the mudline in water depths as great as 130 m. Waves were presumed to impose an oscillatory motion on the soft sediments which would weaken the sediments and lead to downslope motion. Subsequent work has tended to verify Henkel's basic scheme.

Mitchell et al. (1972) evaluated the stability of a bentonite clay under wave action in a laboratory wave tank. It was found that an infinite slope-type failure in the clay occurred where the initial point of failure was at some finite depth below the mudline. A critical wave-induced bottom pressure was defined that caused failure, and that value was found to be about 50 to 70 percent of the value predicted by Henkel (1970). Doyle (1973) presented the results of a laboratory study in which soils were observed under influence of waves. The bottom sediments moved in a wave-like fashion, undergoing an orbital path at the mudline. The vertical component of the orbital motion decayed more rapidly with depth than did the horizontal motion, until at some depth only back-and-forth movement existed. Remolding of the soil took place, and a failure zone was gradually developed. Doyle also found that the wave-induced bottom pressures were less than predicted using rigid-bottom wave theories.

The purpose of this study was to examine, in a field experiment, the characteristics of the orbital wave-like motion of bottom sediments under wave action. The orbital motions are the first reaction of bottom sediments to the waves, apparently occurring before failure of the sediments. Their amplitude and phase relation to the pressure wave are clues to how the sediments are behaving in nature. For a soil behaving elastically, the soil movements should be 180° out of phase with the pressure wave, while for a viscous fluid-like behavior the movements should be nearly in phase with the pressure wave. Such oscillations could have an important effect on the hydrodynamics of the wave motion above the bottom.

The experiments described in this chapter were conducted as a part of a multidisciplinary investigation of the mechanisms responsible for mass movement of sediments in the shelf region of the Mississippi Delta. The project involved geologic, oceanographic, and geochemical studies of a common field site. Geologic studies (Coleman et al., 1974; Garrison, 1974) indicated that there was considerable variability in the characteristics of the mass movement of sediment in the Delta region. The experimental site chosen for the study was a featureless area of East Bay, Louisiana, because a more complex setting would make interpretation of the data too difficult.

Forces on Sediments

Deposited sediments can be affected by a number of natural hydrodynamic forces, several of which occur in a natural shelf area like the upper Mississippi Delta. Bottom pressures induced by surface waves have been identified as a principal force, and bottom forces produced by internal waves are important as well. The magnitudes of the forces, both average and extreme, and their length and time scales are of equal interest. In general, the forces on the bottom sediments can be separated into surface-normal (pressure) forces and shear (drag) forces.

Storm surge, tides, and river floods are all water-level changes having spatial scales of several tens of kilometers. Within the Delta they represent a broad-scale raising and lowering of the mean water level, having an amplitude as large as 0.5 to 1 m offshore (water depth $\sim$ 150 m) and 3 m nearshore. Such depth changes correspond to bottom pressure changes of 300 Pa (or 3.06 g/cm^2) to about 3000 Pa (or 30.6 g/cm^2). High river stages or floods are seasonal, whereas storm surges have time scales of 1 to 2 days. Tides vary on periods of about 12 to 24 hours.

Internal waves can occur in the Delta region because of the presence of several water masses having different densities. Periods for internal waves range from several hours to a few minutes. Well-defined internal waves having periods of 1 and 15 min have been observed emigrating from South Pass. Although knowledge of wave height is meager, typical values would appear to be 1 m, but an extreme value might be as large as 3 m. The pressure force developed by these waves would be smaller than for an equivalent surface wave because the density change across the water-mass boundaries is a small fraction of the density of water. Variations in density of about 2 kg/m^3 are typical. Thus the pressure forces owing to internal waves would be in the range of 170 Pa (or 1.73 g/cm^2) to a maximum of about 570 Pa (or 5.81 g/cm^2). The lengths of those waves range from about 30 m to several kilometers.

Surface waves can produce large changes in bottom pressure. The waves are actually a complex mixture of waves of various periods and heights, which can be separated into long-period waves (infragravity) and wind waves (Munk, 1962). Infragravity waves range in period from 10 min to 30 sec, whereas wind waves and swell have periods of 30 to 1 sec. The infragravity waves will be considered first in some detail.

Infragravity waves result from storm systems, surf action, and underwater landslides. Typically their amplitude is small, about 15 cm; however, under storm conditions, waveheights can reach 0.1 of the storm-wind waveheight. The importance of those long waves can be illustrated by comparing their bottom pressures to surface-storm wave pressures. For example, consider the storm wave to be 13 sec in period and 15 m in height and the infragravity wave to be 100 sec and 1.5 m. The bottom pressures predicted using linear wave theory for several depths of water are shown in table 3–1. It can be seen from the table that the long-wave pressures are relatively constant in water depths less than 300 m. The pressures actually are larger than storm wind-wave pressures for depths greater than about 128 m. The lengths of long waves range from several kilometers in deep water to about 1 km in shallow water.

Waves and water currents can also induce surface drag forces on bottom sediments. Those forces are generally expressed as a function of the square of the current velocity:

$$\tau = \rho C_d U^2 \tag{3.1}$$

where τ is the drag force per unit area (or tangential stress), ρ is the density of water, C_d is a drag coefficient, and U is the water particle velocity away from the boundary layer. There are other forms of the drag law, however, this form will serve to illustrate that drag forces are expected to be quite small generally. The drag coefficient is a function of bottom geometry (roughness), fluid viscosity, and flow condition of the water; however, the values are on the order of 10^{-2}.

Ocean currents on the shelf are quite variable, primarily resulting from density effects, tides, and wind. In the Gulf of Mexico, the most severe currents evidently are associated with hurricanes, during which currents are reported to be up to 5 m/s (11 knots) on the surface and about 1 m/s near the

Table 3–1
Bottom Pressures

Depth (m)	Storm Wave Pressure[a] (kPa)	Infragravity Wave Pressure[b] (kPa)
30	14.10	15.2
60	6.13	15.1
90	3.22	15.0
120	1.69	14.9
150	7.7	14.8
300	0.15	14.4

[a]$H = 15.2$ m; $T = 13$ s; $L_\infty = 164$ m.
[b]$H = 1.52$ m; $T = 100$ s; $L_\infty = 15,610$ m.

bottom in 90 m of water. Using the bottom value, the maximum drag predicted would be about 9.6 Pa (or 98.0 mg/cm^2); that value could be larger near shore, but weaker at the shelf edge.

A maximum value of drag resulting from waves can be estimated using equation 3.1, where the maximum orbital velocity on the bottom is on the order of

$$U_{max} = \frac{\pi H}{T \sinh kh} \qquad (3.2)$$

Here, wave number k is $2\pi/L$, L is wavelength, H is the wave height, T is the wave period, and h is the water depth. For a wave height of 9 m and period of 9 sec, the drag force at a depth of 30 m is 19 Pa (or 193.9 mg/cm^2). In 12 m of water, the force is 96 Pa (or 979.6 mg/cm^2). Therefore, the major force over most of the continental shelf is a wave-derived pressure force. Other pressure forces are generally an order of magnitude smaller. In deep water (7,150 m), long-period infragravity waves tend to dominate. Drag forces, whether produced by waves or currents, are generally quite small.

Field Site

This study was conducted in East Bay, Louisiana. The instrumented site was located in Block 28, near the center of East Bay, in a water depth of about 20 m. The regional bathymetry and location of the field site are shown in figure 3–1. Bottom slope in the area is approximately 0.005. The site is exposed to moderate wave action (wave height $\sim$ 0.5 m) much of the year. However, during winter storms, wave heights increase to about 1 to 2 m, and during the summer hurricanes can generate waveheights in the range of 4 to 6 m. The suitability of the site for an initial and controlled experiment can be seen by considering the regional setting.

Previous studies (e.g., Coleman et al., 1974) have indicated that several deformational or mass-movement processes are actually occurring on the shelf. The techniques used were sediment borings, high-resolution seismic profiles, side-scan sonar tracks, and repeated hydrographic surveys. Deformational features include (1) peripheral faulting and slumping, (2) radial graben and tensional faulting, (3) differential weighting and diapirism, (4) degassing and mass flowage, (5) deep-seated clay flowage (mud noses), (6) shelf-edge rotational slumps and normal faults, and (7) deep-seated normal faults. The distribution of those features within the Mississippi Delta region (figure 3–2) shows that distinctive patterns of occurrence may be correlated with water depth and proximity to the river passes. The different types of mass movements have diverse characteristics and probably do not result

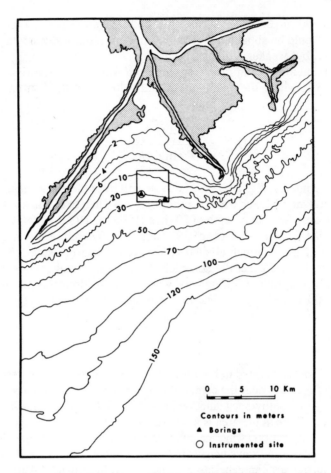

Figure 3-1. Location of field site in East Bay, Louisiana.

from the same causative factors; interpretations of process and form can be difficult. As an initial effort, our study was conducted at a field site where surveys indicated that no bottom feature was present. Absence of features represents the condition of the bottom sediments before deformation (for whatever reason) takes place.

A boring was made at the field site to a depth of 40 m below the mudline, and cores have been analyzed for geological structure, sediment characteristics, and geochemical properties (figure 3–3) (Whelan et al., 1975; Roberts et al., 1976). The following information is taken from Roberts et al. (1976).

The sedimentary sequence of the boring is divided into two very distinct zones: a well-laminated, relatively undisturbed unit accompanied by biotur-

Source: Whelan et al., 1975.

Figure 3-2. Location map showing core locations and the spatial relationship of deformational patterns in the delta platform.

bation features, evidence of secondary mineralization, and only minor gas-related disturbances (0 to −20 m); and a basal zone exhibiting numerous gas-bubble and expansion features, generally massive bedding, and a decrease in evidence of biological activity (−20 to −40 m).

The first 20 m of sediments illustrate characteristics of a sequence that has slowly accumulated by deposition of clays and periodic introductions of silty clay, resulting in a well-stratified unit. Within this upper zone, water content generally decreases from 90 to 100 percent near the mudline to approximately 50 to 60 percent at about −20 m. Shear strength increases from 4.8 kPa (or 49.0 g/cm^2) at the mudline to greater than 23.9 kPa (or

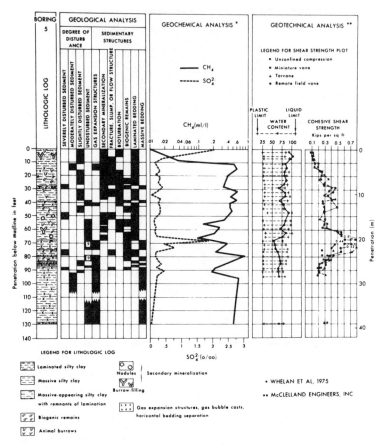

Source: Roberts et al., 1976.

Figure 3-3. Geological, geochemical, and geotechnical data of core taken in East Bay.

243.9 g/cm^2) at the base of the upper unit. Sulfate levels are high at the surface but diminish rapidly to a value of approximately 0.2 ‰ at −3 m, where CH$_4$ values rapidly rise to nearly 0.6 ml/L. A relatively high gas content is characteristic of the remainder of the boring with the exception of the base of the upper zone (∼ −20 m) where SO$_4^{2-}$ levels sharply increase and CH$_4$ content drops to about 0.1 ml/L. Accompanying the rapid geochemical changes is an indication on the x-ray radiographs that silt content of the sediments increases substantially. Most of the upper zone reveals a lack of coring disturbance and a wide variety of thin laminations. bioturbation structures, and delicate diagenetic features, which are generally missing from the more massive underlying unit.

The characteristics of the basal unit (-20 to -40 m), such as massive bedding, numerous gas expansion features, and other disturbances, suggest a pro-delta sedimentary environment. Biogenic remains are sparse, and the evidence of bioturbation is infrequent. Methane content is generally high, and SO_4^{2-} values are expectedly low. Unfortunately, geotechnical data are not available for part of the basal zone. At the upper boundary of the zone, shear strengths up to 33.5 kPa (or 341.8 g/cm^2) define a crust. Below the crust zone, shear strengths quickly decrease with depth to approximately 14.4 kPa (or 146.9 g/cm^2) at the base of the zone. Sedimentation rates for the basal zone appear to have been more rapid than those for the well-laminated and generally less-deformed upper zone.

Therefore, sediments at the field site are nearly uniform clays or silty clays and have very low shear strengths. Methane gas is present within the sediments and is also spread throughout the sediment column. For a natural field site in the Delta, such sediment properties are remarkably simple and uncomplicated. Inclusions of sand lenses or shell layers, large changes in shear strengths with depth, and highly variable gas concentrations within the sediment column are commonly found associated with deformational features (Whelan et al., 1975). Considering the natural variability in sediments, our field site was well suited to initial detailed experiments.

Data Acquisition

Although techniques for measuring wave characteristics are well established, the measurement of bottom movement was complicated by two factors. First, the measurements had to be made away from a platform to ensure that the motion of natural muds would be measured. Second, bottom motions under typically encountered wave conditions were thought to be small, and therefore high resolution (on the order of 0.1 cm) was needed. Both problems were overcome by burying accelerometers in the mud. Though displacements of the bottom were small, accelerations were such that they could be reliably measured and required no fixed reference.

The technique of using accelerometers to measure soil movement is not new; however, most of the previous work has been done on land and has been concerned with frequencies higher than those associated with ocean waves. Our device used a cluster of three Bruel-and-Kjaer-type 8306 accelerometers mounted to measure acceleration in three dimensions. Voltage sensitivity of the accelerometer system was 3.0 V/g from 0.2 Hz to 1,000 Hz and was down 3 db at 0.03 Hz. The maximum sensitivity of the accelerometers to transverse accelerations was typically 3 percent of the sensitivity along the main axis of the accelerometer. The system was designed to allow the measurement of periodic as well as nonperiodic motion.

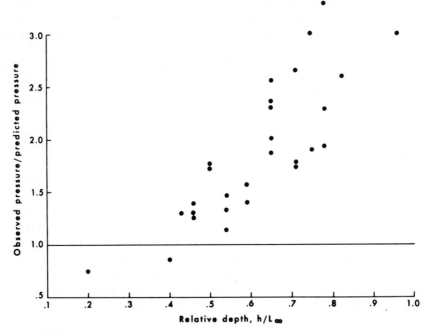

Source: Suhayda, *Marine Geotechnology*, Vol 2: *Marine Slope Stability*, 1977, pp. 135–159. Reprinted with permission.

Figure 3–4. Ratio of the observed and predicted wave pressure shown as a function of relative water depth.

The accelerometers were confined in a waterproof cylindrical package 21 cm in diameter and 63 cm long. Submerged weight was 53 newtons (or 12 lb). It was placed about 45 m from the wave sensors, and the top of the package was about 15 cm below the mudline. The electronic cable was given 5 m of slack and then fixed to taut galvanized cable, which was laid along the bottom between a nearby well jacket and the platform.

Wave characteristics were measured with a wave staff and a pressure sensor. A system of winches and pulleys allowed the pressure sensor to be placed at any depth in the water column on the windward side of the platform. The distance of the wave sensors from the platform legs varied from 5 m at the surface to 2 m at the mudline.

Results and Discussion

Wave-Induced Pressures

Because of the importance of wind-wave–induced bottom pressures in water depths less than 150 m, those pressures were measured during this study.

The wave-induced pressures decrease with depth and are generally taken as given by linear theory (U.S. Army Coastal Engineering Research Center, 1973) and expected to be in phase with the surface waves. Surprisingly, those expectations were not met. The results of several comparisons of observed wave pressures and waveheights showed that pressures can be much larger than predicted. The data are shown in figure 3-4. The ratio of the observed pressure to that predicted by linear theory is given in the figure as a function of the parameter h/L_∞ or the relative depth of the wave, where h is the water depth and L_∞ is the deepwater wavelength. The data indicate that pressures may be lower than predicted for $h/L_\infty \gtrsim 0.4$; however, there were little data in the region. For $h/L_\infty > 0.4$, observed pressures were up to 3 times as large as predicted. Such extreme differences occur where pressures are a very small fraction (~ 1 percent) of the surface wave height. For h/L_∞ in the range 0.4 to 0.6, where pressures are 15 to 5 percent of the surface waveheight, the difference averages 35 percent larger.

Departures from predicted pressures have also been observed in field and laboratory experiments (Homma et al., 1966). In a lab study, it has been noted that measured pressures were about 27 percent of those predicted. The lower measured pressures were attributed to a viscous, fluid-like condition of the bottom sediment (Doyle, 1973). Investigation is now underway into the effect of a non-rigid bottom on wave-induced pressures.

Bottom Movements

Several sets of data involving simultaneous measurements of wave pressure and bottom accelerations were obtained. The conditions ranged from near calm (waveheight < 15 cm) to storm conditions (significant waveheight ~ 1 m and individual wave crests up to ~ 2.5 m). Bottom motions, when visible above background noise, showed a sinusoidal form and had the same general appearance as a wave record. Motions in the vertical direction were always definitely larger than horizontal motions. No results can be reported at this time concerning nonperiodic lateral or vertical movements.

A typical example of the bottom motion observed under moderate wave action (wave height = 1 m, wave period = 5 s) is shown in figure 3-5, which covers 200 seconds of the record of vertical displacement. Aside from the general periodic wave-like appearance of the oscillations, the outstanding feature of the data is the significant time variation of the amplitude of the oscillations. Groups of bottom oscillations occur that involve 2 to 6 cycles having amplitudes 2 to 3 times the amplitudes of the background motion. Those bursts of oscillations occur at intervals of 30 to 100 seconds. A similar pattern was found in both horizontal components of displacement.

The occurrence of groups of bottom oscillations can be explained by the coupling of bottom motion and wave-induced bottom pressures. For example, consider the simultaneous measurements of wave pressure and bottom motion shown in figure 3-6. The pressure sensor was placed directly over the

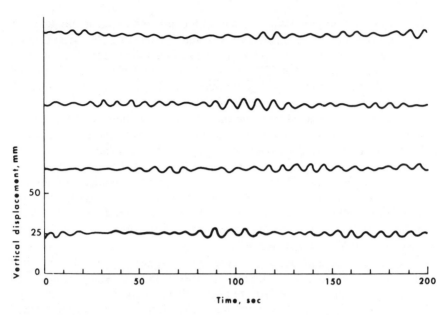

Source: Suhayda, *Marine Geotechnology, 1977.* Reprinted with permission.

Figure 3–5. Record of the vertical displacement of the bottom under wave action.

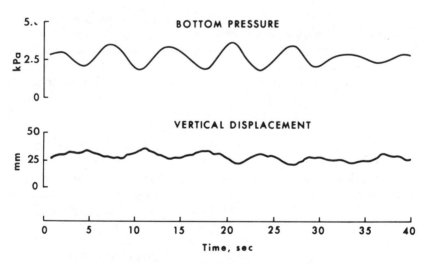

Source: Suhayda, *Marine Geotechnology*, 1977. Reprinted with permission.

Figure 3–6. Simultaneous record of wave pressure and vertical bottom displacement. Note that the crest of the mud wave occurs at a trough in the pressure wave.

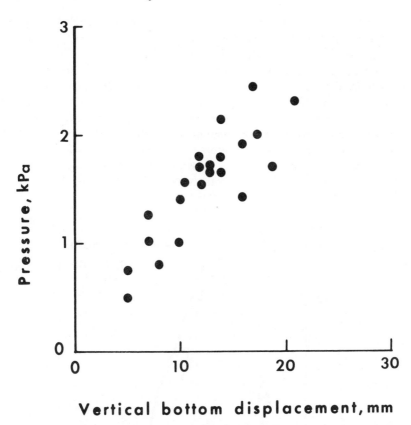

Source: Suhayda, *Marine Geotechnology*, 1977. Reprinted with permission.

Figure 3-7. Relationship of the amplitude of bottom pressure and bottom displacement.

accelerometer. The vertical displacement can be seen to be very nearly 180° out of phase with the wave pressures; i.e., the bottom is forced down under a bottom pressure crest. The correlation of vertical bottom displacement and a bottom pressure change is shown in figure 3-7. Although the range in bottom pressures is limited, there is indication of a linear relationship. Therefore, groups of large bottom oscillations occur because large waves and hence large wave pressures occur in groups. Those larger pressures force a proportionately larger bottom response.

The apparent linear relationship between bottom displacement and wave pressure is interesting; however, being based on a wave-by-wave comparison, it may be misleading. Surface waves and wave-induced pressures are known to have a complex spectrum involving a range of wave frequencies. The nature of the bottom response to waves is more fully described by comparing frequency spectra of wave pressure and bottom motion. Such an example is shown in firgure 3-8. The spectra have band widths of 0.0125 Hz based on sampling records 800 seconds long at 1 sample per second. The spectrum of the vertical displacement and of the wave pressures over the frequency range 0.04 to 0.30 Hz (period range 25 to 3.3 seconds) are shown. The spectra show very similar forms, having a peak at the same frequency (0.14 Hz) and a similar decrease in variance density for higher frequency. However, for frequencies below the peak, bottom displacements are relatively much greater. For those waves, the bottom displacement is about 3 times as great for the same amplitude of bottom pressure change as at the peak frequency.

Three-dimensional bottom motions were also observed. An example of

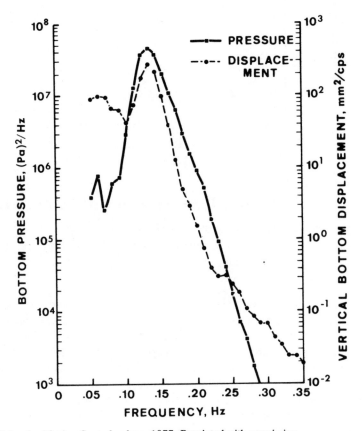

Source: Suhayda, *Marine Geotechnology*, 1977. Reprinted with permission.

Figure 3-8. Spectrum of the frequency of oscillations in bottom pressure compared with the spectrum of frequencies of bottom displacements.

Oscillation Caused by Waves

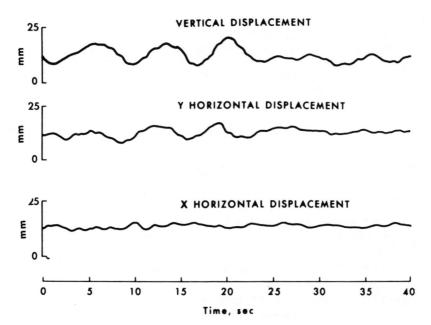

Source: Suhayda, *Marine Geotechnology*, 1977. Reprinted with permission.

Figure 3-9. Simultaneous records of horizontal and vertical displacements of bottom sediments.

the movements recorded on all three channels is shown in figure 3-9. It appears that horizontal motions are nearly 90° out of phase with the vertical motions, but the phase is a lag. A backward horizontal movement, as shown in the Y-component displacement, occurs at the crest of the bottom wave. Thus a retrograde wave-like motion of the bottom similar to a Rayleigh wave seems to be occurring. The ratio of vertical to horizontal displacement over several sets of data averaged approximately 2.0.

Wave Energy Loss

The forcing of a mud wave by the surface-wave–induced pressures results in a net energy loss from the waves. The average energy transmitted through the sea/sediment interface per unit area over one wave cycle (Gade, 1958) is

$$Dm = \frac{1}{T} \int_0^T P \frac{dh}{dt} \, dt \qquad (3.3)$$

where T = wave period
P = wave-induced bottom pressure
dh = an infinitesimal increase in the height of the interface

The general characteristics of the data show that the following functions will accurately describe the motions:

$$\frac{P}{\rho g} = \frac{P_0}{\rho g} + A \cos(kx - \sigma t) \qquad (3.4)$$

$$h = h_0 + MA \cos(kx - \sigma t + \theta) \qquad (3.5)$$

where

P_0 = steady-state bottom pressure
A = amplitude of the wave-induced bottom pressure
h_0 = depth of mud over which motion occurs
M = proportionality constant between the amplitudes of the mud wave and the pressure wave
θ = phase angle between the crest of the bottom pressure wave and the crest of the mud wave

After combining equations 3.4, 3.5, and 3.3 and integrating, and then using linear theory to interpret bottom pressures in terms of surface waveheight, the equation for the rate of average energy loss to the bottom is obtained (Tubman and Suhayda, 1976):

$$Dm = \frac{\pi \rho g M H^2 \sin \phi}{4T \cosh^2 kh} \qquad (3.6)$$

where $\phi = 180° - \theta$

From equation 3.6 it can be seen that the dissipation of wave energy by the soft bottom involves only two important factors, determined by the physics of the sediment movement: (1) the relationship between the pressure force on the sediment and the resultant vertical displacement, given by M, and (2) the phase angle between the crest of the pressure wave and the trough of the mud wave, given by ϕ.

The amount of energy transferred to the bottom sediments can be considerable. At the field site occupied in this study under normal wave action (H = 1 m, T = 8 s) the energy density of the waves is 0.12 J/cm^2, while the energy lost, as given by equation 3.6, equals 0.3 mW/cm^2. That energy-loss rate is about 0.2 percent of the average energy density of the waves. The wave loses about 1.6 percent of the energy each wave period. Under extreme conditions, for example during a hurricane (H = 15.3 m, T = 13 s), the energy input from the waves to the bottom may be as large as

0.2 W/cm², or about equal to the energy input to the atmosphere by the sun (0.14 watts/cm²).

Conclusion

Measurements of accelerations in bottom sediment in the Mississippi Delta indicate the surface waves are, even under moderate wave action, forcing an oscillation of bottom sediments. That wave-like movement appears most closely to resemble a surface wave on an elastic half-space or a Rayleigh wave. The bottom wave is nearly out of phase with the pressure wave, and the vertical motion is larger than the horizontal motion.

Such results provide more support for the role of waves in producing shelf-sediment instability, as proposed by Henkel (1970). Under extreme conditions, storm waves having a height of 15 m and a period of 13 s would produce bottom pressures of 7,600 Pa (or 77.6 g/cm²) in water depths of 150 m. Under those conditions, bottom sediments could be oscillating as much as 0.07 m in 150 m of water depth and 0.20 m in 30 m of water, based on the observations under average conditions.

Henkel (1970), in the conclusion of his paper, states that the submarine stability problem "is complex and that its satisfactory solution depends upon the acquisition of data on the actual bottom pressures found on the sea bed, more details of the shear strength distribution with depths, and the solution of the energy transfer problem between waves and the soft submarine sediments." While some of that work has now been done, the problem of understanding wave/sediment interactions on the shelf is still far from being solved.

It would seem that the most critical problem in need of immediate solution is the identification of the particular sets of processes and soil conditions causing the variety of deformational features observed in the Mississippi Delta. Until that conceptual framework is established, there will be no basis for evaluating the generality of any detailed experimental work or theoretical models that attempt to explain or predict sediment stability on the continental shelf.

Acknowledgments

This study was conducted with the U.S. Geological Survey. Thanks are due to Shell Oil Company for cooperation during the field work. The accelerometer system was built by Michael Tubman, who also provided technical assistance with data acquisition and analysis. CSI salary support was provided by the Geography Programs, Office of Naval Research, Arlington,

Virginia 22217, through a contract with Coastal Studies Institute, Louisiana State University. Parts of this chapter were modified from Suhayda et al. (1976) and are reprinted with permission.

References

Coleman, J.M., Suhayda, J.N., Whelan, T., and Wright, L.D., 1974. Mass movement of the Mississippi River delta sediments. *Transactions of the Gulf Coast Assoc. Geol. Soc.* 24:49–68.

Doyle, .E.H. 1973. Soil wave tank studies of marine soil instability. *Fifth Offshore Technology Conference*, Houston, Preprint 1901, April 29–May 2.

Gade, H.G. 1958. Effects of nonrigid, impermeable bottom on plane surface waves in shallow water. *J. Mar. Res.* 16(2):61–82.

Gade, H.G. 1959. Notes on the effect of elasticity of bottom sediments on the energy dissipation of surface waves in shallow water. *Archiv for Mathematik og Naturridenskab* 3:69–80.

Garrison, L.E. 1974. The instability of surface sediments on parts of the Mississippi Delta front. *U.S. Geological Survey Open File Report, Corpus Christi, Texas* 18:661–673.

Henkel, D.J. 1970. The role of waves in causing submarine landslides. *Geotechnique* 20(1):75–80.

Homma, M., Horikawa, K., and Komori, S. 1966. Response characteristics of underwater wave gauge. *Tenth Conference on Coastal Engineering*, Tokyo, chap. 8, pp. 99–114.

Mallard, W.W., and Dalrymple, R.A. 1977. Water waves propagating over a deformable bottom. *Ninth Offshore Technology Conference*, Houston, Preprint 2895, May 2–5.

Mitchell, R.J., Tsui, K.K., and Sangrey, D.A. 1972. Failure of submarine slopes under wave action. *Proceedings of the Thirteenth Conference on Coastal Engineering*, July 10–14, Vancouver, chap. 84.

Munk, W. 1962. Long ocean waves. In M.M. Hill (Ed.), *The Sea*, Vol. 1. New York: Wiley-Interscience, pp. 647–663.

Roberts, H.H., Cratsley, D.W., Whelan, T., and Coleman, J.M., 1976. Stability of Mississippi Delta sediments as evaluated by analysis of structural features in sediment borings. *Eighth Offshore Technology Conference*, Houston, Preprint 2425, May 3–6.

Suhayda, J.N. 1977. Surface waves and bottom sediment response. *Marine Geotechnology*, Vol. 2: *Marine Slope Stability*, pp. 135–159.

Suhayda, J.N., Whelan, T., Coleman, J., Booth, J., and Garrison, L., 1976. Marine sediment instability: Interaction of hydrodynamic forces and bottom sediments. *Eighth Offshore Technology Conference*, Houston, Preprint 2426, May 3–6.

Tubman, M., and Suhayda, J.N., 1976. Wave action and bottom movements in fine sediments. *Fifteenth Coastal Engineering Conference*, July 11–17, Honolulu, pp. 1168–1183.

U.S. Army Coastal Engineering Research Center. 1973. *Shore Protection Manual*, Vol. 1. Department of the Army, Corps of Engineers, chap. 2, pp. 1–129.

Whelan, T., Coleman, J.M., Suhayda, J.N., and Garrison, L.E., 1975. The geochemistry of recent Mississippi River delta sediments: Gas concentration and sediment stability. *Seventh Offshore Technology Conference*, Houston, Preprint 2342, May 5–8, pp. 71–84.

Wright, S.G. 1976. Analysis for wave induced seafloor movements. *Eighth Offshore Technology Conference*, Houston, Preprint 2427, May 3–6.

Wright, S.G., and Dunham, R.S., 1972. Bottom stability under wave induced leading. *Fourth Offshore Technology Conference*, Houston, Preprint.

4

Current-Induced Sediment Movement in the Deep Florida Straits: Observations

Mark Wimbush, Laszlo Nemeth, and *Barton Birdsall*

Abstract

An apparatus has been built for studying the interaction of sea-floor sediment with overlying current flow. It consists of a recording current meter mounted with a stereo-pair of burst-sampled time-lapse motion picture cameras photographing, from directly above, an obliquely illuminated 1-m patch of the sea bed. A 6-week deployment of the apparatus at 710 m in the eastern Florida Straits produced successful records showing vigorous current/sediment interactions. Sediment analyses of a nearby box-core showed two size modes (1 to 2.75 phi and 4 to 5 phi), indicative of poor sorting. Ripple migrations occurred during periods of higher current speeds (roughly 19 cm/s). Between migrations, the sharpness of the ripples gradually decreased, possibly due to the activity of organisms.

Introduction

Disturbances of sea-floor sediments by bottom currents have been recorded in photographic sequences at shallow and continental shelf depths (Owen et al., 1967; Rhoads, 1970; Miller et al., 1972; Knebel et al., 1976; and papers by Sternberg and Summers and their associates*). Shelf-sediment transport has also been studied with radioisotope tracers (Lavelle et al., 1976). Bioturbation of the sea bed and turbidity fluctuations of the bottom water were filmed for 42 h by Rowe et al. (1974) at 360 m in Hudson Canyon. At greater depths, patches of the sea floor have been photographed in sequences lasting many hours (Schick et al., 1968; Sternberg, 1969, 1970; Thorpe, 1972; Smith et al., 1976). But the only deep record in which significant sea bed disturbances were recorded was a 202-day sequence taken in a manganese nodule field (Paul et al., 1978), and here the disturbances were all due to bioturbations.

---

*Sternberg and Creager (1965), Sternberg (1967, 1971*a* and *b*), Sternberg et al. (1973), Sternberg and Larsen (1975), Summers (1967), Palmer 1969), and Summers et al. (1971).

Therefore, our objective was to investigate the relationships between current patterns and the history of smaller-scale features of the sediment/water interface. To do this we constructed a device that would sit on the sea floor for many weeks and make a stereoscopic motion picture of a small portion of the underlying sediments. Simultaneously, it would obtain a record of the currents affecting those sediments. Our plan was also to take samples of nearby sediment for grain-size analysis.

Equipment Details

In early 1976, a stero time-lapse motion picture camera system (with strobe light) was designed at Nova University for operation in the deep sea (depth < 6 km), and a slightly modified version of that system was constructed for deployment in the Florida Straits (depth < 1 km). The photographic components of our system were incorporated in a specially designed 1.6-m aluminum tripod together with a current meter and dual acoustic releases. Mushroom-shaped lead anchors and a spherical float provided, respectively, the means for initial descent and final ascent of the apparatus (figure 4–1). Figure 4–2 is a close-up view of the camera system, which consists of a pair of Geodyne 16-mm cameras as used in the Richardson current meter. Principal modifications were the use of Leitz 6.5-mm water contact lenses and a common electronically controlled film drive. Camera axes were 10 cm apart. Illumination of the seafloor was provided by the strobe light, which was controlled by the same electronics as the camera. Thirty alkaline D cells mounted within the camera housing provided all the electrical power needed by the strobe light, camera motor, and electronics during the 6-week experiment. Acoustic releases and Vector Averaging Current Meters (VACMs) were commercially manufactured by AMF Corporation. Instrument elevations above the sea floor were: camera, (N_1) 113 cm; strobe light discharge tube, 52 cm; VACM (rotor-vane separator plate), 82 cm.

Figure 4–1. Camera/current meter apparatus. Black housing in frame center foreground contains stereocamera pair. Strobe light with "barn-door" light concentrator is attached to rear leg. Current meter is just behind camera housing. Acoustic releases are on either side; firing either release causes separation of the three lead anchors (here roughly simulated in styrofoam), and the syntactic float (with radios and flashing light to aid in recovery) returns apparatus to surface. Wimbush (right) is 1.72 m tall.

Figure 4-2. Stereocamera pair and controlling electronics, withdrawn from pressure housing.

The Experiment

During a 6-week period in late summer 1976, our apparatus was deployed close to the deepest part of the Florida Straits. On the basis of past records, a region 58 km due east of Port Everglades, Florida was selected as a promising location for observation of sediment ripples at depth. The region is on the eastern flank of the Florida Straits at 710 m, with a bottom slope downward to the west of slightly less than 1°. In May 1976 a Decca Hi-Fix precision navigation system was set up to provide good coverage of the region, and in early June a survey of the bottom at our sampling site was conducted with a free-fall camera. The resulting photographs showed that 75 percent of the bottom appeared to be covered by a layer of rippled sand. The rest of the bottom was free of ripples.

Despite the patchiness of the ripples, in midsummer we decided to go ahead with the experiment. The cameras were loaded with Ektachrome 7256 film and set to take one stereo-photograph every ¼ h. The current meter was set to record ⅛-h averages of current and temperature. The apparatus was dropped at 26°06′N, 79°31′W on June 29 and successfully recovered by acoustic command on July 15. Although all instruments functioned well during the first attempt, the patch of bottom in the field of view of the cameras was devoid of ripples. Grain motions were occasionally visible and did correspond with current speed peaks in the VACM record. On July 16, an attempt was made to box-core the area, but despite several lowerings, only a single rather shallow sample was obtained. Presumably the bottom was too hard to allow significant penetration.

For our second attempt, the apparatus was prepared as before except that programming controls on the camera electronics were set to take sequences of three stereophotographs at 10-second intervals once every hour. On July 23, it was dropped within 700 m of the first launch position. A month later, acoustic release commands were given, but because of excessive friction and inadequate buoyancy, the apparatus did not leave the bottom. On September 4 it was located with an underwater television camera lowered from the U.S. Navy R/V *A.B. Wood*. When nudged by the camera, the apparatus broke loose and returned to the surface.

Results and Discussion

The stereoscopic color films from the second attempt contain an interesting record of sediment suspension and a pattern of ripple formation, migration, and decay related to distinct features in the current record. In this chapter we present black and white reproductions of selected still photographs from that film along with the record from the current meter and the grain size of a nearby sediment sample.

Sediment Sample

The surficial sediment sample was 30 times 20 times 4 cm and was collected with a modified Navy Electronics Laboratory box corer (Bouma, 1969). The sample was homogeneous with no apparent stratification. Analysis of sediment included sieving at ¼-phi intervals (Ingram, 1971), pipetting of the silt/clay fraction (less than 62 μm) at 1-phi intervals (Folk, 1974), computing hydraulic equivalence from settling velocities (Felix, 1969; Cook, 1969), and the determination of percent carbonates by acidification (Ireland, 1946).

The total sediment sample was 98 percent carbonate by weight. Work by Gassaway (1970) showed that the carbonate constituents in this area are aragonite, low-magnesium calcite, and high-magnesium calcite, in decreasing order of abundance. Mean grain density was estimated to be 2.85 g/cm^3. Hydraulic equivalences of the sand-sized fraction were not notably different from grain-diameter equivalences. A histogram of the size analysis (figure 4–3) shows that there were two major sediment size modes: from 1 to 2.75 phi and from 4 to 5 phi. The bimodal histogram suggests poor sorting, which is indicative of several sediment sources. Mean phi was $M = 2.48$, and the

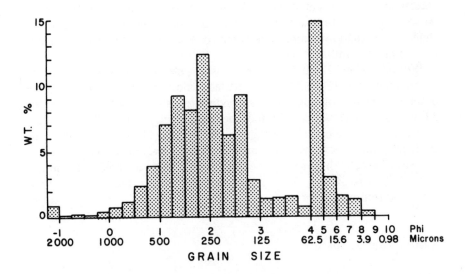

Figure 4–3. Particle-size distribution of the surface of a box-cored sediment sample from the Florida Straits near the July 23 launch position of the camera/current meter apparatus. Principal mode is at 1.875 phi (bar at 4.5 phi is taller only because the interval changes from ¼ to 1 phi at 4 phi).

Current-Induced Sediment Movement 83

standard deviation was $\sigma = 1.63$. Phi skewness and kurtosis showed departures from normal at $Sk = 0.46$ and $K = 1.25$. Here Sk and K are defined as in Krumbein and Pettijohn (1938), and all statistics were computed by the method of Schlee and Webster (1967).

Neumann and Ball (1970) described a sample taken 55 km to the south of our sampling site as having "a bimodal distribution of particle sizes with peaks in the 250 to 62 μm (2 to 4 phi) and 8 to 4 μm (7 to 8 phi) ranges." Hathaway (1971) reported that a sediment sample from 20 km to the east (his station 2442) was bimodal with peaks at 1.6 and 4.0 phi and with $M = 3.17$, $\sigma = 2.06$, $Sk = 0.67$, and $K = 3.45$. Hathaway also found that a sample from 12 km to the south-southwest (his station 2441) had a single mode at 1.6 phi with $M = 2.11$, $\sigma = 1.12$, $Sk = 0.29$, and $K = 1.20$. Note that for each of the four statistics—M, σ, Sk, and K—the values for our sample lie between the values for Hathaway's stations 2441 and 2442.

Camera and Current Meter Records

According to the current meter record (figure 4-4), the current near the bottom was typically northward at 5 to 15 cm/s with occasional "storms" of up to 40 cm/s. Still photographs of the bottom corresponding to major events in the current record are in figure 4-5. (a — o). Transverse starved sediment ripples of 15-cm wavelength were always present, except at the very beginning of the film (figure 4-5a), where the pre-existing bedform had presumably been eliminated by the disturbance of the apparatus landing on the seabed. The sediment ripples we found appear similar to ripples photographed at 815 m in the Straits of Gibraltar (Kelling and Stanley, 1972). During the two most violent "storms" on July 25-27 and August 14, turbidity observed in the photographs (figures 4-5c, h, and i, respectively) showed that sediment resuspension was occurring either locally or upstream. The durations of those storms were 65 and 7 h, respectively.

The film shows sporadic sediment ripple migrations, separated by quiescent periods during which there was a gradual rounding and smoothing of the ripple crests, perhaps by biogenic disturbances (figure 4-5d). For example, by August 1 the sand ripples formed 5 days previously had lost some of their initial sharpness due to gradual erosion of the ripple peaks (compare figures 4-5b and e). Then the current speed rose to 34 cm/s (⅛-h average) causing the ripples to move downstream and re-sharpen their profiles at the crest (compare figures, 4-5e and f). Between August 1 and August 14, the existing ripples again lost much of their initial sharpness (compare figures 4-5f and g). No further ripple migration was observed until after the turbidity caused by the rapid onset of the August 14 "storm" (figures 4-5h and i) had dissipated. The sequence of ripple migration

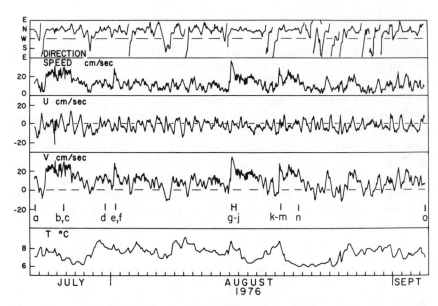

Figure 4–4. Current-meter record giving current directions, speed, eastward (u) and northward (v) velocity components and temperature during the 6-week experiment. Data lopassed to eliminate frequencies above 0.5 c/h. Tic marks on the time axis are at 0000 GMT. Marks on the northward component record labeled $a, b, \ldots$ correspond in time to photos in figure 4–5a, b, ..., respectively.

occurring after the storm is illustrated in figures 4–5j through o. Ripple migrations seemed to be more frequent during the second half of the film.

Joint analysis of the film and current meter records (Wimbush and Lesht, 1979) indicates that mean critical speeds (at the 70-cm rotor elevation) for sediment ripple migration and suspension were 19 and 22 cm/s, respectively.

Acknowledgments

The U.S. Navy provided support for this project under contract N00014-75-C-0165 from the Office of Naval Research and by making the R/V *A.B. Wood* available for a 1-day emergency recovery of the apparatus.

The box corer was loaned to this project by the Bureau of Land Management.

Current-Induced Sediment Movement 85

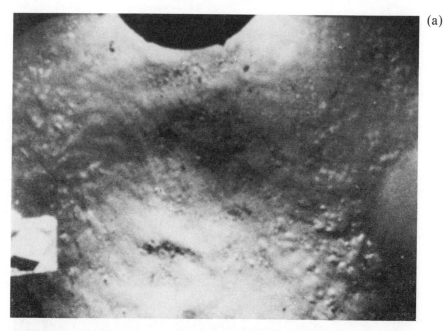
(a)

Figure 4-5. Selected bottom photographs at various times during the 6-week experiment. Bottom area shown is 143 × 104 cm. Short edges of photos are oriented 016°T (toward top of page). The dark object at the top of each picture is part of the baseplate of the current meter. The light object on the left is an Accuquartz watch (overexposed). Behind the watch, part of a lead anchor is visible. Another anchor is similarly visible on the right. Stereopairs may be viewed with the viewer supplied in Hersey (1967) or in Cravat and Glaser (1971). Photographs b, d, j, n, and o are stereopairs. Dates and Greenwich mean times of the photographs are: (a) July 23, 1300; (b) July 26, 1800; (c) July 26, 1900; (d) July 31, 1000; (e) Aug. 1, 1100; (f) Aug. 1, 1200; (g) Aug. 14, 0500; (h) Aug. 14, 0600; (i) Aug. 14, 0900; (j) Aug. 14, 2100; (k) Aug. 19, 1300; (l) Aug. 19, 1400; (m) Aug. 19, 1500; (n) Aug. 21, 1200; (o) Sept. 4, 1500.

(b)
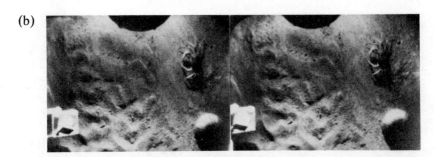

(c)
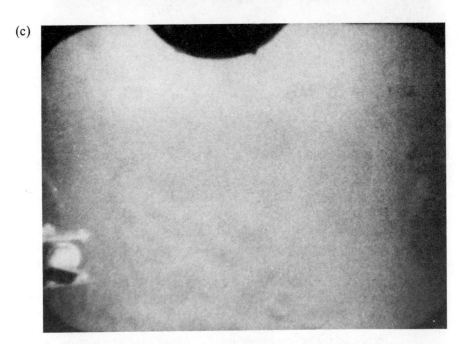

(d)
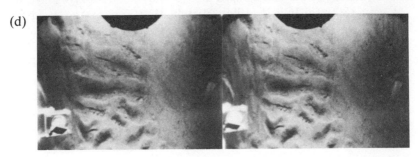

Current-Induced Sediment Movement 87

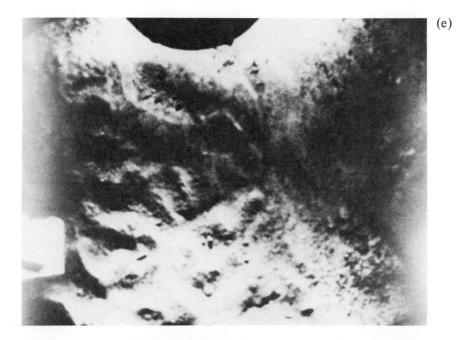
(e)

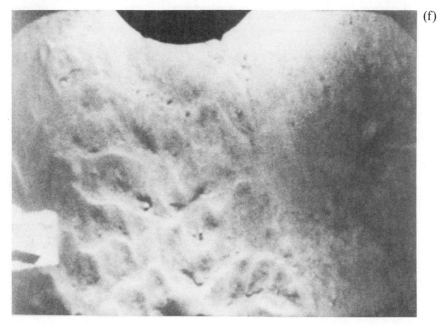
(f)

(g)

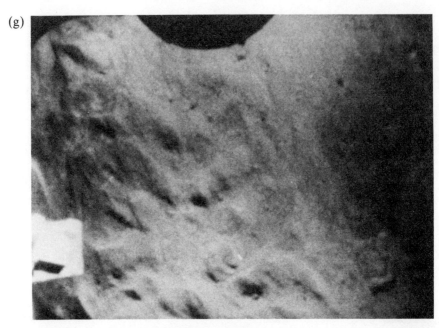

(h)

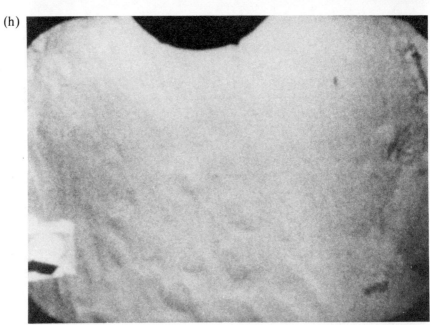

Current-Induced Sediment Movement

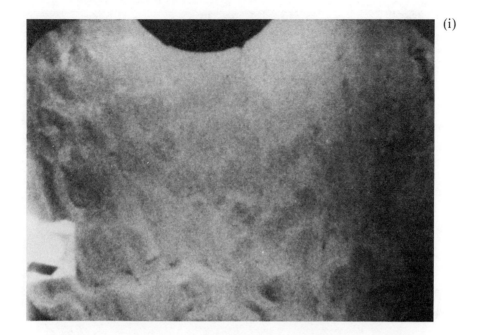

(i)

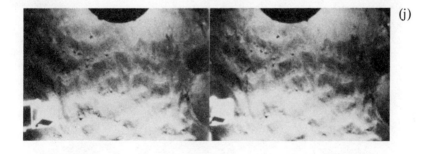

(j)

90 Dynamic Environment of the Ocean Floor

(k)

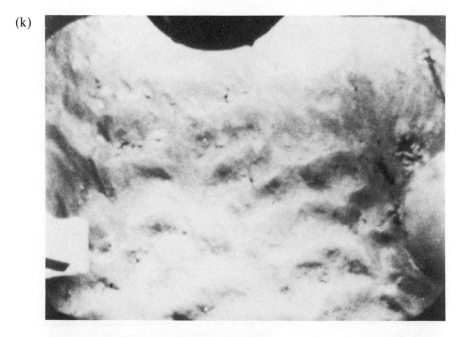

(l)

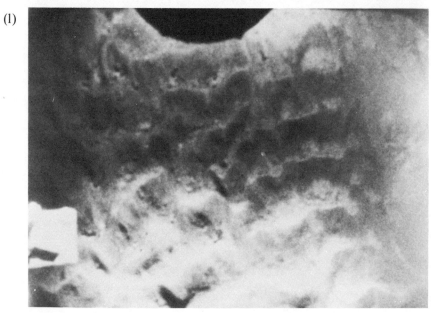

Current-Induced Sediment Movement

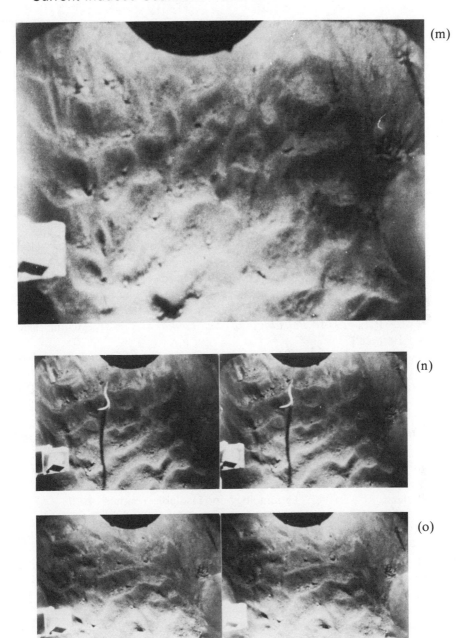

Wilford Gardner and Roger Flood suggested the site for the experiment. Alan Carr designed the camera electronics. Nemeth was responsible for the rest of the camera system and the overall apparatus. Also, Nemeth developed the free-fall camera. Philip Bedard supervised the preparation of acoustic releases and current meters. Birdsall was responsible for the sediment collection and analysis.

References

Bouma, A.H. 1969. *Methods for the Study of Sedimentary Structures.* New York: Wiley.
Cook, D.O. 1969. Calibration of the University of Southern California automatically recording settling tube. *J. Sed. Petrol.* 39:781–786.
Cravat, H.R., and Glaser, R. 1971. Color aerial stereograms of selected coastal areas of the United States. National Ocean Survey, NOAA, U.S. Dept. of Commerce, Rockville, Maryland
Felix, D.W. 1969. An inexpensive recording settling tube for analysis of sand. *J. Sed. Petrol.* 39:777–780.
Folk, R.L. 1974. *Petrology of Sedimentary Rocks.* Austin, Texas: Hemphill.
Gassaway, J.D. 1970. Mineral and chemical composition of sediments from the Straits of Florida. *J. Sed. Petrol.* 40:1136–1146.
Hathaway, J.C. 1971. Data file, continental margin program, Atlantic Coast of the United States. Vol. 2, Sample Collection and Analytical Data. *WHOI Technical Report 71-15.*
Hersey, J.B., (Ed.). 1967. *Deep-Sea Photography—Johns Hopkins Oceanographic Studies, 3.* Baltimore: Johns Hopkins Univ. Press.
Ingram, R.L. 1971. Sieve analysis. In R.E. Carver, (Ed.), *Procedures in Sedimentary Petrology.* New York: Wiley, pp. 49–69.
Ireland, H.A. 1946. Terminology for insoluble residues. *Bull. American Association of Petroleum Geologists.* 31:1479–1490.
Kelling, G., and Stanley, D.J. 1972. Sedimentary evidence of bottom current activity, Strait of Gibraltar Region. *Marine Geology* 13:M51–60.
Knebel, H.J., Butman, B., Folger, D.W., Cousins, P.W., and McGirr, R.R. 1976. Maps and graphic data related to geologic hazards in the Baltimore Canyon trough area. *U.S. Geological Survey Miscellaneous Field Studies Map*, Mf-828.
Krumbein, W.C., and Pettijohn, F.J. 1938. *Manual of Sedimentary Petrography.* New York: Appleton-Century-Crofts.
Lavelle, J.W., Gadd, P.E., Han, G.C., Mayer, D.A., Stubblefield, W.L., and Swift, D.J.P. 1976. Preliminary results of coincident current meter

and sediment transport observations for wintertime conditions on the Long Island inner shelf. *Geophys. Res. Lett.* 3:97–100.
Miller, R.L., Albro, C., Cohen, J.M., and O'Sullivan, J.F. 1972. A preliminary study of tidal erosion in Great Harbor at Woods Hole, Massachusetts. *WHOI Technical Report 72-12*.
Neumann, A.C., and Ball, M.M. 1970. Submersible observations in the Straits of Florida: Geology and bottom currents. *Bull. Geological Soc. Am.* 81:2861–2874.
Owen, D.M., Emery, K.O., and Hoadley, L.D. 1967. Effects of tidal currents on the sea floor shown by underwater time-lapse photography. In J.B. Hersey (Ed.), *Deep-Sea Photography—Johns Hopkins Oceanographic Studies 3*, pp. 159–166.
Palmer, H.D. 1969. Wave-induced scour on the sea floor. *Proc. Civil Engineering in the Oceans. II.* Miami Beach: American Society of Civil Engineers, pp. 703–716.
Paul, A.Z., Thorndike, E.M., Sullivan, L.G., Heezen, B.C., and Gerard, R.D. 1978. Observations of the deep-sea floor from 202 days of time-lapse photography. *Nature* 272:812–814.
Rhoads, D.C. 1970. Mass properties, stability and ecology of marine muds related to burrowing activity. In T.P. Crimes and J.C. Harper (Eds.), *Trace Fossils*. Liverpool: Seel House Press, pp. 391–406.
Rhoads, D.C., and Young, D.K. 1970. The influence of deposit-feeding organisms on sediment stability and community tropic structure. *J. Mar. Res.* 28:150–178.
Rowe, G.T., Keller, G., Edgerton, H., Staresinic, N., and MacIlvaine, J. 1974. Time-lapse photography of the biological reworking of sediments in Hudson Submarine Canyon. *J. Sed. Petrol.* 44:549–552.
Schick, G.B., Isaacs, J.D., and Sessions, M.H. 1968. Autonomous instruments in oceanographic research. *Proc. 4th National Instrument Society of America Marine Sciences Instrumentation Symposium, Cocoa Beach*, pp. 203–230.
Schlee, J., and Webster, J. 1967. A computer program for grain-size data. *Sedimentol.* 8:45–53.
Smith, K.L., Clifford, C.H., Eliason, A.H., Walden B., Rowe, G.T., and Teal, J.M. 1976. A free vehicle for measuring benthic community metabolism. *Limnol. Oceanogr.* 21:164–170.
Sternberg, R.W., and Creager, J.S. 1965. An instrument system to measure boundary-layer conditions at the sea floor. *Marine Geology* 3:475–482.
Sternberg, R.W. 1967. Measurements of sediment movement and ripple migration in a shallow marine environment. *Marine Geology* 5:195–205.
Sternberg, R.W. 1969. Camera and dye-pulser system to measure bottom boundary-layer flow in the deep-sea. *Deep-Sea Res.* 16:549–554.

Sternberg, R.W. 1970. Field measurements of the hydrodynamic roughness of the deep-sea boundary. *Deep-Sea Res.* 17:413–420.

Sternberg, R.W. 1971a. Boundary layer observations in a tidal current. *J. Geophys. Res.* 71:2175–2178.

Sternberg, R.W. 1971b. Measurements of incipient motion of sediment particles in the marine environment. *Marine Geology* 10:113–119.

Sternberg, R.W., Morrison, D.R., and Trimble, J.A. 1973. An instrumentation system to measure near-bottom conditions on the continental shelf. *Marine Geology* 15:181–189.

Sternberg, R.W., and Larsen, L.H. 1975. Threshold of sediment movement by open ocean waves: Observations. *Deep-Sea Res.* 22:299–309.

Summers, H.J. 1967. Time-lapse photography used in the study of sand ripples. *Coastal Research Notes, Geology Department, Florida State University* 2(6):6–7.

Summers, H.J., Palmer, H.D., and Stone, R.O. 1971. Underwater time-lapse motion picture systems. *Marine Geology* 11:M51–57.

Thorpe, S.A. 1972. A sediment cloud below the Mediterranean outflow. *Nature* 239:326–327.

Wimbush, M., and Lesht, B. 1979. Current induced sediment movement in the deep Florida Straits: critical parameters. *J. Geophys. Res.* 84:2495–2502.

5

The Benthic Interface of Deep-Sea Carbonates: A Three-Tiered Sequence Controlled by Depth of Deposition

W.H. Berger

Abstract

The study of box-core profiles from the equatorial Pacific suggests that the uppermost zone of sediment reacting with the bottom water consists of three "tiers": a 6-cm Mixed Layer, homogenized by intense fine burrowing; a 6-cm Mixed Layer Transition, characterized by lumpy mixing; and a deglaciation record, which reflects major changes in conditions of sedimentation. Sediment properties are related to depth of deposition and result primarily from carbonate dissolution and winnowing processes.

Introduction

In many respects, the uppermost layer of deep-sea sediments is as much a part of the overlying ocean as it is a part of the underlying deposits. The *interface*, therefore, is here understood as the transition zone between the ocean and its sediment. It is a product of various biological, chemical, and physical processes, as well as of the more recent history of the ocean. The various factors influencing conditions at the interface and fluxes across it have received considerable attention (see McCave, 1976; and Andersen and Malahoff, 1977).

In this preliminary report on box cores from the equatorial Pacific, I propose the working hypothesis that the interface is a composite of three "tiers." At the surface, there is a mixed layer, about 5 to 7 cm thick, within which homogenization is geologically instantaneous; i.e., there is essentially no depth gradient between particle ages. Below that layer is a transitional zone of the same thickness, within which mixing is slow relative to the deposition rate and is performed by rather large burrowers able to displace lumps of sediment. An underlying third zone represents the glacial-postglacial transition, an interval of drastic change between 16,000 and 10,000 years ago. The sediment properties in the interface zone are largely controlled by factors dependent on depth of deposition. That carbonate content and fossil composition are related to depth is well known (e.g., Sliter et al., 1975); our box-core data show that the effects of water depth are influenced by many other effects as well. Mixing by small and large burrowers, the effects of the deglaciation event, and dissolution and associated depth-related processes result in complex stratigraphies which are difficult to model correctly. The hypothesis is being tested by serial radioisotope dating in collaboration with T. H. Peng and W. S. Broecker of Lamont and with S. Somayajulu and D. Lal of the Tata Institute.

Field Work

Short of in situ study, box-coring is the most reliable way to recover an intact position of oceanic sediment. Large-diameter box cores (50 times 50 cm, about 40 cm deep) were raised in the western and eastern equatorial Pacific in 1975 and 1976, respectively (figure 5-1). About 160 kg of sediment was recovered in each core, so that there was plenty of material for the study of physical and chemical properties, fossil composition, radioisotopes, stable isotopes, and burrowing.

Each core was treated in the following way immediately after recovery. First, subsamples were taken using wide-diameter piston core liners. Then special samples were taken for acoustic velocity determinations and for interstitial water analyses. After subsampling, with 40 percent of the core untouched (except for the scraped-off surface layer), the core was cut vertically and the exposed face was carefully washed with running water to clearly expose the burrow structures and their textures and colors. The artificial outcrops were then described and photographed.

Features of the Interface Profiles

Box-core faces usually show three more or less distinct zones: an uppermost mixed layer, an intermediate zone with large burrows and maximum color contrasts, and a layer with fewer contrasts near the bottom (figure 5-2).

There are marked differences between the cores, depending on depth of deposition. In the east equatorial box-cores (figure 5-2), those from shallower than 4,000 m show abundant vertical burrows (Box 70, 74), while such structures are rare or absent in the deeper cores (Box 90, 77). Apparently, open vertical burrows close more rapidly in the deeper cores, preventing their being filled with sediment; or they are destroyed by shear after being filled; or both. Similar depth-related differences also occur in the western equatorial Pacific (Berger and Johnson, 1976), although the depth of change in mottling is shallower. One possible reason for that discrepancy could be a greater

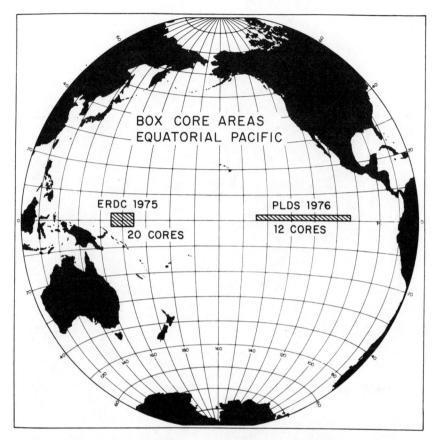

Figure 5-1. Locations of box cores discussed. ERDC: Eurydice Expedition R/V *Washington*. PLDS Pleiades Expedition, R/V *Melville*. Depth ranges: ERDC, 1,600 to 4,400 m; PLDS, 3,500 to 4,900 m.

influence of earthquakes in the western set of cores because of proximity to the Solomon Trench. The patterns of burrows and mottles in the box cores were interpreted in terms of a generalized burrow stratigraphy (figure 5-3).

The Mixed Layer (ML) is intensely burrowed by small organisms. Small-diameter ($\sim$ 1 mm) open ducts are abundant, facilitating the chemical exchange between overlying and interstitial water. Large-diameter ($\sim$ 1 cm) burrows are present, although their lifetime is probably short because of the softness and mobility of the surface sediment.

The Mixed Layer Transition (MLT) is somewhat similar to the ML, but it does not have the abundant open fine ducts. It is stiffer than the ML and

Figure 5–2. The face of box-core profiles from the Pleiades Expedition. All cores about 1°N. Box 70: 107°13'W, 3694 m; box 74: 113°39'W, 3945 m; box 90; 135°04'W, 4296 m; box 77: 119°47'W, 4366 m. Width of box: 50 cm. Core tops at left.

Benthic Interface of Deep-Sea Carbonates

GENERALIZED BURROW STRATIGRAPHY

cm

0

MIXED LAYER
Uniform color. Soft. Intensely burrowed. Abundant open ducts. Homogenous.

5-7

MIXED LAYER TRANSITION
Similar to ML but much fewer organisms and open ducts. Blebs of older sediments in ML matrix.

10-14

TRANSITION ZONE
Maximum color contrast. Firm sediment. Virtually no macroscopic organisms. Burrowed by (rare) large animals over a long time period.

20-40

HISTORICAL LAYER
Fading color contrast. Firm sediment. Burrows and associated reduction haloes grade into "mottles".

Figure 5-3. Generalized burrow stratigraphy based on gross aspects of burrow abundances and coloration.

usually has some open coarse ducts. Mixing in the MLT appears to be of the "lumpy" kind, by rather large organisms displacing pebble-sized blebs of sediment. Thus, not only does mixing decrease with depth (as postulated by the Guinasso-Schink model, 1975), but the nature of the mixing process also appears to change.

The Transition Zone (TZ) is marked by maximum color contrasts, presumably due to critical redox conditions with respect to iron and manganese combined with heterogeneities in the concentration of organic matter and in the rates of redox reactions. Despite its depth, this zone is still subject to burrowing, as witnessed by occasional open ducts. Most of the burrows which will eventually be preserved are formed in the Transition Zone.

Benthic Interface of Deep-Sea Carbonates 103

The Historical Layer (HL) has fading color contrasts, presumably due to chemical reduction and upward migration of pigment. There are essentially no open burrows. In general, burrows are being slowly destroyed by strain and by the fading of color contrasts.

The organisms that made the burrows are unknown, although many burrows can be classified within existing "ichno" taxonomy (Ekdale and Berger, 1978). Much time is available (10,000 years or more) to generate the observed distinctive burrowing, so it is not a problem that the burrowers are, in fact, rare. In thirty-two box cores we found only one large burrower, a worm of uncertain taxonomy (length 9.5 cm, width 8 mm). It was found near the boundary between ML and MLT, in a horizontal position, at the end of an open burrow lined with green slime. The worm resembles echiurids in having a spoon-like proboscis. Although it was retracted, in some echiurids the proboscis can extend a considerable distance and may reach up to the sediment/water interface to collect food. It is clear that reactive organic matter is transferred from the sediment surface to within it by the feeding activity of burrowing worms.

Sediment Properties as a Function of Depth of Deposition

The difference in burrow assemblages between shallow-water and deep-water box cores is rather striking. We have earlier suggested (Berger and Johnson, 1976) the existence of some type of flow within the sediment which affects those assemblages differentially. One might argue that food supply on the sea floor should decrease with water depth and that such a decrease should affect the distribution and activity of burrowers. However, that type of primary effect seems unlikely since the overall pattern of depth change is similar in both east and west equatorial Pacific, areas of greatly differing fertility. In the shallow-water cores, vertical burrows are much more abundant than horizontal ones. That is expected; for every horizontal burrow there should be at least one vertical one. However, the reverse is not true; the deeper-water cores are clearly impoverished in vertical burrows. Thus flowage increases with depth of deposition.

Various sediment properties have been analyzed in the box cores from the west equatorial Pacific (figures 5-4 through 5-6). The most interesting one in the context of burrow preservation is shear strength (figure 5-4). The relationship between depositional depth and shear strength shows considerable scatter, but the trend is clear. The cores from deeper water have the lower strength. In the box cores from the eastern Pacific, vane shear measurements by Homa Lee (personal communication) showed subsurface

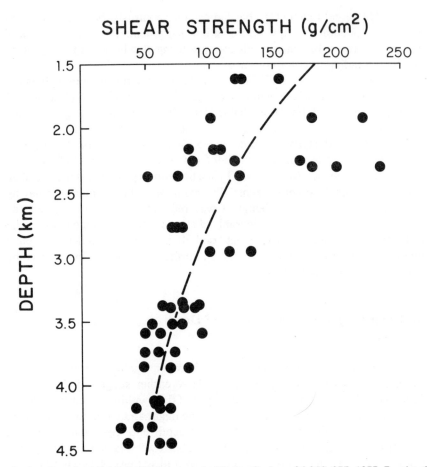

Source: Data from Table I of Johnson et al., *Marine Geology*, 24:259–277, 1977. Reprinted with permission.

Figure 5–4. Shear-strength distribution with water depth in ERDC box cores (three measurements per core).

values near 50 g/cm^2 for depths between 3,500 and 4,800 m, which are in good agreement with the values from the western Pacific. Surface values in the east are near 25 g/cm^2, demonstrating the softness of the upper layer.

Shear strength is greatly influenced by grain size: the more sandy sediments show greater strength (Johnson et al., 1977). The excellent correlation of grain size with depth is rather important in this respect (figure 5–5).

At depths shallower than 3 km, effects of dissolution on physical properties of the sediment become progressively less important, while the

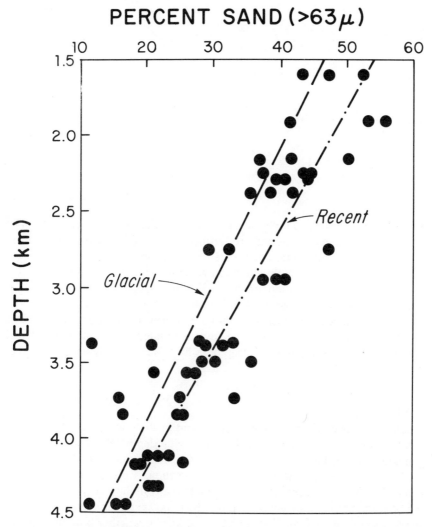

Source: Data from Johnson et al., *Marine Geology* 24:259–277, 1977. Reprinted with permission.

Figure 5-5. Grain-size trends with water depth in ERDC box cores. Three measurements per core: top = "recent," bottom = "glacial" regression lines.

effects of winnowing increase. Two end-member cases of the relationship between winnowing and grain size can be envisaged, as follows:

1. *A decrease of bottom current activity with depth of water.* The question is what type of currents and which part of their spectrum might

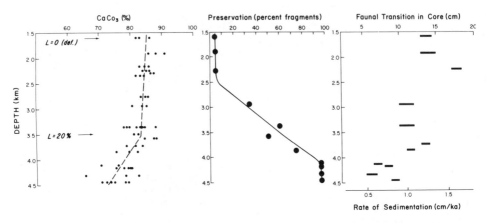

Figure 5–6. Various sediment properties as a function of water depth. The faunal transitions are fertility dominated in cores above 3 km and dissolution dominated below 3 km. The respective changes are not exactly synchronous (the fertility change is later than the dissolution change). For the sedimentation-rate scale a transition age of 11,000 years has been adopted, which is considered correct to within 15 percent of the true values. Note that the rates will be underestimated for shallow cores and overestimated for deep ones by this commonly used method of dating.

produce the observed coarsening of sediment in shallow depths. Apparently bottom currents in general do decrease with depth (R. Knox, personal communication). However, it is doubtful that the average depth gradient is sufficient to produce the observed phenomenon. Episodic currents from tsunamis may merit consideration. Tsunamis are shallow-water waves, and the bottom currents they generate might be important over a shallow feature such as the Ontong-Java Plateau. It is not known whether bottom currents generated from that effect would be sufficient to winnow sediment.

2. *Steady-state winnowing.* If there is any winnowing at all for whatever reason, each depositional site loses fine material, but it also receives fines from upslope sites. The grain size at any one site therefore reflects a balance between the amount of fines coming in and being lost. A shallow site has few upslope sources. Thus the balance at burial will show a low content of fines. That content may be reached asymptotically; a low proportion of fines will decrease the loss rate. The reverse is true for a deep site. Suspension of fines, which allows down-slope transport, could be by bottom currents or by the activity of organisms. Evidence for such activity is

Benthic Interface of Deep-Sea Carbonates 107

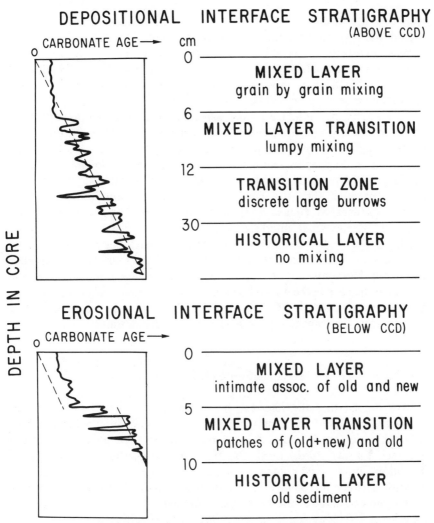

Figure 5–7. Conceptual model of depositional and erosional interface stratigraphy. The graphs of "carbonate age" are hypothetical and denote expected coarse fraction age distributions for very small samples.

considerable, e.g., from the abundance of small mounds surrounding open "exhaust" burrows (Heezen and Hollister, 1971; Ekdale and Berger, 1978).

Below 3-km depth the grain-size distribution is strongly influenced by dissolution processes; excess fines are derived from fragmentation of foraminifera. In addition to any redistribution of fines *between* sites, there is, therefore, a redistribution of mass between size classes *within* sites. The coarser fractions feed the finer ones by fragmentation, which is balanced by increasing losses to dissolution in the finer fractions. Such transfer of material to finer fractions during dissolution is the reason why forminiferal preservation is a much more sensitive indicator of dissolution than either carbonate percentage or sedimentation rate (figure 5-6).

The overriding influence of increasing carbonate dissolution with depth on all sediment properties below 3 km is apparent in the depth profiles of shear strength, percent sand, percent carbonate, percent foraminiferal fragmentation, and apparent sedimentation rates. As the carbonate compensation depth (CCD) is approached and sedimentation rates slow to zero or become negative through chemical-mechanical erosion, the generalized burrow stratigraphy (figure 5-3) needs to be modified. In addition to a "normal" depositional interface stratigraphy, one needs to consider an erosional one (figure 5-7). The point is that even though there is no net accumulation of carbonate, the mixed layer will tend to incorporate modern carbonate. Thus, unless erosion is rapid, a hiatus resulting from dissolution is formed at the base of the mixed layer. Clearly, such erosional conditions may be widespread, since the CCD rose considerably over wide areas of the world ocean during the transition from glacial to postglacial time.

The Third Tier: The Deglaciation Event

I have argued that the interface has at least two "tiers" or layers, the homogenized Mixed Layer and the Mixed-Layer Transition below it, where heterogeneities are introduced by large burrowers. We have already seen evidence for the third "tier," the transition from glacial to postglacial conditions, in the sand content (figure 5-5). It appears from the great color contrast between mixed layer and underlying sediment that older carbonate is being eroded in the deepest core shown in figure 5-2. That observation is in agreement with evidence from other parts of the Pacific.

A rapid rise of the CCD at the end of the last glacial event appears well established (Berger, 1970; Broecker and Broecker, 1974; Adelseck, 1977). Similarly, the lysocline and other dissolution levels rose during deglaciation. presumably in response to sequestration of carbonates on the growing Holocene shelves, as well as other reasons (Berger, 1977). The compensation depth of pink *Globigerina rubescens*, a very delicate species, rose from near 3,800 m during the last glacial episode to about 2,700 m during the postglacial one (figure 5-8).

Shoaling of the CCD was probably less than that amount. However, the determination of its rise is difficult, since the Mixed Layer interferes with the deglaciation record at water depths close to the CCD. Modern shells could

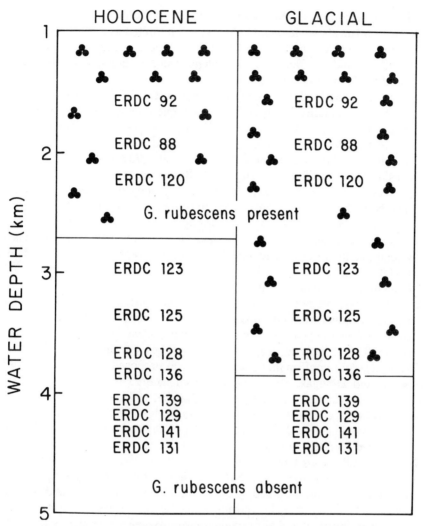

Figure 5–8. The rise of the dissolution levels during deglaciation, as shown by the change in the compensation depth of pink *Globigerina rubescens*, a highly susceptible species. The labels ERDC 92, ERDC 88, etc. are at the water depth from which the box cores were raised.

easily be mixed into the dissolving assemblage of older ones. Thus the age of carbonate particles as well as the faunal aspects would tend to show Holocene characteristics throughout the Mixed Layer, and the base of the Mixed Layer becomes a pseudo-transition between glacial and postglacial periods. According to that argument, the lowest sedimentation rates between 4 and 4.5 km, shown in figure 5-6, can be interpreted as artifacts of mixing in a situation of zero net accumulation. That would put the postglacial CCD between 4 and 4.5 km. There is evidence from oxygen isotope stratigraphy that the inference is correct.

When exactly did dissolution effects intensify? We investigated this question by detailed analysis of the isotopic composition of foraminifera shells (Berger and Killingley, 1977). To illustrate the method, two cores from the western Pacific are compared, one from a depth of 1,598 m (ERDC 92) with no dissolution and one from a depth of 3,732 m (ERDC 128) with considerable dissolution (figure 5-9). Based on preservation stratigraphy (Berger, 1977), the two cores have approximately the same sedimentation rate. Note that the variations in oxygen isotopic composition of *Pulleniatina obliquiloculata* in the two cores closely parallel each other. The oxygen composition is heaviest between 32 and 33 cm in both cores, that which is the glacial maximum usually assumed to be 18,000 years ago. (The ^{14}C date in ERDC 92 at that level is 19,000 years, according to measurements by T.H. Peng in the laboratory of W.S. Broecker).

The two isotope curves for *Globigerinoides sacculifer* also show similar trends. However within the postglacial the curve for the deeper core is shifted to heavier isotopic values. The reason is that partial dissolution concentrates the shells that grow in deeper depths, and they are thus rich in $\delta^{18}O$. The present results confirm that effect, which was proposed earlier (Berger, 1971; Savin and Douglas, 1973) but has been ignored in the interpretation of isotopic fluctuations (Shackleton and Opdyke, 1973; Emiliani and Shackleton, 1974) until recently (Shackleton and Opdyke 1976). Such confirmation dispels any lingering doubts that the data of Savin and Douglas (1973), who compared core-top samples from various depths, can only be interpreted in terms of admixtures of glacial material at the deeper sites. Such glacial carbonate appears to be exposed (or present near the surface) at and below 4-km depth in the area studied by Savin and Douglas (Broecker and Broecker, 1974). The effect of such exposure would be that the oxygen-isotope values of conspecific planktonic Foraminifera would show the depth trend described by Savin and Douglas, independent of any effects of dissolution.

The shift to heavier values in the postglacial *G. sacculifer* starts at about 22 cm in the box cores. In ERDC 92, that depth corresponds to a ^{14}C age of 12,000 years (T.H. Peng and W.S. Broecker, personal communication). The maximal faunal change is slightly later at about 14 cm, or approximately 9000 years ago (see Berger, *in* Andersen and Malahoff, 1977). There is an indication, then, that the onset of appreciable dissolution preceded the maximum drop in fertility in this area by several thousand years. That an increase in corrosiveness of the bottom water preceded a decrease in surface productivity in the equatorial Pacific has previously been proposed by Pisias

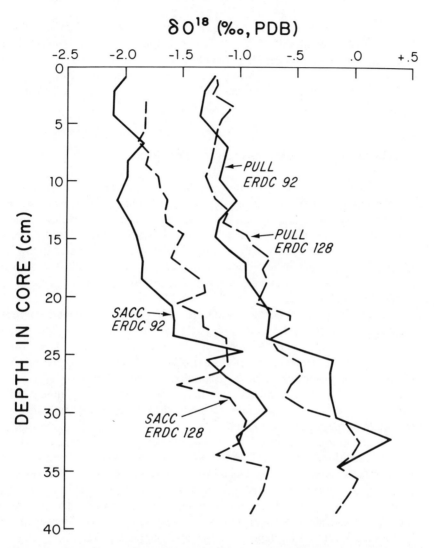

Figure 5-9. Oxygen isotopic composition of *Globigerinoides sacculifer* (SACC) and of *Pulleniatina obliquiloculata* (PULL) in box cores ERDC 92 (from 1598 m) and ERDC 128 (from 3732 m). The *Pulleniatina* values are similar at both depths, while the *G. sacculifer* values diverge at 20 cm (about 14,000 years ago). Isotopically "light" *G. sacculifer* are being removed by dissolution in the deeper core. For additional data and discussion, see Berger and Killingley (1977).

et al., (1975). However, their suggestion that those events preceded deglaciation (as indicated by $\delta^{18}O$ shift) is not borne out in the present data. The sequence indicated here is (1) melting of glaciers, (2) sharply increased dissolution, and (3) drop in fertility. Further study of those intriguing "leads and lags" (Luz and Shackleton, 1975; Moore et al., in Andersen and Malahoff, 1977) is necessary.

Conclusions and Implications for Interface Studies

Short of in situ study, box-coring is the most reliable way to recover a piece of sedimentary sea floor. The distribution and character of burrows indicates that the benthic interface is not a surface but is instead a zone which includes the surficial layer of the sediment. The most reactive part of that surficial layer is the Mixed Layer, which is about 6-cm thick. The sediment below the Mixed Layer is not at rest but is still being burrowed, and it moves in response to shear stress and dewatering. The resulting strain depends on the depth of deposition.

Depth of deposition controls a number of important sediment properties, including shear strength, grain size, carbonate content, and sedimentation rate. The overriding cause is increasing dissolution with depth, as shown in the deteriorating preservation of Foraminifera. Above 3000 m, winnowing effects are very important. Both dissolution and winnowing strongly affect grain size, which in turn affects other properties. The depth gradients in the various properties are considerable and should markedly influence all exchange of matter across the interface.

The zone immediately below the homogenous Mixed Layer, the Mixed Layer Transition, is characterized by lumpy mixing. The nature of the two-tiered mixed-layer system is crucial for interface processes. Open burrows (and their lifetime) are of special interest. Open ducts near the sea floor facilitate exchange processes between sediment and water. Deeper ducts provide for exchange between the mixed-layer system and older sediment. The mixed layer can be depositional or erosional in nature.

The deglaciation event introduced additional complications into the mixed-layer system, especially if sedimentation rates were low. Deglaciation produced drastic changes in various paramenters of sedimentation between 9,000 and 16,000 years ago. The corresponding depths below the sea floor are between 5 and 25 cm for deep-sea carbonates. The event complicates steady-state models of interface chemistry because of important changes in supply, preservation, and rates of accumulation of carbonate, silica, and organic matter.

Acknowledgments

The physical properties were determined in collaboration with T.C. Johnson and E.L. Hamilton. Oxygen isotope measurements were done at the Isotope Sediment Lab at S.I.O., by J. S. Killingley. The ^{14}C data mentioned in the text were supplied by T. H. Peng and W.S. Broecker, Lamont. J.P. Kennett critically read the manuscript and made helpful suggestions for improvement. The research was supported by the National Science Foundation (Oceanography), OCE 75-04335 A01, and by the Office of Naval Research, USN N 00014-69-A-0200-6049.

References

Adelseck, C.G. 1977. Recent and late Pleistocene sediments from the eastern equatorial Pacific Ocean: Sedimentation and dissolution. Ph.D. thesis, University of California, San Diego.

Andersen, N., and Malahoff, A. (Eds.) 1977. *The Fate of Fossil Fuel CO_2 in the Ocean,* Vol. 6: *Marine Science Series.* New York: Plenum.

Berger, W.H. 1970. Planktonic Foraminifera: Selective solution and lysocline. *Marine Geol.* 8:111–138.

Berger, W.H. 1971. Sedimentation of planktonic Foraminifera. *Marine Geol.* 5:325–358.

Berger, W.H. 1977. Deep-sea carbonate and the deglaciation preservation spike in pteropods and planktonic Foraminifera. *Nature* 269:301–304.

Berger, W.H., and Johnson, T.C. 1976. Deep sea carbonates: Dissolution and mass wasting on Ontong Java Plateau. *Science* 192:785–787.

Berger, W.H., and Killingley, J.S. 1977. Glacial-Holocene transition in deep-sea carbonates: Selective dissolution and the stable isotope signal. *Science* 197:563–566.

Broecker, W.S., and Broecker, S. 1974. Carbonate dissolution on the western flank of the East Pacific Rise. In W.W. Hay (Ed.), *Studies in Paleoceanography: Soc. Econ. Paleont. Mineral. Spec. Pub.* 20:44–57.

Ekdale, A.A., and Berger, W.H. 1978. Deep-sea ichnofacies: Modern organism traces on and in pelagic carbonates of the western equatorial Pacific. *Palaeogeogr., Palaeoclimatol., Palaeoecol.* 23:263–278.

Emiliani, C., and Shackleton, N. 1974. The Brunhes epoch: Isotopic paleotemperatures and geochronology. *Science* 183:511–514.

Guinasso, N.L., Jr., and Schink, D. 1975. Quantitative estimates of biological mixing rates in abyssal sediments. *J. Geophys. Res.* 80:3032–3043.

Heezen, B.C., and Hollister, C.D. 1971. *The Face of the Deep*. New York: Oxford Univ. Press.

Johnson, T.C., Hamilton, E.L., and Berger, W.H. 1977. Physical properties of calcareous ooze: Controlled by dissolution at depth. *Marine Geol.* 24:259–277.

Luz, B., and Shackleton, N.J. 1975. $CaCO_3$ solution in the tropical east Pacific during the past 130,000 years. In W.V. Sliter, A.W.H. Be, and W.H. Berger (Eds.), *Dissolution of deep-sea carbonates: Cushman Found. Foram. Res.* 13:142–150.

McCave, I.N. (Ed.). 1976. *The Benthic Boundary Layer*. New York: Plenum.

Pisias, N.G., Heath, G.R., and Moore, T.C. 1975. Lag times for oceanic responses to climatic change. *Nature* 256:716–717.

Savin, S.M., and Douglas, R.G. 1973. Stable isotope and magnesium geochemistry of recent planktonic Foraminifera from the South Pacific. *Geol. Soc. Am. Bull.* 84:2327–2342.

Shackleton, N.J., and Opdyke, N.D. 1973. Oxygen isotope and paleomagnetic stratigraphy of equatorial Pacific core V28-238: Oxygen isotope temperatures and ice volumes on a 10^5 year and 10^6 year scale. *Quat. Res.* 3(1):39–55.

Shackleton, N.F., and Opdyke, N.D. 1976. Oxygen-isotope and paleomagnetic stratigraphy of Pacific core V28-239: Late Pliocene to latest Pleistocene. *Geol. Soc. Am. Memoir* 145:449–464.

Sliter, W.V., Be, A.W.H., and Berger, W.H. (Eds.). 1975. Dissolution of deep-sea carbonates. *Cushman Found. Foram. Res.* (Spec. Publ.) 13.

6

The Influence of a Diffusive Sublayer on Accretion, Dissolution, and Diagenesis at the Sea Floor

Bernard P. Boudreau and
Norman L. Guinasso, Jr.

Abstract

Fluxes of dissolved materials across the sea-floor boundary play an important part in determining what sediment dissolves, what sediment accumulates, and what diagenetic processes occur. Above the sea floor, turbulent diffusion is the dominant transport process for vertical fluxes, but in a thin layer immediately adjacent to the sediment/seawater interface, turbulence is suppressed by viscous forces to such an extent that the process of molecular diffusion is the dominant mode of mass transport. That layer is the diffusive sublayer. The flux of mass across the diffusive sublayer can be described in terms of a mass transfer coefficient β, which has the dimensions of velocity. If the concentration of the bulk fluid is C_∞ and the concentration at an interface is C_0, the flux J across the diffusive sublayer and toward the interface is given by an expression analogous to Ohm's Law:

$$J = \beta(C_\infty - C_0)$$

The mass transfer coefficient β is a function of turbulence, viscosity, and the molecular diffusion coefficient. Many useful models that can be used to estimate β for sea-floor conditions are available in the chemical engineering literature. By analogy to Ohm's Law, the inverse of mass transfer coefficient can be thought of as the impedance of the diffusive sublayer to mass flow.

Kinetics of reactions at surfaces and the combined effects of reaction kinetics and diffusion in pore water also limit fluxes of mass in the sea-floor region. Expressions are derived for the kinetic impedance to mass flow for surface reactions and for the combination of the kinetic and diffusion impedances (internal impedance) of reactions taking place in pore water. Because the impedances are directly comparable to the impedance of the diffusive sublayer to mass transfer, the effect of a diffusive sublayer on mass transport can be assessed.

When the impedance of the diffusive sublayer is much greater than the kinetic or internal impedance, then reaction rates are controlled by mass transport across the diffusive sublayer. That appears to be the case for the growth of manganese nodules. If the impedance of the diffusive sublayer is much less than the kinetic or internal impedance, then the diffusive sublayer has little effect. Silica fluxes from the sea floor have that property. Dissolution of calcium carbonate at the sea floor is an intermediate case.

Introduction

The chemistry of seawater is regulated by sedimentation, dissolution, precipitation, and associated chemical fluxes. Some authors have suggested that seawater has remained substantially constant in composition over a vast period (Sillen, 1961; Mackenzie and Garrels, 1966). Others have either suggested or even proposed that temporal variations have occurred (Broecker, 1971; Holland, 1974). Whether steady-state composition can be maintained or even exist in the oceans is largely a function of the rate at which excess dissolved materials are removed to sediments and the rate at which deficits are made up by the release of dissolved substances from the sediments. The response of the sediment/pore-water system to oceanic perturbations is determined by at least two mechanisms: (1) interstitial reactions and (2) transport of solutes by diffusion within the sediments. As a consequence, diagenetic equations describing such interactions include terms for diffusion and reaction within the sediment (Anikouchine, 1967; Tzur, 1971; Berner, 1971, 1975, 1976b; Hurd, 1973; Imboden, 1975; Lerman, 1975, 1977, etc.).

Attention has also been drawn to chemical fluxes from deep-sea sediments into overlying waters (Morse, 1974; Schink, Guinasso, and Fanning, 1975; Schink and Guinasso, 1977a and b; Berner, 1976a; Takahashi and Broecker, 1977; Wimbush, 1976). Seawater and pore water represent two chemical systems capable of reacting on an immense scale; models useful in chemical engineering have much to say about how those systems can interact. Laboratory studies of heterogenous reactions occurring on the walls of flow reactors have shown that molecular diffusion of a reactant or product through a thin boundary layer adjacent to the walls often determines the overall rate of reaction (see, e.g., Sherwood, Pigford, and Wilke, 1975). Boundary layers at the sea floor offer a similar impedance to mass flux and thus may play significant role in controlling the rate of chemical exchange between the sediments and seawater.

Mass transfer resistance in boundary layers does not necessarily determine rates of interaction, but it must be considered in any model used to predict such rates. Rates of sediment/seawater interaction depend on the relative magnitudes of the transfer rate across the boundary layer and the rate at which material is released or taken up by the sediments. If the rate-controlling step in sediment/seawater interaction can be identified and quantified, we can predict the response of the sediment/pore-water system to changes in environmental parameters.

Impedance to Transfer in Boundary Layers

Turbulent Transport of Momentum

Turbulence associated with boundary flow dominates the transport of momentum, heat, and mass in the vicinity of the sea floor. But very near the sediment/seawater interface, turbulent processes are suppressed and molecular processes—viscosity, thermal conductivity, and molecular diffusion—become the major mechanisms for the transport of momentum, heat, and mass. Wimbush and Munk (1970), and later Wimbush (1976), pointed out that the lower meter of the benthic boundary layer behaves as a neutrally stable, nonrotating, turbulent boundary layer, much like laboratory systems near flat plates, and that theories of the transfer of mass, heat, and momentum developed for laboratory systems should apply to the sea floor.

To understand how the mechanism of momentum transfer changes from turbulent to viscous transport, we examine the movement of a fluid in turbulent motion flowing across an infinite flat plate. We assume two-dimensional flow with a predominant direction of flow parallel to the plate (x direction). We also assume that the viscous shear stress is significant only in the direction perpendicular to the plate surface (i.e., in the z direction). The first assumption means that the z component of velocity is small compared to the x component (at least far from the plate surface), but that it cannot be ignored. Those simplifications are known as the *boundary layer approximation* and, for our purpose, describe conditions of flow in the benthic boundary layer. Ignoring contributions from body forces (gravity) and pressure and assuming density to be constant, we can write a conservation equation for the instantaneous horizontal component of momentum as

$$\frac{\partial u}{\partial t} + u\frac{\partial u}{\partial x} + w\frac{\partial u}{\partial z} = \nu\left(\frac{\partial^2 u}{\partial z^2}\right) \qquad (6.1)$$

where x = directional coordinate parallel to the interface (cm)
z = directional coordinate perpendicular to the interface (cm)
u = velocity component in the x direction (cm/s)
w = velocity component in the z direction (cm/s)
t = time (s)
ν = kinematic viscosity (cm^2/s)

In turbulent flows we can describe the velocity at any given instant as the sum of a time-independent mean velocity and a fluctuating component (Reynolds, 1895),

$$u = \bar{u} + u'$$
$$w = \bar{w} + w'$$
(6.2)

where $\bar{u}$ = mean velocity in x direction (cm/s)
u' = fluctuation of u component (cm/s)
$\bar{w}$ = mean velocity in z direction (cm/s)
w' = fluctuation motion of z component (cm/s)

If the instantaneous velocities (equation 6.2) are substituted for u and w in equation 6.1 and the resulting equations are averaged over a sufficiently long period of time so that the average of the velocity fluctuations is zero, an expression involving the mean velocities can be written (Sideman and Pinczewski, 1975, p. 58):

$$\bar{u}\frac{\partial \bar{u}}{\partial x} + \bar{w}\frac{\partial \bar{u}}{\partial z} = \frac{\partial}{\partial z}\left[\nu \frac{\partial \bar{u}}{\partial z} - (\overline{u'w'})\right]$$
(6.3)

We have used the fact that the time-averaged fluctuations are equal to zero ($\bar{u}' = 0$, $\bar{v}' = 0$) to eliminate some of the terms generated in the expansion from equation 6.1 to equation 6.3. The reader is referred to standard texts for a detailed account of the time-averaging procedure (e.g., Schlichting, 1968, pp. 525–530; or Parker, et al., 1974, pp. 221–224).

The term $-(\overline{u'w'})$ can be interpreted as the average shearing stress due to the turbulent fluctuations of velocity. It is generally called the *Reynolds stress* and in boundary layers acts like viscous stresses in transferring momentum from one layer of the fluid to another.

Boussinesq (1877) first suggested that turbulence acts to increase the apparent kinematic viscosity and that the turbulent transport term could be expressed as the product of a mean velocity gradient and an eddy (or turbulent) viscosity coefficient E:

$$-(\overline{u'w'}) = E\frac{\partial \bar{u}}{\partial z}$$

That allows equation 6.3 to be written as

$$\bar{u}\frac{\partial \bar{u}}{\partial x} + \bar{w}\frac{\partial \bar{u}}{\partial z} = \frac{\partial}{\partial z}\left[(\nu + E)\frac{\partial \bar{u}}{\partial z}\right]$$
(6.4)

Influence of a Diffusive Sublayer

The term in brackets represents the total shearing stress; E has the same units as kinematic viscosity; and their sum can be considered to be the effective viscosity. In the turbulent stream above an infinite plate, eddy viscosity E is many orders of magnitude larger than the viscous transport coefficient. For example, near the sea floor, E is of the order of 1 to 100 cm^2/s while molecular viscosity is about 1.8×10^{-2} cm^2/s.

Very near the sea floor, frictional forces retard the motion of the fluid. Intermolecular attractions cause the fluid to adhere to the surface, and as a consequence the velocity of the fluid must be zero at the interface. That is often referred to as the "no-slip condition." The forces that act to retard the fluid near the wall also dampen fluctuations in the velocity components so that u' and w' become zero at the interface. As a result, the contribution of the Reynolds stresses $-(\overline{u'w'})$ to the total momentum flux must decrease, becoming zero at the surface. At some point near the wall, the Reynolds stresses will be suppressed to such an extent that the eddy viscosity E will equal the kinematic viscosity ν. Above that point, momentum transfer is dominated by turbulence; below that point, momentum transport is dominated by viscosity. The region where viscosity dominates is referred to as the *viscous sublayer* (figure 6-1).

Numerous laboratory studies of turbulent flow (Taylor, 1916; Prandtl. 1928; Nikuradse, 1933; Reichardt, 1951; Laufer, 1953; Deissler, 1954) have shown that, for different fluids and different conditions of flow, mean-velocity profiles near a wall become similar when certain dimensionless parameters are used to describe distance and velocity. Figure 6-2, adopted from Deissler (1954), shows such a velocity profile above a flat plate in terms of those dimensionless parameters. Velocity is scaled by dividing it by the square root of the turbulent shearing stress. The scaled velocity is called the *friction velocity* and is a measure of the intensity of turbulence. Distance is scaled by the ratio of the friction velocity and kinematic viscosity. The scaled quantities are given by

$$u^+ = u/u_* \quad \text{and} \quad z^+ = zu_*/\nu \qquad (6.5)$$

where u^+ = dimensionless velocity
z^+ = dimensionless distance above the surface
u_* = the friction velocity (cm/s)
ν = the kinematic viscosity (cm^2/s)

The inflection in the curve at $z^+ = 26$ in figure 6-2 indicates the point at which $E \approx \nu$. In the region above that point, momentum is transferred mostly by the turbulent Reynolds stresses; the velocity profile in that region can be described using Prandtl's mixing-length hypothesis, which leads to a logarithmic velocity profile (Schlichting, 1968; Wimbush and Munk, 1970). The viscous sublayer is the region below $z^+ = 26$, where viscosity dominates

REGIONS NEAR THE SEA FLOOR AFFECTING DIFFUSIVE TRANSPORT

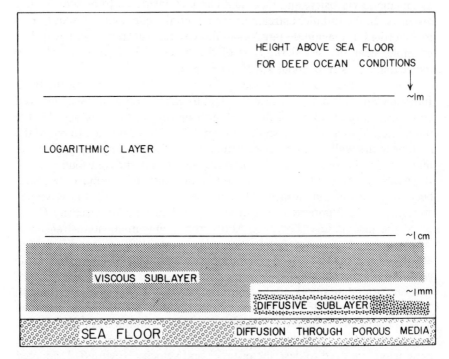

Figure 6–1. A schematic diagram of regions that play a roll in mass transfer across the sediment/seawater interface.

momentum transport and where most of the velocity change takes place. The velocity profile in the lower portion of the viscous sublayer is closely approximated by a laminar profile where u^+ equals z^+ (Schlichting, 1968); however, the flow is not laminar. The velocity profile can be described by a linear mathematical expression because the Reynolds stresses are so much smaller than the viscous stresses.

Turbulent Transport of Chemical Species

When dissolved material is transferred to or from a wall, a conservation-of-mass equation similar to the equation for conservation of momentum can be written. We here take *conservation of mass* to mean the concentration of a chemical species, such as an ion or dissolved molecule. We also assume that the masses of the ions do not appreciably change the density of the fluid. The

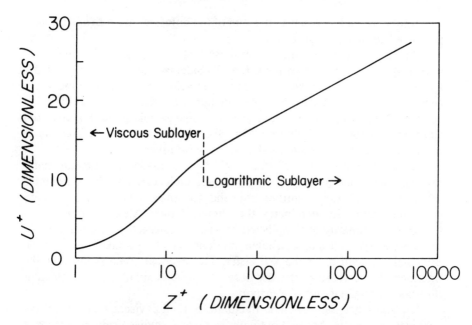

Figure 6–2. The "universal dimensionless velocity profile" plates (adopted from Deissler, 1954). Dimensionless quantities for distance and velocity are defined by equation 6.5. The curve is based on many experiments with different media.

time-averaged equation analogous to equation 6.3 is (Sideman and Pinczewski, 1975, p. 59):

$$\bar{u}\frac{\partial \bar{c}}{\partial x} + \bar{w}\frac{\partial \bar{c}}{\partial z} = \frac{\partial}{\partial z}\left[D\frac{\partial \bar{c}}{\partial z} - \overline{(w'c')}\right] \qquad (6.6)$$

Convective   Diffusive
transport   transport

where $\bar{c}$ is the mean concentration, c' is the fluctuations, and D is the molecular diffusion coefficient (cm²/s). By analogy to momentum, the term $\overline{(w'c')}$ is the turbulent contribution to the flux. That term is often called the *Reynolds flux*. Following Boussinesq's hypothesis, the Reynolds flux can be expressed in terms of the product of an eddy diffusion coefficient K and the mean gradient:

$$-\overline{(w'c')} = K\frac{\partial \bar{c}}{\partial z} \qquad (6.7)$$

The total mass flux is equal to the sum of the diffusive flux $-D(\partial \bar{c}/\partial z)$, and the turbulent flux, $-K(\partial \bar{c}/\partial z)$ or $(\overline{w'c'})$.

The eddy diffusion coefficient for mass K is usually assumed to be of the same order as that for momentum, $E \approx K$ (Sideman and Pinczewski, 1975). Concentration profiles near a wall resemble velocity profiles and include a sublayer in which K has become smaller than the molecular diffusion coefficient D. Most of the concentration change takes place across the sublayer (Chilton and Colburn, 1934; Sherwood and Pigford, 1952; Lin et at., 1953), which we denote as the *diffusive sublayer*.

In water the molecular diffusion coefficient is smaller than the kinematic viscosity, and consequently the region where molecular viscosity dominates momentum flux extends further from the wall than does the region where molecular diffusivity dominates the chemical flux. Although momentum transport is primarily accomplished by viscous interactions throughout the viscous sublayer, the small turbulent motions in that layer continue to be the principal mechanism of mass transfer. However, at some point near the interface, the decreasing intensity of turbulent fluctuations allows molecular diffusion to dominate mass transport.

The existence of turbulent fluctuations in the viscous and diffusive sublayers has been observed optically in several studies using microscopes and high-speed cameras to examine the flow near walls (Kline et al., 1967; Corino and Brodkey, 1969). The observations have given rise to more sophisticated models of mass transfer, such as "penetration" and "surface renewal" theories (again see Sideman and Pinczewski, 1975). However, as presented here, classical models of eddy or turbulent diffusivity serve equally well to predict fluxes across boundary layers (Sideman and Pinczewski, 1975; Sherwood et al. 1975).

Relative thickness of the diffusive sublayer and the viscous sublayer is determined by the ratio of kinematic viscosity to molecular diffusivity. That ratio is called the *Schmidt Number*, $Sc = v/D$. In fluids where the Sc is of order 1 (air, for example), the diffusive sublayer occupies the entire viscous sublayer. As Sc increases, resistance to mass transfer is restricted to a proportionally smaller part of the viscous sublayer. When Sc is in the range of 10^3 to 10^4, as it is for most dissolved species in seawater (table 6–1), the thickness of the diffusive sublayer is approximately $z^+ = 1$ (Sideman and Pinczewski, 1975), whereas the viscous sublayer extends to $z^+ \approx 26$.

Because most of the resistance to mass transfer takes place in the diffusive sublayer, most of the concentration change between a surface and the bulk fluid also occurs there. That has been demonstrated in an elegant manner by Lin et al. (1953, p. 653), who used the change of refractive index of a liquid due to changes in concentration to obtain photographic evidence that the concentration gradient is indeed restricted to a thin diffusive sublayer adjacent to a reacting surface. Their results are reproduced in figure 6–3,

Table 6–1
Molecular Diffusion Coefficients (D) and Schmidt Numbers (Sc) for a Number of Chemical Species of Interest to Oceanographers

Ion	$D\ (10^{-6}\ cm^2 s)$[a]	Sc
H^+	56.1	320
OH^-	25.6	700
Na^+	6.3	2,900
K^+	9.9	1,800
NH_4^+	9.8	1,800
Mg^{2+}	3.6	5,000
Ca^{2+}	3.7	4,900
Ra^{2+}	4.0	4,500
Mn^{2+}	3.0	6,000
Cu^{2+}	3.4	5,300
SO_4^{2-}	5.0	3,600
HCO_3^-	5.0	3,600
NO_3^-	9.8	1,800
CO_3^{2-}	4.4	4,100
Gas	$D\ (10^{-6}\ cm^2/s)$[b]	Sc
CO_2	17.3	1,000
CH_4	16.0	1,100
H_2	37.9	500

Source: Ionic Data from Li and Gregory, 1974; gas data from Sherwood, Pigford and Wilke, 1975.
[a] at 0° C.
[b] at 25° C.

which shows clearly that 90 percent of the concentration gradient is confined to a layer less than 0.15-mm thick or, in dimensionless terms, a layer of thickness $z^+ \leq 3$ for $Sc = 900$ and $u_* \approx 1.0$.

Because the concentration gradient is confined to a thin diffusive sublayer, a simple model useful for predicting mass flux across a boundary layer can be constructed by assuming that the flux in the diffusive sublayer is given by the First Law of Diffusion (Fick, 1855). If the diffusive sublayer has an effective thickness δ, then the concentration gradient is given by the difference between concentrations in the bulk fluid and at the interface divided by the thickness of the diffusive sublayer. Accordingly, the flux can be written

$$J = \frac{D}{\delta}(C_\infty - C_0) \tag{6.8}$$

where J = the flux from fluid and toward the interface [g/(cm²·s)]
C_∞ = the bulk fluid concentration (g/cm³)
C_0 = the concentation at the interface (g/cm³)
δ = the thickness of the diffusive sublayer (cm)

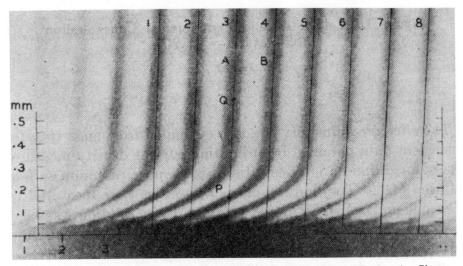

Source: Lin et al., 1953. Reprinted with permission from *Industrial and Engineering Chemistry* 45(3):6. Copyright by the American Chemical Society.

Figure 6–3. Interference fringe photograph showing concentration versus height in a diffusive sublayer. The dark bands (1–8) are adjacent interference fringes formed by monochromatic light passing through two identical paths. In each path light passes through a liquid cell filled with 0.01 M cadmium sulfate. One cell is a reference cell; the other contains a mercury electrode. When a current passes through this electrode, cadmium metal is electroplated from the solution. The flow of water through the cells gives rise to a boundary layer that affects the concentration profile. In the absence of electrode current, the fringes would be located along the drawn lines. When current flows, the displacement of each fringe from that line is proportional to the decrease in concentration; hence the dark lines represent concentration profiles. The figure shows that the concentration gradient near the surface is confined to a thin layer less than 0.03 cm thick for $Sc = 900$ and $u_* = 1.0$ cm/s.

The ratio D/δ defines the relationship between the flux and the difference between the bulk concentration and the concentration at the interface. It forms the basis of the phenomenological mass transfer coefficient β, where $\beta = D/\delta$. It has the dimensions of velocity and is sometimes called the *piston velocity*.

The mass transfer coefficient β is a function of the molecular diffusion coefficient D and the chemical and hydrodynamic parameters which determine the effective thickness of the diffusive sublayer, that is, ν, u_*, and Sc.

Table 6–2
Some Formulas for the Coefficient β that Are Applicable to Mass Transfer at the Sea Floor

Function for Mass Transfer Coefficient	Range of Schmidt Number	Reference
$\beta \approx \dfrac{Du_*}{\nu}$	$Sc \approx 1{,}000$	Sideman and Pinczewski (1975)
$\beta = \dfrac{0.11 u_*}{Sc^{3/4}}$	$Sc > 200$	Deissler (1954)
$\beta = \dfrac{Du_* Sc^{1/3}}{12\nu}$	$Sc > 1$	Wimbush (1976)
$\beta = \dfrac{Du_* Sc^{1/7}}{4.0\nu}$	$Sc > 1$	Katsibas and Gordon (1974)
$\dfrac{1}{\beta} = \dfrac{1}{u_*}\left(\dfrac{1}{\dfrac{1}{1.6 Sc} + 0.062 Sc^{-2/3}} + \dfrac{1}{0.0615 Sc^{-1/2}} \right)$	$Sc > 500$	Hughmark (1975)
$\beta = \dfrac{Du_* Sc^{1/3}}{24\nu}$	$Sc > 1$	Schink and Guinasso (1977b)

Much has been written concerning models that can be used to estimate the value of β (see reviews by Kestin and Richardson, 1963; Leont'ev, 1966; Scriven, 1969a and b; Sideman and Pinczewski, 1975). Table 6–2 presents several expressions for β that have proved successful in predicting mass transfer from smooth surfaces. Values predicted by the equations are within a factor of two of each other (figure 6–4) for a Schmidt Number of 4,000 (typical of the sea floor).

Mixing-length theory permits calculation of shear stress and, consequently, the friction velocity u_* from current measurements made near the sea floor. Wimbush and Munk (1970) calculated values of the friction velocity from their current meter records taken at a spot on the Pacific Ocean floor. They determined values for u_* that ranged from about 0.05 to 0.3 cm/s. Since Schmidt Numbers are about 4,000 for most species in seawater (table 6–1), the thickness of the diffusive sublayer in the deep ocean should be of the order of 1 mm. Weatherly (1972) calculated values of u_* from current meter records taken beneath the Florida Current. In this more

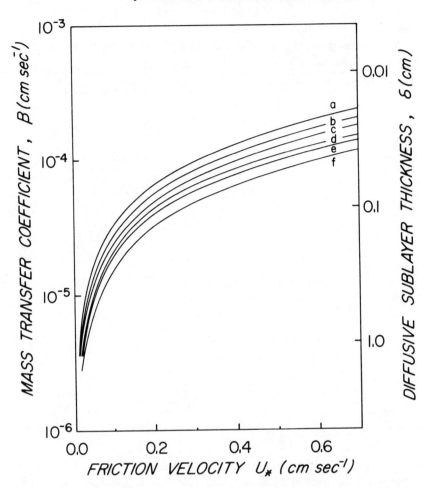

Figure 6-4. Plot of the mass-transfer coefficient (β) and the thickness of the diffusive sublayer (δ_D) as a function of the friction velocity (u_*) calculated using the equations of table 6-2: (a) Wimbush (1976); (b) Hughmark (1975); (c) Sideman and Pinczewski (1975); (d) Deissler (1954); (e) Katsibas and Gordan (1974); (f) Schink and Guinasso (1977b). A Schmidt Number of 4,000 was used, corresponding to a diffusion coefficient about 4.3×10^{-6} cm²/s. All the functions predict a diffusive sublayer about 1-mm thick for a shear velocity typical of the sea floor ($u_* = 0.1$ cm/s, Wimbush, 1976).

energetic region, u_* varied from about 0.1 to 1.0 cm/s. Measurements by Heathershaw (1976) in the shallow waters of the Irish Sea gave u_* values ranging from 0.4 to 2.6 cm/s. Based on the limited number of studies now available, we conclude that in shallow waters the thickness of the diffusive sublayer is almost an order of magnitude smaller than diffusive sublayers in the deep sea. As in laboratory boundary layers, turbulent mixing in the benthic boundary layer should ensure homogeneity above the diffusive sublayer. Continuous profiling of salinity and temperature in the boundary layer of the deep Atlantic by Armi and Millard (1976) and Armi (1977) has shown that those conservative scalar qualities are indeed well mixed to a height of at least 10 m above the bottom. Similarly, Schink et al. (1975) failed to detect any systematic gradients of dissolved silica in the benthic boundary layer immediately above the diffusive sublayer.

Effect of Surface Roughness

Ripples and other sedimentary features; burrows, tubes, and mounds of benthic organisms; manganese nodules; and intrinsic irregularities due to the finite grain size of sediment all contribute to sea-floor roughness. If the heights of the roughness elements are smaller than the "laminar" portion of the viscous sublayer, then the presence of the roughness will have only a negligible effect on the flow and transport characteristics (Schlichting, 1968; Dawson and Trass, 1972; Hinze, 1975). When roughness elements are submerged within the layer, they are not capable of generating sufficient turbulence to disrupt the diffusive sublayer. The thickness of the "laminar" layer is approximately $5\,\nu/u_*$ (Schlichting, 1968; Sternberg, 1970) or 0.9 cm for typical deep-sea conditions ($u_* = 0.1$ cm/s, $\nu = 1.8 \times 10^{-2}$ cm^2/s). Given the fine grain size of deep-sea sediments, roughness due to grain size should have minimal effect.

Studies of the transfer of scalar quantities (heat or chemical species) across boundary layers where roughness elements protrude through the "laminar" sublayer often disagree either partly or wholly on the effect of the roughness. For example, investigations by Nunner (1956), Dipprey and Sabersky (1963), and Dawson and Trass (1972) indicate an increase in the transfer rate with increasing roughness, while studies by Levich (1962), Mahato and Schemilt (1968), and Hughmark (1972, 1975) show the opposite effect. However, in all of those studies, transfer rates from rough surfaces differed from those for smooth surfaces by less than a factor of 3.

Although the diffusive sublayer at the sea floor is very thin, it cannot be ignored when considering exchange of dissolved material between seawater and the sediment/pore-water system or between seawater and solid surfaces such as those of manganese nodules. Sharp gradients within the diffusive sublayer would indicate that diffusion across the layer controls the rates of chemical exchange. In the next section, we investigate conditions under which the diffusive sublayer will affect sediment/seawater interaction.

Modeling the Influence of Boundary-Layer Impedance on Sediment/Seawater Interactions

Reactions at Interfaces

Heterogeneous reactions such as adsorption, desorption, dissolution, precipitation, and microbiological activity dominate interactions at the interface between seawater and solid surface. To maintain those processes, transport of dissolved material through the benthic boundary layer is necessary, and gradients are established because the rate of transfer is finite. As discussed in the previous section, the greatest part of the concentration change takes place across a thin diffusive sublayer. Assuming steady state, the fluxes across the sublayer and toward the interface can be described by equation 6.8.

The formulation is limited to situations where (1) only small mass fluxes occur, i.e., no appreciable density change, (2) there are no homogeneous reactions involving product or reactant within the boundary layer, and (3) the surface is equi-accessible. The first condition is usually readily met at the sediment/seawater interface. Modifications can be made to accommodate systems that do not fulfill the second requirement; however, those will not be considered here (see Vieth et al., 1963; Hershey, 1973; Sherwood et al., 1975). The last condition requires that all points on the surface of the reacting medium be equally accessible with respect to diffusive transport. For the case of a porous material, the last condition can only be approximated.

The rate of exchange across the interface is constrained not only by the diffusive sublayer, but by reaction kinetics as well. The effect of kinetic impedance on the rate of accretion or uptake at an interface can be expressed as a rate law of the form

$$R_s = k(C_0 - C_s)^n \qquad (6.9)$$

where R_s = surface rate of reaction [g/(cm^2 · s)]
k = apparent kinetic rate constant for the reactant under consideration [cm · (cm^3)$^{n-1}$/(g^{n-1} · s)]
n = order of reaction for the reactant under consideration (dimensionless)
C_s = saturation concentration (g/cm^3)

Influence of a Diffusive Sublayer

For the release of a soluble product, the terms in parentheses on the right side of equation 6.9 become $(C_s - C_0)$ (Konak, 1974b; Plummer and Wigley, 1976). Because the relative concentrations of chemical species in seawater are generally non-stoichiometric for any given precipitation reaction, the observed rate of consumption of one of the reactants, often the least abundant, controls the overall rate. At steady state, the rate of reaction must equal the rate at which the limiting reactant reaches the surface through the diffusive sublayer, which, considering equations 6.8 and 6.9, implies that

$$k(C_0 - C_s)^n = \beta(C_\infty - C_0) \tag{6.10}$$

Expressions similar to equation 6.10 have been obtained by Bircumshaw and Riddiford (1952), Levich (1962), Frank-Kamenetskii (1969), Konak (1974a), Plummer (1972), and Plummer and Wigley (1976).

Although k represents an apparent rate constant, dimensional analysis shows that $k(C_0 - C_s)^{n-1}$ has the same units as the mass transfer coefficient (cm/s), and consequently, the terms are directly comparable (Plummer and Wigley, 1976). If we call $k(C_0 - C_s)^{n-1}$ the equivalent rate constant k', equation 6.10 can be written

$$\beta(C_\infty - C_0) = k'(C_0 - C_s) \tag{6.11}$$

Equation 6.11 can be rearranged to solve for concentration of the limiting reactant at the solid interface:

$$C_0 = \frac{\beta C_\infty + k' C_s}{\beta + k'} \tag{6.12}$$

Using equation 6.12 and the fact that the surface reaction rate is given by $k'(C_0 - C_s)$, we can derive a new expression for the surface reaction rate:

$$R_s = \frac{\beta k'(C_\infty - C_s)}{\beta + k'} = k^*(C_\infty - C_s) \tag{6.13}$$

In equation 6.13, the observed reaction-rate coefficient k^* is given by

$$k^* = \frac{\beta k'}{\beta + k'} \tag{6.14}$$

Rearranging equation 6.14 allows us to express the reciprocal of the observed rate constant as

$$\frac{1}{k^*} = \frac{1}{k'} + \frac{1}{\beta} \qquad (6.15)$$

Equation 6.15 is well known in chemical engineering (Frank-Kamenetskii, 1969; Roberts, 1973; Konak, 1974a; Strickland-Constable, (1968) and is often called "the rule of additive resistances." The boundary-layer resistance and kinetic resistance are $1/\beta$ and $1/k'$, respectively.

If the mass transfer coefficient β is much larger than the equivalent rate constant ($\beta \gg k'$), the resistance of the diffusive sublayer can be ignored, and the observed rate is determined only by the kinetics of the surface reaction ($k^* \approx k'$). That is known as the *kinetically controlled regime*. Under such conditions, $C_0 \approx C_\infty$, and the concentration of the limiting reactant at the interface is the same as the concentration in the fluid.

However, if $\beta \ll k'$, then $k^* \approx \beta$, $C_s \ll C_\infty$, and the observed rate of reaction is determined by the rate of mass transfer through the diffusive sublayer:

$$R_s = \beta C_\infty \qquad (6.16)$$

That is known as the *diffusion-controlled regime*. The kinetics of the reaction at the surface do not influence the observed rate since β is primarily a function of the hydrodynamics of the flow. Under those conditions, the reaction will always appear to be of first order regardless of the order of the surface reaction.

If $\beta \approx k'$, the reaction does not fall into either of the limiting regimes, and equation 6.14 must be used to explain the observed rate of reaction (intermediate regime). In that situation, changes in hydrodynamic parameters like flow velocity can cause a transition to either of the two "limiting regimes."

A reaction can occur no faster than reactants can be supplied to and soluble products transported away from the interface. Consequently, reaction rates at surfaces can be no greater than those allowed by the transport of material across the diffusive sublayer. Even if the kinetics at the surface are not known, the impedance of the diffusive sublayer can be used to place an upper limit on actual reaction rates (Plummer, 1972). With that in mind, Boudreau and Scott (1978) have considered the accretion of manganese nodules with the hypothesis that nodule growth is controlled by diffusion of metal ions from seawater. They used equation 6.16 to predict rates of metal transfer through the diffusive sublayer to the nodule surface. The predicted rates were compared to accumulation rates calculated using radiometric techniques. Their results indicated that the impedance of the boundary layer either partially or totally controls the rate of manganese incorporation into nodules and consequently the rate of nodule growth. They further concluded

Influence of a Diffusive Sublayer

that dissolved metals in seawater could supply (by diffusion) all the trace metals found in nodules and that the incorporation of iron is probably kinetically controlled.

The dissolution of biogenic carbonate and silica tests at the sea floor is an important step in both the carbon and silica cycles of the ocean (Broecker and Broecker, 1974: Schink et al. 1975). Diffusion and reactions in the sediment coupled with the transport of dissolution products into the overlaying water figure prominently in dissolution of this material (Schink and Guinasso, 1977b; Takahashi and Broecker, 1977). The kinetics of reaction and mass transfer across the diffusive sublayer can play equally important roles in determining the rate of dissolution.

Reactions Occurring within Sediment

The interaction between sediment and seawater extends into the sediment, as shown by the sharp interstitial gradients of dissolved O_2, NO_3^-, NO_2^-, Mn^{2+}, $Si(OH)_4$, HCO_3^-, NH_4^+, CH_4, etc. in the first few meters of pore water (Berner, 1976a). To determine the flux of dissolved constituents between pore water and the overlying seawater, the pore water concentration gradients must be calculated. To that end, a one-dimensional mathematical model, called *the diagenetic equation* (Berner, 1964, 1971, 1975, 1976a and b; Anikouchine, 1967; Tzur, 1971; Lerman, 1975; Schink et al., 1975; Schink and Guinasso, 1977a and b; etc.), has been developed. One of the more general forms of the equation is (Berner, 1975, 1976a)

$$\frac{\partial(\phi C)}{\partial t} = \frac{\partial[\phi D_c(\partial C/\partial z)]}{\partial z} - \frac{\partial(\phi w C)}{\partial z} + \phi R(C) \qquad (6.17)$$

where
ϕ = porosity (dimensionless)
C = concentration of the dissolved species in the pore water (g/cm^3)
t = time (s)
D_c = diffusion coefficient (cm^2/s)
z = distance from the sediment/seawater interface with reference to the interface and increasing downward (cm)
w = advective velocity of pore water (cm/s)
$R(C)$ = the rate of all chemical reactions affecting the value of C[g/(cm$^3 \cdot$s)] per unit volume of pore water

A number of assumptions can be made to simplify equation 6.17. Such simplifications reduce the general applicability of the equation. However, the conclusions drawn from the following analysis should be relevant to most situations.

The porosity will be considered constant with depth and time for the sediment section of interest—the top meter. That appears to be valid for abyssal sediments (Schink and Guinasso, 1977a). Then,

$$\frac{\partial \phi}{\partial t} = 0 \quad \text{and} \quad \frac{\partial \phi}{\partial z} = 0.$$

We also assume that

$$\frac{\partial w}{\partial z} = 0 \quad \text{and} \quad \frac{\partial D_c}{\partial z} = 0,$$

and thus the rate of advection is equal to the rate of deposition (Berner, 1975). If the rate of deposition is considered small compared to diffusion, which is usually the case for abyssal sediments, then the advection term can be neglected. In addition, equation 6.17 ignores the effects of adsorption (Berner, 1976b; Schink and Guinasso, 1978) and "ventilation" (Goldhaber et al., 1977). Ventilation of pore water does not obliterate the effects of a diffusive sublayer at the sediment/seawater interface. Ventilation primarily affects vertical transport; the horizontal velocity would still go to zero at the interface. The following development could be modified to accommodate the process; however, irrigation is thought to be a relatively minor process in the deep sea (Schink and Guinasso, 1977a), where the results of this discussion are intended to apply.

With the preceding assumptions, equation 6.17 can be rewritten as

$$\frac{\partial C}{\partial t} = D_c \frac{\partial^2 C}{\partial z^2} + R(C) \tag{6.18}$$

The diffusion coefficient D_c takes into account the effects of tortuosity only (Petersen, 1965; Strieder and Aris, 1973; van Brakel and Heertjes, 1974).

The exact form of the reaction term on the right side of equation 6.18 has not been specified. The term represents the production or removal of solute per unit volume of pore water. It will be assumed that the reaction takes place throughout the volume occupied by the pore water; in other words, the reaction can be considered pseudo-homogeneous (Frank-Kamenetskii, 1969). The rate constant associated with the reaction is assumed to be independent of depth; the reaction will, however, depend on the amount of solid reactive material present. Thus, for the release of soluble product, the reaction term has the form

$$R(C) = k_c (C_s - C)^n \frac{B(z)}{B_0} \tag{6.19}$$

where k_c = apparent (pseudo-homogeneous) rate constant
$[(cm^3)^{n-1}/(g^{n-1} \cdot s)]$
n = order of reaction (dimensionless)
C_s = saturation concentration of the soluble product per unit volume of pore water (g/cm^3)
C = concentration of dissolved material at some depth z per unit volume of pore water (g/cm^3)
$B(z)$ = concentration of dissolving solid material at depth z $(g/cm^3$, bulk phase)
B_0 = concentration of dissolving solid material at $z = 0$ $(g/cm^3$, bulk phase)

The pseudo-homogeneous rate constant k_c in equation 6.19 is related to the rate constant of the heterogeneous reaction k by the expression

$$k_c = k \mathcal{A} \, (B_0/\phi) \tag{6.20}$$

where $\mathcal{A}$ is the specific surface area of the reactive material (cm^2/g).

Combining equations 6.18 and 6.19 we can write a simplified diagenetic equation

$$\frac{\partial C}{\partial t} = D_c \frac{\partial^2 C}{\partial z^2} + k_c (C_s - C)^n \frac{B(z)}{B_0}$$

which at steady state becomes

$$D_c \frac{d^2 C}{dz^2} + k_c (C_s - C)^n \frac{B(z)}{B_0} = 0 \tag{6.21a}$$

The boundary conditions associated with equation 6.21a, when there is a surplus of B, i.e., $B(\infty) > 0$, are

$$\begin{aligned} C &= C_0 & \text{at } z = 0 \\ C &= C_s & \text{at } z = \infty \end{aligned} \tag{6.21b}$$

In addition to equation 6.21a, we also have the differential equation describing the distribution of soluble solid material with depth. At steady state and ignoring bioturbation, it has the form

$$w \frac{dB}{dz} + \phi k_c (C_s - C)^n \frac{B(z)}{B_0} = 0. \tag{6.22a}$$

The advective velocity w has been treated as a constant in equation 22a. Although only an approximation in some systems (e.g., carbonate dissolution), the assumption nevertheless greatly reduces the mathematical complexity of the problem and is sufficient for our purposes. The boundary conditions for equation 22a are

$$B(z) = B_0 \quad \text{at } z = 0$$
$$\frac{dB}{dz} = 0 \quad \text{at } z = \infty$$
(6.22b)

The boundary conditions are similar to those used by Schink et al. (1975).

The flux of the dissolved substance into the overlying boundary layer can now be calculated. This flux is given by

$$F_s = -\phi D_c \frac{dC}{dz}\bigg|_{z=0} \quad (6.23)$$

where F_s denotes mass of substance diffusing per unit of interfacial surface area per unit time [g/(gcm$^2 \cdot$s)]. An analytical expression for the flux at the interface ($z = 0$) is found by integrating equations 6.21a and 6.22a and combining the results with equation 6.23 to yield

$$F_s = -\phi A \left[D_c k_c (C_s - C_0)^{n-1} \right]^{1/2} (C_s - C_0) \quad (6.24)$$

where A is a dimensionless constant of order 1. The method of solution is given in the appendix 6A. The term $\phi A [D_c k_c (C_s - C_0)^{n-1}]^{1/2}$ (which we will denote as k_i) has the units of centimeters per second and represents the reciprocal of the combined kinetic and diffusional resistance of the sediment/pore-water system to chemical exchange with seawater. It is analogous to k' in equation 6.11, and we will call it the *internal transfer velocity* (or *coefficient*).

If reactions involving the dissolving species do not occur in the diffusive sublayer, then the diffusive flux through the interface must equal that across the diffusive sublayer. Equating those two fluxes gives

$$\beta(C_0 - C_\infty) = k_i(C_s - C_0) \quad (6.25)$$

An analysis for reactions at a solid surface similar to the analysis in the previous section can be applied to reactions in a porous media. If the resistance of diffusive sublayer is much less than the resistance within the sediment, i.e., if $\beta \gg k_i$, then the concentration at the interface must be

Influence of a Diffusive Sublayer

approximately that of the bulk water ($C_0 \approx C_\infty$) and pore-water concentration will increase from C_∞ at the interface to C_s at some depth. In such a case, the hydrodynamics of the boundary layer do not affect the observed reaction rate, a situation known as an *internally controlled regime*.

If the boundary-layer resistance is much greater than the internal resistance $\beta \ll k_i$, the overall reaction rate is controlled by mass transfer through the diffusive sublayer. Under the *external diffusion regime*, the concentration at the interface must be near C_s. Because the mass transfer coefficient β is primarily a function of the flow characteristics, the reaction rate is determined by the hydrodynamics and not by kinetics of reactions occurring at the interface or within the sediments.

If we divide β into D and k_i into D_c, the results are, respectively, the effective thickness δ of the diffusive sublayer and the scale length δ_i of the diffusion profile in the sediments. When $\beta \gg k_i$, as in the internally controlled regime, the diffusion scale length δ_i is much greater than the thickness of the diffusive sublayer. For the external-diffusion regime, where $\beta \gg k_i$, $\delta_i \ll \delta$. As one might expect, a diffusion profile extending into the sediment to depths much greater than the diffusive sublayer thickness indicates an internal regime.

Intermediate (or transitional) regimes may exist where the rates of reaction and transfer are such that both kinetics and hydrodynamics are important. The properties of the three regimes are summarized in figure 6–5.

The preceding treatment indicates that in order to properly model the dissolution of biogenic debris on and within pelagic sediments, the relative importance of kinetic and transfer resistances must be determined for the hydrodynamic conditions prevailing at the ocean floor. To illustrate, consider the carbonate dissolution studies conducted by Morse and Berner (1972) and Plummer and Wigley (1976). In their experiments, the dissolution of fine suspended calcite particles was controlled by the kinetics of the surface reaction at the levels of undersaturation encountered at the sea floor. If solution kinetics rather than diffusion also controlled dissolution on the sea floor, then we should find virtually no gradient of dissolved products in the benthic boundary layer ($C_0 \approx C_\infty$, figure 6–5a), and current velocities should have essentially no effect on the rate of dissolution. But the geometry and mass transfer characteristics of fine suspended particles do not reflect those of the sea floor. A diffusive sublayer of 1-mm thickness typical of the benthic boundary layer is 100 times thicker than the boundary layer around particles 10 μm in size (as calculated from the mass transfer coefficients for suspended particles in the Plummer and Wigley study). In addition, the fine biogenic calcium particles in deep-sea sediments have a large specific surface area (Honjo, 1977). Both properties enhance dissolution rates and consequently decrease the internal resistance of the sediment.

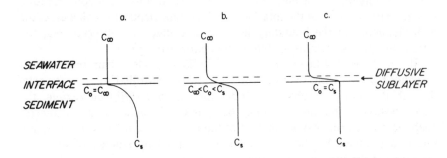

Figure 6–5. Schematic representation of the concentration gradient of a species diffusing from the sediment: (a) internally controlled regime; (b) intermediate regime; (c) external diffusion regime.

Schink and Guinasso (1977b), using a model that includes boundary-layer transfer resistance, concluded that bottom currents do have an influence on carbonate dissolution. Their calculated concentration profiles for pore-water carbonate ion suggest that calcite dissolution occurs within the intermediate regime (figure 6–5b).

Unlike the carbonate ion profiles, dissolved silica exhibits gradients that extend several centimeters into the sediment (Fanning and Pilson, 1971, 1974; Hurd, 1973; Schink et al., 1975). That prompted Schink and Guinasso (1977a) to conclude that the dissolution of biogenic silica is little influenced by the presence of the benthic boundary layer and that dissolution occurs within the internal regime (figure 6–5a). Therefore, little error should be introduced into silica-flux calculations by assuming $C_0 \approx C_\infty$.

Conclusions

Viscous forces acting on seawater near the sea floor suppress turbulence to the extent that a thin layer adjacent to the sediment/seawater interface exists in which molecular diffusion is the dominant mode of transport of dissolved material. Although that diffusive sublayer presents an impedance to chemical exchange between seawater and sediments, its influence on the rate of the interaction depends on the kinetics of the reactions going on at the sea floor.

We have presented simple models for the kinetic impedance of reactions taking place on a surface and for the combined kinetic and diffusive impedance (internal impedance) for reactions taking place in porous media. The comparable impedance of the diffusive sublayer to mass transfer is given by the inverse of the mass transport coefficient, which can be estimated by models available in the literature. A diffusive sublayer has an effect on reaction rates at the sea floor only if its impedance to mass transfer is the same order as or greater than the kinetic impedance or internal impedance.

By assessing the relative magnitudes of the diffusive-sublayer impedance and the kinetic or internal impedances, one can distinguish three types of regimes of chemical reactions that are taking place at the sea floor. A diffusion-controlled regime exists when the boundary-layer impedance limits mass transfer. A kinetically controlled regime arises when the kinetic or internal impedance limits mass transfer. An intermediate case results when the kinetic and diffusive impedances are of the same order.

The ability to distinguish between those regimes is important to understanding the kinetics of chemical reactions at the sea floor. The different regimes respond quite differently to changes in environmental parameters, such as current velocity, and also produce different types of concentration profiles in abyssal sediments.

Acknowledgments

This work is based in part on two papers presented in Edinburgh in the fall of 1976 at the Joint Oceanographic Assembly, Symposium on Benthic Processes and the Geochemistry of Interstitial Waters of Marine Sediments. M.R. Scott's paper, "Accumulation rates of metals along active ocean ridges," described a model for maganese nodule growth that treated metal fluxes to the nodule as processes controlled by diffusion across a diffusive sublayer. A paper by N.L. Guinasso, Jr., K.A. Fanning, and D.R. Schink, "Modelling the dissolution of biogenic tests at the sea floor," presented calculations of silica and calcium carbonate dissolution using a model that treated the effect of the diffusive sublayer.

Appreciation is given to D.R. Schink and A.D. Kirwan, Jr., who reviewed this manuscript and made valuable suggestions. The authors acknowledge useful discussions with A. Nachman and S. Taliaferro.

This work was supported by the Office of Naval Research Contract N00014-75-C-0537 and by National Science Foundation Grant Numbers OCE 75-21275, OCE 77-11372 and OCE 77-21009. Boudreau received support from a National Research Council of Canada Post-Graduate Scholarship.

References

Ames, W.F. 1968. *Nonlinear Ordinary Differential Equations in Transport Process*. New York: Academic Press.
Anikouchine, W.A. 1967. Dissolved chemical substances in compacting marine sediments. *J. Geophys. Res.* 72:505–509.

Armi, L. 1977. The dynamics of the bottom boundary layer in the deep ocean. In J.C.J. Nihoul (Ed.), *Bottom Turbulence*. Amsterdam: Elsevier, pp. 153–164.

Armi, L., and Millard, R.C. 1976. The bottom boundary layer of the deep ocean. *J. Geophys. Res.* 81(27):4983–4990.

Berner, R.A. 1964. An idealized model of dissolved sulfate distribution in recent sediments. *Geochim. Cosmochim. Acta* 28:1497–1503.

Berner, R.A. 1971. *Principles of Chemical Sedimentology*. New York: McGraw-Hill.

Berner, R.A. 1975. Diagenetic models of dissolved species in the interstitial waters of compacting sediments. *Am. J. Sci.* 275:88–96.

Berner, R.A. 1976a. The benthic boundary layer from the viewpoint of a geochemist. In I.N. McCave (Ed.), *The Benthic Boundary Layer*. New York: Plenum, pp. 33–55.

Berner, R.A. 1976b. Inclusion of adsorption in the modelling of early diagenesis. *Earth Planet. Sci. Lett.* 29:333–340.

Bircumshaw, L.L., and Riddiford, A.C. 1952. Transport control in heterogeneous reaction. *Quart. Rev. Chem. Soc. London* 6:157–187.

Boudreau, B.P., and Scott, M.A. 1978. A model for the diffusion controlled growth of deep-sea manganese nodules. *Am. J. Sci.* 278:903–929.

Boussinesq, M.J. 1877. Essai: Sur la théorie des eaux courantes. *Mem. Prés. par Divers Savants, Acad. Sci. Paris* 23:46.

Broecker, W.S. 1971. A kinetic model for the chemical composition of seawater. *Quat. Res.* 1:188–207.

Broecker, W.S., and Broecker, S. 1974. Carbonate dissolution on the western flank of the East Pacific Rise. In W.W. Hay, (Ed.), *Studies in Paleo-Oceanography*. Soc. Econ. Paleo. Mineral., Spec. Publ. No. 20, pp. 44–57.

Chilton, T.H., and Colburn, A.P. 1934. Mass transfer (adsorption) coefficients. *Ind. Eng. Chem.* 26:1183–1187.

Corino, E.R., and Brodkey, R.S. 1969. A visual investigation of the wall region in turbulent flow. *J. Fluid Mech.* 37:1–30.

Dawson, D.A., and Trass, O. 1972. Mass transfer at rough surfaces. *Int. J. Heat Mass Transfer* 15:1317–1336.

Deissler, R.G. 1954. Analysis of turbulent heat transfer, mass transfer and friction in smooth tubes at high Prandtl and Schmidt numbers. *Nat. Adv. Comm. Aero. Tech. note* 3145.

Dipprey, D.F., and Sabersky, R.H. 1963. Heat and momentum transfer in smooth and rough tubes at various Prandtl numbers. *Int. J. Heat Mass Transfer* 6:329–335.

Fanning, K.A., and Pilson, M.E.Q. 1971. Interstitial silica and pH in marine sediments: Some effects of sampling procedures. *Science* 173:1228–1231.

Fanning, K.A., and Pilson, M.E.Q. 1974. The diffusion of dissolved silica out of deep-sea sediments. *J. Geophys. Res.* 79:1293–1297.

Fick, A. 1855. Über diffusion. *Poggendorff's Annalen der Physik* 94:59–86.

Frank-Kamenetskii, D.A. 1969. In J.P. Appleton (Ed.), *Diffusion and Heat Transfer in Chemical Kinetics*, 2d ed. New York: Plenum.

Goldhaber, M.B., Aller, R.C., Cochran, J.K., Rosenfeld, J.K., Martens, C.S., and Berner, R.A. 1977. Sulfate reduction, diffusion, and bioturbation in Long Island Sound sediments: Report of the FOAM group. *Am. J. Sci.* 227:193–237.

Heathershaw, A.D. 1976. Measurements of turbulence in the Irish Sea benthic boundary layer. In I.N. McCave (Ed.), *The Benthic Boundary Layer*. New York: Plenum, pp. 11–31.

Hershey, D. 1973. *Transport Analysis*. New York: Plenum.

Hinze, J.O. 1975. *Turbulence*, 2d ed. New York: McGraw-Hill.

Holland, H.D. 1974. Marine evaporites and the composition of seawater during the Phanerozoic. In W.W. Hay (Ed.), *Studies in Paleo-Oceanography*. Soc. Econ. Paleon. Mineral. Spec. Publ. No. 20, pp. 187–192.

Honjo, S. 1977. Biogenic carbonate particles in the ocean: Do they dissolve in the water column? In N.R. Andersen, and A. Malahoff (Eds.), *The Fate of Fossil Fuel CO_2 in the Oceans*. New York: Plenum, pp. 269–294.

Hughmark, G.A. 1972. Turbulent heat and mass transfer from rough surfaces. *Am. Inst. Chem. Eng. J.* 18(3):667–668.

Hughmark, G.A. 1975. Heat, mass, and momentum transport with turbulent flow in smooth and rough pipe. *Am. Inst. Chem. Eng. J.* 21(5):1033–1035.

Hurd, D.C. 1973. Interactions of biogenic opal, sediments, and seawater in the Central Equatorial Pacific. *Geochim. Cosmochim. Acta* 37:2257–2282.

Imboden, D.M. 1975. Interstitial transport of solutes in non-steady state accumulating and compacting sediments. *Earth Planet. Sci. Lett.* 27:221–228.

Katsibas, P., and Gordon, R.J. 1974. Momentum and energy transfer in turbulent pipe flow: The penetration model revisited. *Am. Inst. Chem. Eng. J.* 20(1):191–193.

Kestin, J., and Richardson, P.D. 1963. Heat transfer across turbulent, incompressible boundary layers. *Int. J. Heat Mass Transfer* 6:147–189.

Kline, S.J., Reynolds, W.C., Schraub, F.A., and Runstadler, P.W. 1967. Structure of turbulent boundary layers. *J. Fluid Mech.* 30:741–773.

Konak, A.R. 1974a. A new model for surface reaction-controlled growth of crystals from solution. *Chem. Eng. Sci.* 29:1537–1543.

Konak, A.R. 1974b. Surface reaction-controlled dissolution of crystals in a solvent or solution. *Chem. Eng. Sci.* 29:1785–1788.

Laufer, J. 1953. The structure of turbulence in fully developed pipe flow. *Nat. Adv. Comm. Aero., Tech. Note* 2954.

Leont'ev, A.I. 1966. Heat and mass transfer in turbulent boundary layers. In T.F. Irvine and J.P. Hartnett (Eds.), *Advances in Heat Transfer.* New York: Academic Press, pp. 33–100..

Lerman, A. 1975. Maintenance of steady state in oceanic sediments. *Am. J. Sci.* 275:609–635.

Lerman, A. 1977. Migrational processes and chemical reactions in interstitial waters. In E.D. Goldberg (Ed.), *The Sea*, Vol. 6. New York: Wiley, pp. 695–738.

Levich, V.G. 1962. In L.E. Scriven (Ed.), *Physiochemical Hydrodynamics.* New Jersey: Prentice-Hall.

Li, Y.-H., and Gregory, S. 1974. Diffusion of ions in sea water and in deep-sea sediments. *Geochim. Cosmochim. Acta* 38:703–714.

Lin, C.S., Moulton, R.W., and Putnam, G.L. 1953. Mass transfer between solid wall and fluid streams. *Ind. Eng. Chem.* 45:636–646.

Mackenzie, F.T., and Garrels, R.M. 1966. Chemical mass balance between rivers and oceans. *Am. J. Sci.* 264:507–525.

Mahato, B.K., and Schemilt, L.W. 1968. Effect of surface roughness on mass transfer. *Chem. Eng. Sci.* 23:183–185.

Morse, J.W. 1974. Calculation of diffusive fluxes across the sediment-water interface. *J. Geophys. Res.* 33:5045–5048.

Morse, J.W., and Berner, R.A. 1972. Dissolution kinetics of calcium carbonate in seawater. II. A kinetic origin for the lysocline. *Am. J. Sci.* 272:840–851.

Nikuradse, J. 1933. Stromungsgesetze in rauhen Rohren. *Forschungsheft* 361:1–22.

Nunner W. 1956. Warmeubergang und Druckaball in rauhen Rohren. *Ver. Deut. Ing. Forschungsh* 455B(22):5–39.

Parker, J.D., Boggs, J.H., and Blick, E.F. 1974. *Introduction to Fluid Mechanics and Heat Transfer.* Reading, Mass.: Addison-Wesley.

Petersen, E.E. 1965. Physical properties of porous catalysts. In *Chemical Reaction Analysis.* New Jersey: Prentice-Hall, pp. 106–128.

Plummer, L.N. 1972. Rates of mineral-aqueous solution reactions. Ph.D. dissertation, Northwestern Univ.

Plummer, L.N., and Wigley, T.M. 1976. The dissolution of calcite in CO_2-saturated solutions at 25°C and 1 atmosphere total pressure. *Geochim. Cosmochim. Acta* 40:191–202.

Prandtl, von L. 1928. Bemerkung uber der Warmeubergang im Rohr. *Physik. Zeitschr.* 29:487–492.

Reichardt, H. 1951. Die Grundlagen des Turbulenten Warmeuberganes. *Arch. Ges. Warmetech.* 2:129.

Reynolds, O. 1895. On the dynamical theory of incompressible viscous fluids and the determination of the criterion. *Phil. Trans. Roy. Soc. Lond.* A186:123–164.

Roberts, J. 1973. Towards a better understanding of high rate biological film flow reactor theory: *Water Res.* 7:1561–1588.

Schink, D.R., Guinasso, N.L., Jr., and Fanning, K.A. 1975. Processes affecting the concentration of silica at the sediment-water interface in the Atlantic Ocean. *J. Geophys. Res.* 80:3013–3031.

Schink, D.R., and Guinasso, N.L., Jr. 1977*a*. Effects of bioturbation on sediment seawater interaction. *Mar. Geol.* 23:133–154.

Schink, D.R., and Guinasso, N.L., Jr. 1977*b*. Modelling the influence of bioturbation and other processes on calcium carbonate dissolution at the sea floor. In N.R. Andersen and A. Malahoff (Eds.), *The Fate of Fossil Fuel CO_2 in the Oceans*. New York: Plenum, pp. 375–399.

Schink, D.R., and Guinasso, N.L., Jr. 1978. Redistribution of dissolved and adsorbed materials in abyssal marine sediments undergoing biological stirring. *Am. J. Sci.* 278:687–702.

Schlichting, H. 1968. *Boundary-Layer Theory*, 6th ed. J. Kestin, trans. New York: McGraw-Hill.

Scriven, L.E. 1969*a*. Flow and transfer at fluid interfaces. II. Models. *Chem. Eng. Educa.* 3:26–29.

Scriven, L.E. 1969*b*. Flow and transfer at fluid interfaces. III. Convective diffusion. *Chem. Eng. Educa.* 3:94–98.

Sherwood, T.K., and Pigford, R.L. 1952. *Adsorption and Extraction*, 2d ed. New York: McGraw-Hill.

Sideman, S., and Pinczewski, W.V. 1975. Turbulent heat and mass transfer: Transport models and mechanisms. In C. Gutfinger (Ed.), *Topics in Transport Phenomena*. New York: Wiley, pp. 47–207.

Sillen, L.G. 1961. The physical chemistry of sea water. In M. Sears (Ed.), *Oceanography*. Washington: Am. Assoc. Adv. Sci. Pub. 67, pp. 549–581.

Sternberg, R.W. 1970. Field measurements of the hydrodynamic roughness of the deep-sea boundary. *Deep-Sea Res.* 17:413–420.

Strickland-Constable, R.F. 1968. *Kinetics and Mechanism of Crystallization*. New York: Academic Press.

Strieder, W., and Aris, R. 1973. Diffusion through a porous media. In *Variational Methods Applied to Problems of Diffusion and Reaction*. New York: Springer-Verlag, pp. 18–41.

Takahaski, T., and Broecker, W. 1977. Mechanism of calcite dissolution on the sea floor. In N. Andersen and A. Malahoff (Eds.), *The Fate of Fossil Fuel CO_2 in the Oceans*. New York: Plenum, pp. 455–478.

Taylor, G.I. 1916. Conditions at the surface of a hot body exposed to the wind. *Brit. Adv. Com. Aero. R. and M.* 272:423–429.

Tzur, Y. 1971. Interstitial diffusion and advection of solute in accumulating sediments. *J. Geophys. Res.* 76:4208–4211.

van Brakel, J., and Heertjes, P.M. 1974. Analysis of diffusion in macroporous media in terms of a porosity, a tortuosity and a constrictivity factor. *Int. J. Heat Mass Transfer* 17:1093–1103.

Vieth, W.R., Porter, J.H., and Sherwood, T.K. 1963. Mass transfer and chemical reaction in a turbulent boundary layer. *Ind. Eng. Chem. Fund.* 2(1):1–3.

Weatherly, G.L. 1972. A study of the bottom boundary layer of the Florida Current. *J. Phys. Ocean.* 2:54–72.

Wimbush, A.H.M.H., and Munk, W. 1970. The benthic boundary layer. In A.E. Maxwell (Ed.), *The Sea*, Vol. 4. New York: Wiley, pp. 731–758.

Wimbush, M. 1976. The physics of the benthic boundary layer. In I.N. McCave (Ed.), *The Benthic Boundary Layer*. New York: Plenum, pp. 3–10.

Appendix 6A: Solution to the Diffusion Equation

In the text it was stated that changes in concentration of a soluble product with depth in a sediment can be described by a second-order nonlinear ordinary differential equation (that is, time independent)

$$D_c \frac{d^2 C}{dz^2} + k_c(C_s - C)^n \frac{B(z)}{B_0} = 0 \qquad (6\text{A}.1)$$

with the boundary conditions

$$C = C_0 \quad \text{at } z = 0$$
$$C = C_s \quad \text{at } z = \infty \qquad (6\text{A}.2)$$

To calcualte the flux out of the sediment, it is necessary to evaluate the first derivative of concentration with respect to depth. We first introduce dimensionless variables,

$$\theta = \frac{(C_s - C)}{(C_s - C_0)} \quad \text{Dimensionless concentration of dissolved material}$$

$$\zeta = z \left[\frac{k_c(C_s - C_0)^{n-1}}{D_c} \right]^{1/2} \quad \text{Dimensionless distance}$$

$$b(\zeta) = \frac{B(z)}{B_0} \quad \text{Dimensionless concentration of dissolvable material}$$

Using these variables, equation 6A.1 becomes

$$\frac{d^2 \theta}{d\zeta^2} = \theta^n b(\zeta) \qquad (6\text{A}.3)$$

To solve equation 6A.3 the dependence of b on z or ζ must be known. We integrate equation 6.22a introduced earlier in the main text.

Using our dimensionless variables, equation 6.22a becomes

$$\frac{d[b(\zeta)]}{b(\zeta)} = -\gamma \theta^n d\zeta \tag{6A.4}$$

where γ is a dimensionless constant given by

$$\gamma = \frac{\phi[D_c k_c (C_s - C_0)^{n+1}]^{\frac{1}{2}}}{w B_0} \tag{6A.5}$$

Equation 6A.4 is readily integrated to yield

$$b(\zeta) = \exp\left[-\gamma \int_0^\zeta \theta^n(\omega) d\omega\right] \tag{6.A6}$$

where ω is a dummy variable.

Substituting equation 6A.6 into equation 6A.3, we arrive at

$$\theta''(\zeta) = \theta^n(\zeta) \cdot \exp\left[-\gamma \int_0^\zeta \theta^n(\omega) d\omega\right] \tag{6A.7}$$

where θ'' is the second derivative of θ with respect to ζ.

Equation 6A.7 can be integrated with respect to ζ to give

$$\theta'(\zeta) - \theta'(0) = -\frac{1}{\gamma}\left\{\exp\left[-\gamma \int_0^\zeta \theta^n(\omega) d\omega\right] - 1\right\} \tag{6A.8}$$

where θ' is the first derivative of θ with respect to ζ.

Now we are interested in the flux produced by the gradient in concentration from $\zeta = 0$ to $\zeta = \infty$, that is, $z = 0$ to $z = \infty$. For those depths, the boundary conditions in equation 6A.2 give that

$$\theta'(\infty) = 0$$

and since $\theta'(0)$ is negative, the right side of equation 6A.8 must also be negative. We will denote the right side of equation 6A.8 as A, so we can simply write

$$\theta'(0) = -A \tag{6A.9}$$

and A is a constant of order 1.

We are now in a position to calculate the flux out of the sediment and across the interface. It is assumed that the flux can be described by Fick's First Law of Diffusion,

Influence of a Diffusive Sublayer

$$F_s = -\phi D_c \frac{dC}{dz}\bigg|_{z=0} \quad (6\text{A}.10)$$

If we introduce our dimensionless variables, the flux can be written

$$F_s = \phi[D_c k_c (C_s - C_0)^{n+1}]^{1/2}\, \theta'(0) \quad (6\text{A}.11)$$

Bringing $(C_s - C_0)^2$ outside the square root brackets and using equation 6A.9 to replace θ', equation 6A.11 can be written as

$$F_s = -\phi A [D_c k_c (C_s - C_0)^{n-1}]^{1/2} (C_s - C_0) \quad (6\text{A}.12)$$

As a final note of interest, consider the solution of the flux for $B(z)$ equal to a constant. Equation 6A.3 then reduces to

$$\frac{d^2\theta}{d\zeta^2} = \theta^n \quad (6\text{A}.13)$$

The method of solution for this equation is given in Ames (1968, pp. 42–43). The result for the given boundary conditions in equation 6A.2 is

$$\theta' = \left(\frac{2}{n+1}\right)^{1/2} \theta^{(n+1)/2} \quad (6\text{A}.14)$$

and the flux is thus

$$F_s = -\phi \left(\frac{2}{n+1}\right)^{1/2} [D_c k_c (C_s - C_0)^{n-1}]^{1/2} (C_s - C_0) \quad (6\text{A}.15)$$

Clearly, from equation 6A.15,

$$A = \sqrt{\frac{2}{n+1}}$$

which is of order 1 for any reasonable value of n.

Part III
Biological Interactions at the Sea Floor

7 Microbiological Studies of Decompressed and Undecompressed Water Samples Collected with a Deep-Ocean Sampler

Rita R. Colwell and
Paul S. Tabor

Abstract

A sampler has been developed which permits collection of deep-ocean-water samples for study of the ecology and systematics of deep-sea bacteria. In situ conditions of pressure and temperature are maintained during recovery of samples and subsequent incubation. The sampler has been successfully deployed in the Puerto Rico Trench of the Atlantic Ocean. Morphology of the deep-sea bacteria, using scanning electron microscopy, and rates of conversion of ^{14}C-labeled substrates by the bacteria have been determined under in situ conditions maintained in the sampler. Low rates of respiration of labeled substrate were observed for Puerto Rico Trench water samples maintained under in situ temperature and pressure.

Introduction

Microbial activity in the deep ocean is only just beginning to be examined using the more critical tools of modern microbiology. Relatively few studies on microorganisms isolated from the deepest regions of the world oceans have been done, and those few are not definitive because it is quite difficult to maintain and control the fundamental environmental parameters of temperature and hydrostatic pressure (> 300 atm) during recovery of samples. Since temperature and pressure act synergistically on the metabolic activity of microorganisms, those parameters cannot be considered independently if one is to study the composition and activities of hyperbaric communities. Although respiration rates and oxygen debts have been measured for deep-sea sediments (Pamatmat, 1973; Smith and Teal, 1973), the role of bacteria in carbon, nitrogen, and phosphorous cycling in deep-sea sediment and overlying water has been difficult to elucidate because of problems in maintaining in situ pressure and temperature of bacteria recovered from the deep sea. Information available in the literature suggests that since the nutrients present in water and sediment can be sufficient to support microbiological populations in many regions of the deep sea, parameters other than nutrient concentration may act to control microbial activities.

Low temperatures alone do not necessarily result in low rates of microbial activity, as shown by Gillespie et al. (1976) who described significant heterotrophic activity for amino acids in the Antarctic, where the water temperature is in the range of -3 to $-1°C$. However, the combined effect of low temperature and high hydrostatic pressure does appear to restrict microbial activity. Data collected to date indicate that heterotrophic activity proceeds relatively slowly under such conditions compared with identical microbial activities at ambient pressure (Jannasch and Wirsen, 1976, 1977).

Comparative studies of barophobes and slightly barophilic bacteria have not yet been done, but will be required if morphological or physiological characteristics peculiar to deep-sea microorganisms exist and are to be recognized.

The microbial ecology of the deep ocean commands interest because of increased activity toward economic exploitation of the deep ocean for mining, energy, and dumping of industrial and domestic wastes. The consequences of such activities in the deep sea, particularly the responses of native microbial populations, are not well known.

A sampler developed in our laboratory permits collection of deep-ocean water for analysis of the ecology and systematics of deep-sea bacteria so that in situ conditions of pressure and temperature can be maintained during recovery and incubation. Jannasch and Wirsen (1976, 1977) constructed a sampler which has a purpose similar to that of our sampler; however, ours has a different and somewhat simpler design.

Materials and Methods

The Department of Microbiology at the University of Maryland, in a collaborative effort with the National Bureau of Standards, Gaithersburg, Maryland, designed and constructed a deep-sea sampling device to collect seawater without decompression. The sampler also has a transfer system to maintain pressure during subsampling and incubation aboard ship or in the laboratory.

The Deep-Ocean Sampler or DOS consists of two interlocking stainless steel sub-units (figure 7–1.) A programmed, cam-driven triggering mechanism and a spring-driven pump are integrated into a single subunit; the sequence of timed events during sampling is messenger-activated through a spring-wound triggering mechanism. The second subunit consists of a thick-walled cylinder with two main valves of the same diameter as the cylinder

Microbiological Studies

bore at either end and two smaller valves, a check valve and a sampling valve, on the wall of the chamber. The 400-ml cylinder serves as the sample receiving chamber, and its inner surfaces are Teflon-coated. The chamber can be used for continuous incubation of samples at deep-ocean pressures and temperatures.

Prior to deployment, the sampler chamber is filled with distilled water containing a coloring agent, pressurized slightly above ambient pressure, and autoclaved. After cooling to 5°C, a sterilized Enerpac hydraulic pumping system of the type commonly employed in laboratory pressure work is used to pressurize the chamber to the approximate hydrostatic pressure at the desired sampling depth in the ocean. The colored solution provides a means for adequately determining the extent of the flushing of the chamber during sampling. The two main end valves are spring-loaded in the open position; however, the valves remain closed because of the hydrostatic pressure in the chamber (figure 7–1.1). Similarly, the check valve cannot be accidentally released because of increased pressure in the chamber.

With the main valve springs loaded, the pump and programmer subunit are joined to the sample chamber, and a Teflon tube links the pump intake to one of the end valves. During sampling, seawater is drawn from the lower end into the chamber and through the pump, thus eliminating the possibility of contamination from the pump. The flow is unrestricted through the 4.8-cm bore of the chamber when the biconically tapered main end valves are open (figure 7–1.2). The flow rate generated by the tandem spring-driven pump averages 4 liters per minute, while the total flow is 8 liters, or 20 times the volume of the chamber. Three cam-activated triggering arms are connected to the triggering mechanism. The arms release and close the check valve at the beginning of the sampling sequence and release the large springs at each of the two main end valves at the end of the sampling sequence.

Aboard ship the sampler is shackled to the end of a hydrographic wire, and a 50-kg weight is attached below the sampler to maintain it in a vertical position during deployment. A self-contained pinger and an alcohol-washed Niskin bottle fitted with reversing thermometers are also hung on the wire for each cast. Completely assembled, including shrouds and tie-rods, the Deep-Ocean Sampler weighs 36 kg, stands 1 meter tall, and is 20 cm in diameter. The decrease in sample pressure arising from structural distortion during the time the sampler traverses from the deep sea to the surface has been calculated to be 7 to 10 percent of the in situ pressure at the sampling depth. Details of sampler handling, testing, and specifications have been reported elsewhere (Tabor and Colwell, 1976; Waxman, et al., *in preparation*).

The transfer system permits subsampling of 100-ml volumes without decompression. The system has two latex bladders: one within the sampler to receive the displacement volume from a pumping system, and a second fitted to a pressure reactor (HIP) to receive the subsample. The void volume

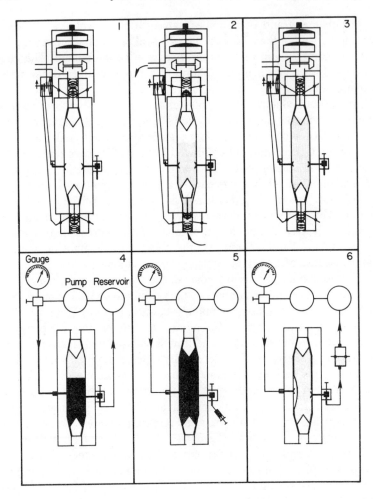

surrounding the second bladder is displaced through a pressure generator-rectifier system to allow precise flow control. A special valve in the pressure-reactor vessel permits undecompressed transfer of sample volumes into the bladder of the pressure reactor and transfer of the displacement volume to the pressure generator-rectifier. The bladder in the subsampling vessel can be preloaded with substrate and the subsampler pressurized via the additional valve system before transfer from the Deep-Ocean Sampler. During the cruise reported here, the subsampling transfer system was not employed. Thus pressure maintained on the first sample successfully obtained from 6,800 m was sacrificed in order to re-use the sampler.

Sterile plastic 3-ml syringes were filled by releasing pressure in the sample chamber at a rate of 150 atm/min through the flame-sterilized

Figure 7-1. Schematic diagram of sampling events and sample handling: (1) sterilized chamber and liquid contents are pressurized to the desired in situ pressure and the sampler is then deployed; (2) the check valve releases any pressure differential in situ, and large springs, located in juxtaposition to the valves, open the chamber at equivalent deep-sea pressure. The pump causes the chamber to be flushed and the sample to be drawn into the sampler. Large springs release, allowing a smaller set of springs to seat the valves, thereby trapping the sample; (3) pressure inside the chamber is retained on return to the surface, with the chamber being hydrostatically sealed; (4) by means of a sterilized, pressurized system, labeled substrate is introduced into the sample chamber through preloaded capillary tubing. The sample is then incubated with continual mixing, the latter being accomplished by Teflon balls in the chamber; (5) subsamples are taken for observation under the microscope and for assay of heterotrophic activity by replacing the sample volume with sterile seawater in order to maintain pressure in the chamber; (6) improvements in the sampler design include location of a bladder in the chamber to prevent contamination and dilution of subsampling and construction of a transfer system to permit withdrawal of undecompressed subsamples.

sampling valve (figure 7-1.5). The syringes were repressurized, at the same rate, to 690 atm at 5°C in conventional stainless steel pressure reactors. Modified seawater yeast extract (MSWYE) agar plates were inoculated with 0.2-ml subsamples bled directly from the sampling valve (Schwarz and Colwell, 1975). Additional subsamples of 2.5-ml volume were drawn into syringes from a main valve opened at one end of the sample chamber (figure 7-1.2). The subsamples were inoculated into MSWYE broth and either incubated at 5°C or immediately repressurized to 690 atm at 5°C.

The repressurized samples were decompressed after incubation for 7 and 10 weeks at 4 to 5°C, and 0.3-ml aliquots were inoculated onto MSWYE plates.

A second sample of seawater was obtained at a depth of 3,450 m and maintained under in situ pressure and temperature. The pressure of the sample chamber was 345 atm upon recovery, and it remained at that pressure during subsequent incubation. Uniformly labeled ^{14}C-glutamate was introduced into a high-pressure capillary tube (0.46-mm ID) by syringe. The tube was connected to the check valve of the sample chamber and to the sterilized hydrostatic pressurizing system. Backflow of the radioactive substrate was prevented by a valve through which the ^{14}C-glutamate passed before it entered the sample chamber. The final activity of the labeled substrate within the chamber was 0.18 μCi/ml, and the final concentration of glutamate added was 0.12 μg/ml.

After incubation of the sample on a shaker for 12 weeks, three 1-ml volumes were brought to atmospheric pressure from the sampler by bleeding from the valve on the transfer system. In order to maintain pressure, a solution of sterile salts was added to replace the volume of the sample withdrawn for scanning electron microscopy. Five replicate subsamples (4.5 ml) were collected for determination of heterotrophic activity at 23 and 43 weeks after addition of ^{14}C-glutamate. Subsamples were bled directly into prepared serum-capped bottles to prevent loss of CO_2, and the sample volume again replaced with a salts solution. The method used for heterotrophic assay was that of Hobbie and Crawford (1969), as modified by Paul and Morita (1971) and Schwarz and Colwell (1975).

Cells were collected by filtration through 0.4-μm Nuclepore filters. The filters were desiccated in a series of ethanol solutions of increasing strength, critical-point dried, and coated for scanning electron microscopy (SEM).

Results

The two successful hydrocasts using the Deep-Ocean Sampler (see above) were accomplished in the Puerto Rico Trench during February 1976. As mentioned, the shallower sample was examined in the laboratory for microbial growth and activity under in situ conditions using the sampling chamber of the DOS as an incubator. The first cast was at 6,800 m depth and the second at 3,450 m.

For both casts, deep-sea reversing thermometers were placed on a Niskin bottle 5 m above the Deep-Ocean Sampler to provide an independent measure of the in situ sampling depth. Thermometric depth calculations, in conjunction with results of calculations derived from measured pressures within the retrieved sample chambers, provided good estimates of the depth at which the water samples were taken. The results of both depth determinations were in close agreement in view of the distortion factor of 7 to 10 percent for the sample volume during retrieval. In addition, when the calculated sampling depth for the 3,450-m cast was compared to the depth of the bottom using a precision depth recorder, we concluded that the 3,450-m bacterial sample was taken within 50 meters of the ocean floor.

A decompressed subsample of the 6,800-m sample plated out immediately upon retrieval was used to determine the total viable count of the whole water sample. The plates were incubated for 5 weeks at 5°C. A total of twenty colonies, eight bacteria and twelve fungi, grew on the seven MSWYE

Microbiological Studies 155

plates, which were inoculated with 0.2 aliquots of seawater bled from the sampler. Thus the total viable count was calculated to be 14.3 organisms per milliliter in non-repressurized subsamples. Five subsamples repressurized to 690 atm and incubated for 7 weeks at 4 to 5°C yielded seventy-nine colonies per 11.9 ml of subsample. Fifty of the fifty-four colonies observed on the plates from one 2.5-ml subsample were similar in morphology. The average, aerobic, heterotrophic, colony-forming population of the repressurized subsamples incubated for 7 weeks was calculated to be 6.6 microorganisms per milliliter. Other repressurized subsamples were incubated for 10 weeks at 690 atm. Eleven bacteria and five fungi were observed on the plates inoculated with 10 milliliters of those repressurized subsamples. A total viable count of 1.6 microorganisms per milliliter was calculated.

The population of bacteria present in the second sample of deep-ocean water (3,450 m) was calculated by examining measured areas of sample grids with a scanning electron microscope. The population present in the water sample after 12 weeks of incubation was 150 cells per milliliter by direct count. At 23 weeks the total count did not show further increase.

Broth cultures of strains isolated from the repressurized subsamples of seawater collected in the first hydrocast were tested for barotolerance by repressurization. They proved to be baroduric, i.e., capable of growth at the pressure equivalent to the depth from which they were obtained.

Morphology of the cells from 3,450 m observed under the SEM was that of rod-shaped, but wrinkled, bacteria. Some of the cells appeared to have ruptured, probably during decompression. Appending structures on the surface of the cells were noted and were prevalent on rod-shaped bacteria, as seen in the micrographs shown in figure 7–2. They were regularly shaped, small, uniformly raised extensions of the cell surfaces and were approximately 0.25 μm. In general, three morphological cell types were most common in the samples examined under SEM, and those three morphological types were the organisms found to grow when incubated in the Deep-Ocean Sampler. However, at 43 weeks, a much lower density of organisms was observed on the filters and a significant amount of cellular debris was noted. Colony-forming units obtained after incubation for 23 weeks were eleven per milliliter; at 43 weeks, five per milliliter were counted. The population size estimated by direct count was always more than an order of magnitude greater than that estimated by colony-forming units, i.e., total, viable, aerobic, heterotrophic plate count.

In the 3450-m sample, the rate of conversion of uniformly labeled ^{14}C-glutamate to $^{14}CO_2$, under simulated in situ conditions, i.e., incubation in the sampler at in situ pressure and temperature, was 2.4×10^{-6} μg/(ml·wk) at 23 weeks. Heterotrophic activity measured at 43 weeks showed CO_2 respiration to be virtually unchanged from that calculated at 23 weeks.

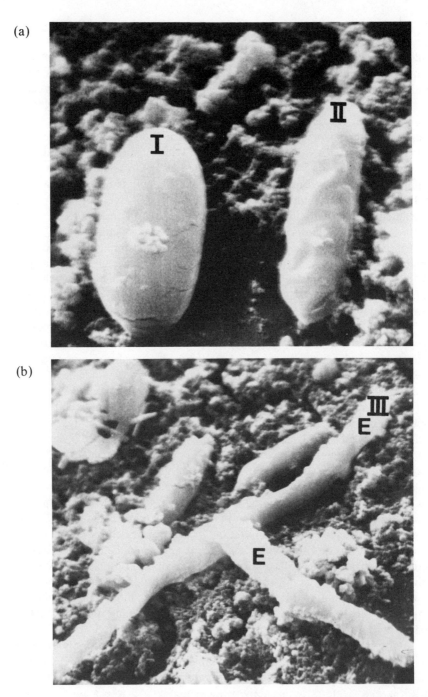

Figure 7–2. (a and b) Scannng electron micrographs prepared from seawater samples incubated under pressure for 12 weeks reveal three distinct morphological types (I, II, III). Elongated forms (E) were observed. (c and d) The surface morphology of many

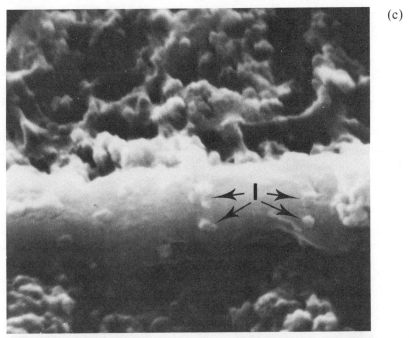

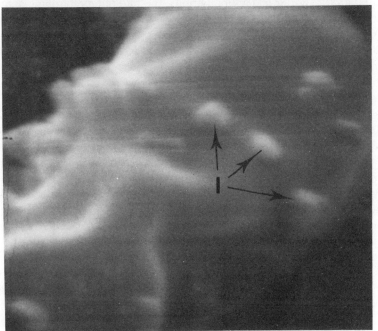

cells revealed small, regularly spaced inclusions (i). Magnification: (a) × 1,800; (b) × 9,600; (c) × 31,000; (d) × 90,000.

Discussion

Decompression and repressurization of barotolerant or barophilic microorganisms retrieved with conventional methods of water sampling and sediment coring may decrease the activity and viability of those microorganisms. Few samplers designed to overcome the problem of maintaining pressure during sample retrieval have been developed. Maintaining in situ hydrostatic pressure during and after sampling is, of course, the major obstacle. Jannasch and Wirsen (1976) reported incorporation and respiration activities of undecompressed microbial populations at depths of 3,100 m, with a maximum reduction of 50 and 11 times the rate of activity noted for the pressurized samples compared with decompressed samples. From their results, Jannasch and Wirsen suggest that free-living microorganisms in the deep sea are not better adapted to conditions of elevated hydrostatic pressure than microorganisms growing at atmospheric pressure. Results from this preliminary study and results of earlier studies in our laboratory using conventional techniques (Schwarz and Colwell, 1975) support that view. Exceedingly low rates of respiration of labeled substrate were observed, and, although direct comparison is difficult because of the lack of data for shorter incubation periods, the rate of heterotrophic activity observed in this study was significantly lower than that reported by Jannasch et al. (1976). A single-point measurement was used to calculate the respiration rate of glutamate by the undecompressed deep-ocean sample. A second end-point measurement showed no increase during extended sample incubation. The question must be asked, however, whether respiration ceased prior to 23 weeks. In their experiments, Jannasch and Wirsen (1976) show no increase in microbial activity after one week in some cases. Further studies are in progress which should shed some light on this vexing question, as well as determine how long any samples, whether of natural populations or pure cultures, can be maintained under elevated pressure.

Decreased viability was detected during extended incubation of mixed cultures of repressurized samples of seawater, with the exception of one plate count. In the latter instance, there may have been a bacterial aggregate present in the sample or growth under pressure. That culture, and all others isolated from the repressurized subsamples of seawater, was found to be baroduric when grown under elevated pressure.

As mentioned, for the 3,450-m sample, a decrease in total count in the undecompressed sample was observed by SEM after incubation for 23 and 43 weeks. The increased cell debris observed at 43 weeks may have arisen from lysis of cells after prolonged incubation, since the sample was not bled through the sampling valve system but rather was transferred after release of pressure on the entire sampler. Hence a "French Pressure Cell" effect was quite unlikely even though the wrinkled and ruptured appearance of the cells observed by SEM was initially thought to be due to a French press effect, i.e., a result of forcing cells through the 0.46-mm ID capillary tubing under a large pressure drop (345 atm). However, an improved SEM preparation technique yielded samples that were also wrinkled compared to cells grown at 1 atm pressure but no longer showed a rupturing of cells. No French press effect has been observed to occur in undecompressed cultures transferred without decompression, as measured by SEM and total viable count. In the transfer from one sampler to another, there was essentially no pressure drop created in the displacing volume; also, minimal lengths of tubing were used. Such precautions were designed to minimize the possibility of any French press effect.

Thus the Deep-Ocean Sampler has proved successful in retrieving samples of seawater from the depths of the ocean while maintaining in situ pressure during retrieval and subsequent incubation. As indicated, a successful deep-ocean sampler has been constructed and used by Jannasch and his colleagues, but there are significant differences between our sampler and theirs. First, no gas-accumulation chamber is employed in our Deep-Ocean Sampler. Decompression by material distortion is 10 percent at most, occurs at a very slow rate during retrieval of the sampler, and can be corrected by slight repressurization of the sample chamber after retrieval. Second, the operation of our Deep-Ocean Sampler is much less complicated than theirs and does not require sophisticated support gear.

Further improvements in the design of the sampler are in progress. One is the adaptation of the pressurizable transfer system for shipboard use, whereby a bladder is placed in the sample chamber so that replacement volumes do not mix with seawater samples after subsamples are taken. Another is the construction of a pressure-sensitive safety mechanism to prevent activation of the sampler at other than a pre-determined hydrostatic pressure.

Acknowledgments

This work was supported by National Science Foundation Grant No. OCE 75-02635 A01. The authors acknowledge the excellent cooperation and assistance of Mr. Meyer Waxman and Mr. Harry Davis of the National Bureau of Standards. Dr. Max Klein provided guidance and advice helpful in designing and fabricating the sampler, for which the authors are most grateful. Funds for completion of the sampler were provided by the National Bureau of Standards through the office of Dr. William Kirchoff.

References

Gillespie, P.A., Morita, R.Y., and Jones, L.P. 1976. The heterotrophic activity for amino acids, glucose, and acetate in Antarctic waters. *J. Oceanogr. Soc. Jap.* 32:74–82.

Hobbie, J.E., and Crawford, C.C. 1969. Respiration corrections for bacterial uptake of dissolved organic compounds in natural waters. *Limnol. Oceanogr.* 14:528–532.

Jannasch, H.W., and Wirsen, C.O. 1976. Substrate conversion by undecompressed microbial deep-sea populations. Presented at the 76th Annual Meeting of the American Society for Microbiology, May 1976, Atlantic City, N.J.

Jannasch, H.W., and Wirsen, C.O. 1977. Retrieval of concentrated and undecompressed microbial populations from the deep sea. *Appl. Environmental Microbiol.* 33:642–646.

Jannasch, H.W., Wirsen, C.O., and Taylor, C.D. 1976. Undecompressed microbial populations from the deep sea. *Appl. Environmental Microbiol.* 32:360–367.

Pamatmat, M.M. 1973. Benthic community metabolism on the continental terrace and in the deep sea in the North Pacific. *Int. Revue Ges. Hydrobiol.* 58:345–368.

Paul, K.L., and Morita, R.Y. 1971. Effects of hydrostatic pressure and temperature on the uptake and respiration of amino acids by a facultatively psychrophilic marine bacterium. *J. Bacteriol.* 108:835–843.

Schwarz, J.R., and Colwell, R.R. 1975. Heterotrophic activity of deep-sea sediment bacteria. *Appl. Microbiol.* 30:639–649.

Smith, K.L., Jr., and Teal, J.M. 1973. Deep-sea benthic community respiration: An in situ study at 1850 meters. *Science* 179:282–283.

Tabor, P.S., and Colwell, R.R. 1976. Initial investigations with a deep ocean in situ sampler. Presented at Oceans '76 the Second Annual MTS-IEEE Conference, September 13–15, 1976, Washington, D.C.

8 Direct Calorimetry of Benthic Metabolism

Mario M. Pamatmat

Abstract

Developments in instrumentation have enhanced the usefulness and feasibility of direct calorimetry in studies of benthic metabolism. A twin calorimeter, with semiconductor thermopile as heat flux sensor, has been used to measure the metabolism of whole sediment as well as macrofauna. Some sediments poisoned with mercuric chloride, formaldehyde, and glutaraldehyde showed a great increase, followed by a slow decrease, in heat production rate. Burrowing beach animals, *Donax variabilis*, *Emerita talpoida*, and Haustoriid amphipods, have much lower metabolic rates when buried in sand than when in water only.

Introduction

Benthic organisms affect sediment structure and distribution of chemical properties of the benthic boundary layer (Berner, 1976; Suess, 1976; and working group reports in McCave, 1976). Further study of benthic metabolism should lead to better understanding of the overall impact of the benthos on the sea bed. Total metabolism of the benthos, however, is difficult to measure because of the complex associations of different metabolic types and the lack of a suitable instrument for dealing with all aerobes and anaerobes combined (Pamatmat, 1975).

Direct calorimetry seemed to be a promising method for measuring total-energy metabolism in complex anaerobic microbial communities such as those in the rumen of ruminants (Walker and Forrest, 1964), in soils (Mortensen et al., 1973), and in sediments (Doyle, 1963; Pamatmat and Bhagwat, 1973). Highly sensitive instruments are available commercially (Forrest, 1972), but those tried in benthic research were unstable, difficult to use, and did not appear promising for shipboard use (Pamatmat, 1975). Furthermore, they are expensive and, with their low reliability and apparently limited applicability, did not seem worthwhile. Significant improvements in the instrument were obviously necessary.

Description of the Calorimeter

Among many calorimeter designs on the market (Skinner, 1969), the twin calorimeter (Calvet, 1956) appears to offer the best solution to the problem of baseline instability. It consists of two chambers hollowed out of a single metal block which serves as the heat sink. Earlier calorimeters were built with a bimetallic thermopile (Atwater and Benedict, 1905; Calvet, 1956), but semiconductor thermopiles are superior (Calvet and Guillaud, 1965) and have been used to make twin microcalorimeters (Ross and Goldberg, 1974; Wadsö, 1974). Wadsö's instrument is a wattmeter whose steady-state voltage is directly proportional to rate of heat production. When the thermopile voltage varies with time, as thermograms of higher animals show (Prat, 1956), the integrated curve represents the total heat liberated during a period.

The thermoelectric modules are attached to the inside walls of the chambers so that heat generated by the sample flows through the sensors and is dissipated by the metal block. Thermoelectric modules within each chamber are wired in series, and the modules of one chamber are connected in series opposition with those of the other chamber. Small, uncontrollable, ambient temperature fluctuations, which cause baseline drift in single-chambered calorimeters, affect the sensors of both chambers equally and simultaneously. Thus the two signals cancel out, and a stable baseline results. Finally, ambient temperature fluctuations are further damped, and uniformity of temperature of the calorimeter block is achieved by enclosing it with concentric layers of insulating and conductive materials (Evans, 1969).

My calorimeter was built along the same design but with variably sized chambers (figure 8–1) to accommodate different species ranging in size from bacteria to thick-shelled bivalves. The opposing pairs of aluminum blocks are parted when inserting or removing samples by pulling on a monofilament line. The opposing thermoelectric modules of each chamber are in contact with sample containers up to $10 \times 10 \times 10$ cm in size for the most efficient heat transfer to the heat sink. The ideal sample container should be as thin and flat as possible. The instrument's response time increases with increasing sample size.

The calorimeter is contained inside a fiberglass-coated plywood box within a larger, similarly waterproofed plywood box. The outer box is lined with 7.5-cm-thick fiberglass wool insulator which surrounds the inner box; the latter is lined with 5-cm-thick Styrofoam which completely surrounds the calorimeter. For use at below-ambient temperatures, the outer box is immersed in a water bath to within 5 cm of the top and the water temperature regulated to within $\pm\ 0.01°C$.

Direct Calorimetry

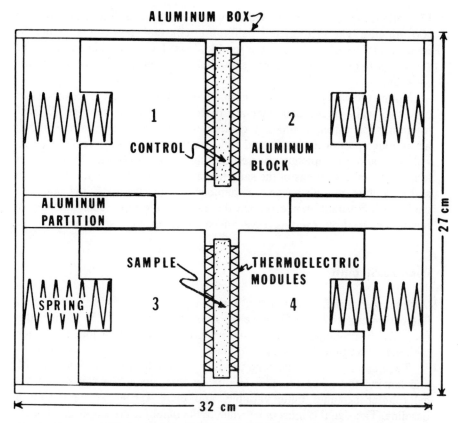

Source: Pamatmat, *Marine Biology* 48:317–325, 1978. Reprinted with permission.

Figure 8–1. Horizontal cross section of the twin calorimeter designed to accommodate sample containers of different thicknesses. Sample is in contact with sensors on two opposite surfaces only. Heat loss through other surfaces exposed to air is accounted for by the calibration factor.

Calibration of the Calorimeter

The calorimeter is calibrated by passing a known current through a known resistance heater (Berger, 1969). Each chamber has a sensitivity of 1.35×10^{-5} watt (3.2×10^{-6} cal/s) per microvolt (μV). The time constant depends on the mass of the sample and the conductivity of both sample and container. With a bare resistor placed between the sensors, full response is attained in

17.5 minutes. With 70 ml of sediment in a plastic cell, full response takes 100 minutes. The instrument's calibration constant, however, remains the same regardless of the size of the sample, from 1 to 100 ml in volume. The absolute size of the sample per se should not materially affect the calibration constant.

Not all the heat generated by the sample passes through the sensors, since only two walls of the container are in contact with the sensors. A small fraction is lost through radiation, conduction, and convection to surrounding air although part of the heat could still flow through the sensors if the sensor wall area is not completely covered by the sample container. Therefore, as long as the relative areas of the sample container in contact with the sensors and in contact with air inside the chamber remain reasonably the same, the calibration constant should remain the same. Calibration constants can be readily checked by inserting a resistance heater inside a sample.

Baseline Stability

With temperature fluctuation of $\pm$ 1.1°C, the calorimeter's baseline shows a fluctuation of $\pm$ 0.2 μV. The baseline stabilizes at about +0.3 μV instead of zero. The instrument has shown that order of stability for as long as 1 week, the longest continuous observation made.

Twin calorimeters have been used by placing a sample in one chamber and leaving the other chamber empty. In that mode, the signal from the calorimeter becomes quite variable, and there appears a slight shift in baseline. The signal is most quiet when both chambers are empty and the gap between sensors is fully closed; however, the apparent baseline in that mode is -0.6 μV, versus +0.3 μV when both chambers contain identical empty aluminum cannisters. Evidently there is a tiny undefined heat flux which changes according to the different modes but remains steady at each mode for long periods.

It seems that +0.3 μV is the reasonable baseline to use. Such a small deviation from zero is negligible when measuring heat-output rates equivalent to tens of microvolts or more. Moreover, the calorimeter is unavoidably disturbed when placing and removing samples. Reproducibility of successive measurements after removing and returning the same sample is within $\pm$ 1 μV.

In a fluctuating thermal environment, if the sample in one chamber and a control in the other are not identical in conductivity and thermal mass (e.g., an empty container in one chamber and an identical container filled with distilled water in the other), the calorimeter signal could show long-term fluctuations of ± 2 to 20 μV. The amplitude of fluctuation appears to increase with increasing discrepancy in thermal mass between sample and

Direct Calorimetry

control in the two chambers. Evidently, as the temperature of the heat sink changes ever so slightly, the heat flow to or from the sample and control changes accordingly. If the two are not matched in conductivity and thermal mass, a fluctuating signal results. It is obviously desirable not only to minimize the thermal mass of the sample but also to minimize the difference in thermal mass between sample and control. Since it is difficult to match a sample with a thermally inert control of the same thermal mass and conductivity, it is desirable to operate the calorimeter in the most stable environment possible. With the water-bath arrangement described previously, an error of about 10 percent or less could result by determining heat-output rates of about 30 μcal/s.

Calorimetry of Sediments

My primary interest has been to measure metabolic heat production of the anaerobic community in sediments. Many sediment samples, however, both marine and freshwater, have not produced heat; instead they absorbed heat.

One endothermic mud sample from a freshwater reservoir was treated with glucose, and the container was provided with a fine-bore Tygon tubing leading outside the instrument (figure 8–2). The tubing was initially clamped shut. About 5 hours after being placed inside the calorimeter, the signal reading was beginning to stabilize at -32 μV indicating heat absorption at the rate of 10^{-4} cal/s. Evidently, sediment bacteria were already fermenting the added glucose because gas pressure had built up inside the container, and when the tubing was unclamped, 0.5 ml of gas was released with instantaneous absorption of heat. The tubing was left open during the next 3 h, during which 0.3 ml more of gas was released. Each time a tiny bubble of gas emerged from the end of the tubing, a small amount of heat was absorbed. For over 9 h, from the time the sample was first put inside the calorimeter, the sample was endothermic despite the metabolic activity that was definitely going on. The level of endothermy was, however, decreasing. Nearly 10 hours after putting the sample in, it became exothermic and increasingly so with time. The signal peaked offscale, was already falling at 17 hours, but remained slightly exothermic until the twelfth day (10 μcal/s). When measured 2 weeks later, the sample was again endothermic.

Leakage and evaporative loss of water have been ruled out, and no obvious explanation for endothermy can be offered at this time. It is clear, however, that wherever gas bubbles are released from sediments, concomitant heat absorption may occur.

A main difficulty, then, in measuring sediment metabolism involves the problem of distinguishing metabolic heat production from heat effects of other physical and chemical processes going on simultaneously. The addition

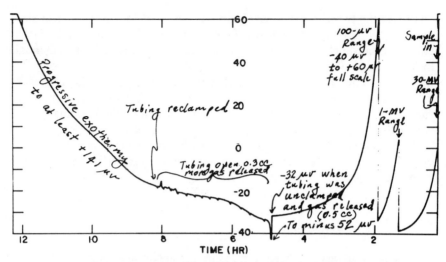

Figure 8–2. Thermogram of freshwater sediment treated with glucose. Sample was warmer than calorimeter when placed inside. Thermopile voltage crossed zero and tended to level off at -32 μV as the sample absorbed heat. The sample cell was equipped with Tygon tubing. Release of gas resulted in cooling and rapid absorption of heat. With tubing left open, periodic evolution of tiny gas bubbles were accompanied by heat absorption. Sample became increasingly exothermic after the tubing was clamped. Thermopile voltage was $+141$ μV at the end of 17 h, decreased gradually with some fluctuation to $+15$ μV another 24 h later. Twelve days later the sample was still slightly exothermic ($+2.5$ μV). Another 14 days later it had become endothermic again (-23 μV).

of poison to reduced sediment generates heat for surprisingly long periods, presumably from the oxidation of sulfides and other reduced substances (figure 8–3). After many days, the rate of heat production drops below the original rate shown before poisoning. It is likely that other heat-producing or heat-absorbing processes could have been altered by poisoning; if not, the final difference represents metabolic heat production.

Calorimetry of Macrofauna

Macrofauna generally have rapidly fluctuating thermograms that look suspiciously "noisy." A specimen of *Uca pugilator* contained in a respir-

Direct Calorimetry

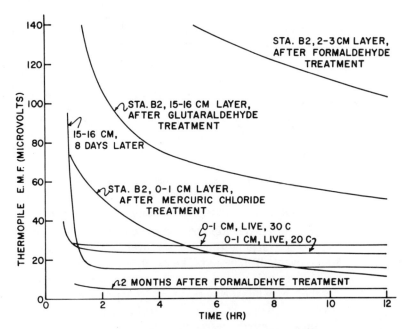

Figure 8-3. Effect of various poisons on rate of heat output of sediments from Mobile Bay, Alabama. Samples were placed inside the calorimeter at time zero; the instrument took about 2 h to return to thermal equilibrium. The 0- to 1-cm layer gave a steady thermopile voltage of 23 μV at 20°C, 27 μV at 20°C. When treated with mercuric chloride, thermopile voltage increased and then decreased exponentially from 52 μV at the end of 2 h to 11 μV at the end of 12 h, and zero by the fourth day. The 2- to 3-cm layer (treated with 10 ml formaldehyde) showed an even greater increase in thermopile voltage, which was still 4 μV 2 months later. Likewise, the 15- to 16-cm layer (treated with glutaraldehyde) also showed increased heat production, which was still substantially above zero after 8 days.

ation chamber with a pair of fine Tygon tubings leading to the outside of the calorimeter shows such a fluctuating signal (figure 8–4). The animal showed a uniform basal level of metabolic heat production (32 μV = 0.37 cal/h) for nearly 30 hours; the fluctuations above that basal level probably reflected active metabolism. Temporary disturbances in the calorimeter signals resulted from injecting formaldehyde into the respiration chamber. Following formaldehyde injection and the animal's death, heat production not only ceased, as indicated by the zero reading, but signal fluctuations also

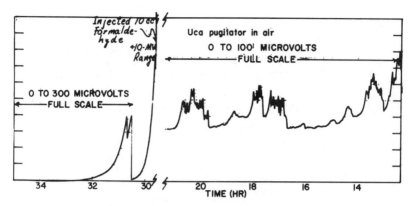

Source: Pamatmat. *Marine Biology* 48:317–325, 1978. Reprinted with permission.

Figure 8–4. Variable heat production of *Uca pugilator* in air, with thermopile voltage ranging from a basal level of 32 μV to as high as 80 μV over a 30-h period. The basal level, which was the same during the whole period, probably represents the animal's basal metabolism, while the increases above it probably resulted from physical activity. At the end of 30 h, 10 ml of formaldehyde was injected into the respiration chamber. Thermal disturbance caused thermopile emf to exceed 10 μV. On regaining thermal equilibrium 2.5 h later, and with the animal dead, thermal emf dropped to a steady zero.

disappeared. In this case, one is able to ascertain metabolic heat production readily.

Other animals, however, like the estuarine clam *Polymesoda caroliniana*, when injected with mercuric chloride showed a decrease in heat production, but the rate of heat production continued to decrease for days. In that case, I used the thermopile signal from long-poisoned bivalves as zero baseline and measured the area of the curve above that baseline to estimate bivalve anaerobic metabolism (figure 8–5).

The calorimeter is also suitable for comparing the metabolic activity of infauna when buried in sediment with that of animals deprived of sediment. Respiration of infauna has been commonly measured while the animals are in water only in order to eliminate complication from the sediment's oxygen uptake. Experiments with three beach-burrowing species, a Haustoriid amphipod, a mole crab (*Emerita talpoida*), and small clams (*Donax variabilis*), show that all of them have higher metabolic activity when in water without sand than when buried in sand (table 8–1).

In the first experiment with the amphipod, twenty-eight specimens were placed in each of two respiration cells: one with sand from the beach

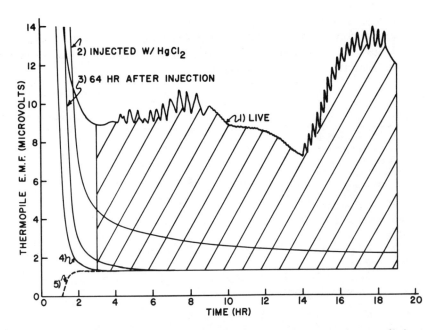

Figure 8-5. Thermogram of an estuarine clam *Polymesoda caroliniana* and the effect of mercuric chloride injected into the clam tissue. The rapid drop in voltage reflects cessation of metabolism, but after 19 h there was still a significant heat production from some chemical reaction with the poison. After 2.5 days, the poisoned clam showed a steady voltage of 1.3 μV, which was exactly the same as that shown by another clam (curves 4 and 5) poisoned 2 weeks before. Curves 4 and 5 show that it is immaterial whether the samples placed inside the calorimeter are initially cooler or warmer than the instrument.

collection site and whose metabolic activity had just been measured, the other with seawater only. The twenty-eight amphipods in water had an average heat production of 0.045 cal/h per animal, while the twenty-eight amphipods able to burrow in sand averaged 0.0082 cal/h per animal, a fivefold difference. The experiment was repeated with only fifteen animals per cell. The amphipods in sand averaged 0.012 cal/h per animal in contrast to 0.022 cal/h for the animals without sand. When the group of fifteen animals with sand was later transferred to water only, their heat production doubled to 0.024 cal/h per animal. The group of fifteen amphipods in sand had slightly higher heat production per animal than the group of twenty-eight animals in sand, a difference that is probably due to animal size only. The two groups of fifteen animals in water, however, had only half the heat

Table 8–1
Metabolic Activity (Calorie of Heat Released per Individual Animal per Hour) of Sandy-Beach–Burrowing Animals When Buried in Sand versus Activity When in Water Only

Species	Number of Animals in Respiration Cell	Metabolic Heat Production	
		In Water Only	Buried in Sand
Amphipod (Haustoriidae)	28	0.045[a]	0.0082[b]
Amphipod (Haustoriidae)	15	0.022[c]	0.012[d]
Amphipod (Haustoriidae)	15	0.024[d]	
Emerita talpoida	1	0.52[e]	0.22[e]
Donax variabilis	10	0.046[f]	0.030[f]

[a] 0.0139 g wet wt/animal.
[b] 0.0131 g wet wt/animal.
[c] 0.0126 g wet wt/animal.
[d] 0.0171 g wet wt/animal, same group of fifteen first in sand, then transferred to water only.
[e] 0.191 g dry wt.
[f] 0.252 g dry wt/animal including shell. Average shell length is 11 mm. In water only first, then sand added to respiration chamber.

production of the group of twenty-eight amphipods in water. It appears that crowding had less effect, if any, on metabolic activity of buried animals than on swimming animals.

Emerita talpoida had metabolic heat production of 0.22 cal/h when buried in sand and more than twice as much (0.52 cal/h) when in water only. *Donax variabilis* had 1.5 times as much heat production in water as in sand (0.046 cal/h per animal versus 0.030 cal/h).

The determination of metabolic activity of animals buried in sediment, of course, requires a separate determination of the metabolic activity of all microorganisms associated with the sediment. Whether the latter is the same in the presence or absence of ventilating macrofauna remains to be shown. In the absence of burrowing animals, the sediment could rapidly go anaerobic with subsequent decline in metabolic activity. However, no such decline was observed during the period of 4 to 8 hours that the experiments lasted. Furthermore, if such a decline occurs, it would tend to show a higher value of apparent heat production for the burrowing macrofauna.

Acknowledgments

This research has been supported by NSF Grant OCE77–08634 and ERDA Grant EE-77-S-05-5465.

References

Atwater, W.O., and Benedict, F.G. 1905. A respiration calorimeter with appliances for the direct determination of oxygen. Carnegie Institution of Washington, Publication 42, Washington, D.C.

Berger, R.L. 1969. Calibration and test reactions for microcalorimetry. In H.D. Brown (Ed.), *Biochemical Microcalorimetry*. New York: Academic Press, pp. 221–234.

Berner, R.A. 1976. The benthic boundary layer from the viewpoint of a geochemist. In I.N. McCave (Ed.), *The Benthic Boundary Layer*. New York: Plenum, pp.33–55.

Calvet, E. 1956. Appareils et méthods. In E. Calvet and H. Prat (Eds.), *Microcalorimétrie: applications Physico-Chimiques et Biologiques*. Paris: Mason, pp. 5–136.

Calvet, E., and Guillaud, C. 1965. Étude d'un microcalorimètre équipé avec des thermoéléments semi-conducteurs. *C.R. Acad. Sci., Paris* 260:525–528.

Doyle, R.W. 1963. Calorimetric measurements of the anaerobic metabolism of marine sediments and sedimentary bacteria. Masters thesis, Dalhousie University, Halifax.

Evans, W.J. 1969. The conduction-type microcalorimeter. In H.D. Brown (Ed.), *Biochemical Microcalorimetry*. New York: Academic Press, pp. 257–273.

Forrest, W.W. 1972. Microcalorimetry. In J.R. Norris and D.W. Ribbons (Eds.), *Methods in Microbiology*, Vol. 6B. London: Academic Press, pp. 285–318.

McCave, I.N. (Ed.). 1976. *The Benthic Boundary Layer*. New York: Plenum.

Mortensen, U., Noren, B., and Wadsö, I. 1973. Microcalorimetry in the study of the activity of microorganisms. *Bull. Ecol. Res. Comm. (Stockholm)* 17:189–196.

Pamatmat, M.M. 1975. *In situ* metabolism of benthic communities. *Cah. Biol. Mar.* 16:613–633.

Pamatmat, M.M. 1978. Oxygen uptake and heat production in a metabolic conformer (*Littorina irrorata*) and a metabolic regulator (*Uca pugnax*). *Mar. Biol.* 48:317–325.

Pamatmat, M.M., and Bhagwat, A.M. 1973. Anaerobic metabolism in Lake Washington sediments. *Limnol. Oceanogr.* 18:611–627.

Prat, H. 1956. Applications biologiques. In E. Calvet and H. Prat (Eds.), *Microcalorimétrie: Applications Physico-Chimiques et Biologiques*. Paris: Mason, pp. 279–377.

Ross, P.D., and Goldberg, R.N. 1974. A scanning microcalorimeter for thermally induced transitions in solution. *Thermochim. Acta* 10:143–151.

Skinner, H.A. 1969. Theory, scope, and accuracy of calorimetric measurements. In H.D. Brown (Ed.), *Biochemical Microcalorimetry* New York: Academic Press, pp. 1–32.

Suess, E. 1976. Nutrients near the depositional interface. In I.N. McCave (Ed.), *The Benthic Boundary Layer.* New York: Plenum, pp. 57–79.

Wadsö, I. 1974. A microcalorimeter for biological analysis. *Science Tools* 21:18–21.

Walker, D.J., and Forrest, W.W. 1964. The application of calorimetry to the study of ruminal fermentation *in vitro. Aust. J. Agric. Res.* 15:299–315.

Part IV
Gases in Sediments

9 Inert Gas Gradients and Concentration Anomalies in Pacific Ocean Sediments

Ross O. Barnes

Abstract

Measured excess helium concentration gradients in marine sedimentary pore waters can be attributed to upward diffusion of radiogenic helium but are influenced by sedimentation rate. The depression of surface concentration gradients in rapidly deposited sediment can explain the observed absence of excess helium gradients in such locations. A one-dimensional model based on heat-flow models can be used to explain the effects of sedimentation rate on helium gradients. No satisfactory explanation has been found for the unexpectedly large concentration variations exhibited by the three heavier inert gases, Ne, Ar, and Kr, presenting an intriguing problem for further study.

Introduction

Helium[a] is generated in the earth's crust by the slow radioactive disintegration of uranium and thorium and their daughter products. A significant percentage diffuses slowly through the crust into the ocean-atmosphere system and is eventually lost to space. The common occurrence of low helium-uranium ages for rocks and minerals is a result of helium loss from the crust. The helium flux should produce helium concentration gradients increasing with depth in sediments porous enough to permit bulk-mode diffusion as long as pore-water advection does not significantly alter the diffusion gradients. Deep-ocean sediments should provide the best medium for measuring such gradients because the high clay content, slow sedimentation rate, and horizontally and temporally uniform temperature field all combine to suppress advective water movements in the sediment column.

We have been investigating the helium flux through marine sediments by measuring the helium concentration gradients in sedimentary pore waters. Measurements have been obtained with the aid of a specially designed in situ pore-water sampler (Barnes, 1973a). Initial results demonstrated the presence of excess ^{4}He gradients that can be attributed to a flux of helium from

---

[a]All references to helium in this chapter are to ^{4}He.

the sediment and underlying crust into the ocean (Barnes and Bieri, 1976). In that paper the measured profiles are discussed in terms of crustal and sedimentary radiogenic helium production, but the discussion will not be repeated here. Instead, this chapter focuses on the lack of excess helium in some sediments and the significant unexplained variations in Ne, Ar, and Kr concentrations that were also detected. Additional in situ pore-water sampling and isotope-dilution mass-spectometric analyses reported in this chapter have confirmed the presence of these gas-concentration anomalies.

Results

Inert gas concentration profiles have been measured at four stations in the northeastern Pacific Ocean. The data collected subsequent to Barnes and Bieri (1976) are listed in table 9–1. Figure 9–1a, b, d and e shows the concentration profiles for the four northeastern Pacific stations. Those results suggest that pore-water inert gas concentrations in this area are influenced by sedimentation rate or some other variable related to sedimentation rate, such as total sediment thickness. The gas concentration profiles in figure 9–1a, b, and d–e were measured in areas of low ($\sim 10^{-3}$ cm/yr), medium (~ 0.01 to 0.08 cm/yr), and high (~ 0.2 to 0.5 cm/yr) sedimentation rate, respectively (Koide et al., 1972; Emery 1960).

The pore-water concentrations of the heavier inert gases at the pelagic station (figure 9–1a) show relatively small variations from overlying seawater values; the variations do not correlate with the monotonically increasing He concentrations that are attributed to crustal He generation (Barnes and Bieri, 1976).

In shallow (~ 900 m) nearshore Santa Monica and San Pedro Basins (SMB and SPB) of figure 9–1d and e, all inert gas concentrations show large correlated changes. There is no evidence of excess radiogenic helium in the profiles, since the variations in normalized Ne concentration are nearly the same as the variations in normalized He concentration. The Ar and Ne isotope ratios showed no significant deviation from atmospheric values within analytical accuracy, and there are no known nuclear production mechanisms that would lead to Ar, Ne, and Kr increases comparable to He generation from alpha-particle decay in the crust and sediments.

Effect of Sedimentation Rate

A lack of excess He relative to Ne in pore waters of the San Pedro and Santa Monica Basins can be explained by the high sedimentation rates at those locations. The effective pore-water diffusion coefficient of He is about

Inert Gas Gradients

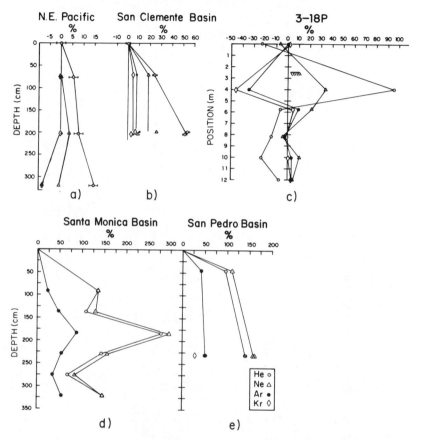

Figure 9-1. Inert gas concentration profiles in Pacific sediments. The concentrations are plotted as percentage deviations from bottom-water gas concentrations. The error bars on the He concentrations in (a) and (b) show the estimated sampling and analytical precision for the He determination. Precisions are better for heavier inert gases. Stations locations are (a) 31°19′N, 119°49′W; (b) 31°59′N, 117°56′W; (c) 7°14′S, 168°30′W, S.E. Pacific; (d) 33°45′N, 118°50′W; and (e) 33°33′N, 118°24′W. Duplicate samples in (e) at 231 cm, collected and analyzed independently, show precisions for all gases of ± 2 percent or better. The relative sample spacings in (c) through (e) are exact. However, absolute depth in the sediment for (c) is uncertain, and the estimated location of the sediment/water interface is shown by the slanted hachures.

Table 9-1
Noble Gas Concentrations in the Pore Waters of Santa Monica and San Pedro Basins Off Southern California

Sample	He (ml/L at STP) $\times 10^{-5}$	Percent[a]	Ne (ml/L at STP) $\times 10^{-4}$	Percent	Ar (ml/L at STP) $\times 10^{-1}$	Percent	Kr (ml/L at STP) $\times 10^{-5}$	Percent
Santa Monica Basin Probe: September 13, 1973 Bottom: 899 m								
Eastern North Pacific water[b]	4.25	0	1.80	0	3.50	0	8.40	0
Seawater 600 m	4.38	+3	2.01	+12	3.57	+2	—	
Probe 91 cm	10.06	+137	4.27	+137	4.33	+24		
Probe 137 cm	8.86	+108	4.14	+130	5.18	+48		
Probe 183 cm	16.13	+280	7.17	+298	6.63	+89		
Probe 229 cm	10.40	+145	4.65	+158	5.47	+56		
Probe 275 cm	7.17	+69	3.34	+86	4.76	+36		
Probe 321 cm	10.56	+148	4.47	+148	5.50	+57		
San Pedro Basin Probe: September 14, 1973 Bottom: 893 m								
Eastern North Pacific water[b]	4.25	0	1.80	0	3.50	0	8.40	0
Probe 45 cm	8.31	+96	3.80	+111	4.97	+42		
Probe 231 cm[c]	10.25	+141	4.65	+158	5.27	+51	10.72	+28
Probe 231 cm[c]	10.30	+142	4.71	+162	5.36	+53	10.74	+28

[a] Percentage deviation from representative values of gas concentrations determined by mass spectrometry on North Pacific waters of the same temperature and salinity characteristics as basin bottom waters (505°C, 34.3‰) (Bieri, Koide, and Goldberg, 1968; Bieri, personal communication).
[b] Gas concentrations.
[c] Duplicate samples collected by strapping two in situ samplers together.

2.5 × 10^{-5} cm^2/s or less (Barnes and Bieri, 1976). A finite sedimentation rate decreases the surface concentration gradient from the equilibrium value at zero sedimentation rate because the slow upward diffusion of helium from a deep source cannot "keep up" with the moving sediment surface. That effect increases with increasing sedimentation rate and sediment accumulation time. A similar effect of sedimentation rate on surface gradients is well known from heat-flow work. However, for a given sedimentation rate, chemical gradients are altered to a greater degree than thermal gradients because chemical diffusion coefficients in pore waters are at least 10^3 times smaller than the thermal diffusion coefficients.

The influence of sedimentation rate would also depend on the depth distribution of the rate of helium release from the solid phase to the pore waters. With no information about that factor, I have chosen a simplified one-dimensional model that calculates the helium concentration gradient at the sediment/water interface—g_t ($z = 0$)—for any arbitrary time $t > 0$, sedimentation rate U, and radiogenic helium release rate R. The model is taken from Jaeger (1965) and is modified for the inclusion of a helium production term R after Carslaw and Jaeger (1959, p. 388). The model assumes that (1) the pore-water helium concentration gradient (g_0) is linear at some arbitrary time $t = 0$; (2) the bottom-water helium concentration remains constant with time; (3) sediment compaction below the upper several meters and vertical advective pore-water movements do not significantly affect a diffusive model; (4) after time $t = 0$, sediment is deposited at a constant rate of U cm/yr; and (5) the release rate of helium to the pore waters from uranium and thorium decay in sediments deposited after time $t = 0$ is a constant R. The resultant equation is

$$g_t(z=0) = g_0 \left[1 - F(p)\left(1 - \frac{R}{Ug_0}\right)\right] \quad (9.1)$$

where

$$F(p) = -\tfrac{1}{2}p^2 + (1 + \tfrac{1}{2}p^2)\,\text{erf}(\tfrac{1}{2}p) + (1/\pi^{1/2})p\,\exp(-p^2/4),$$

p is the dimensionless quantity $Ut/(D'_{He}t)^{1/2}$, and z is the vertical distance below the sediment/water interface. $D'_{He} = \phi D_{He}/2$, where ϕ is sediment porosity (assumed constant at 70 percent), and D_{He} is the liquid-phase diffusion coefficient taken as 5×10^{-5} cm^2/s. D'_{He} is thus 1.75×10^{-5} cm^2/s. Assuming a constant D'_{He} will not materially affect the predictions of this model. $F(p)$ tends to 1 for p large and positive (high sedimentation rate and/or large time t).

Equation 9.1 shows that the original gradient g_0 is modified by two terms.

One depends on the sedimentation rate $[-F(p)]$, and the other term reflects the ratio of helium release rate to the sedimentation rate $[F(p)R/Ug_0]$.

If $R/Ug_0 \ll 1$, then the effect of helium production and release in the sediment is insignificant compared to the sedimentation-rate effect. If $R/Ug_0 \gg 1$, then the sedimentation-rate effect will be minimal and $g_i(z=0)$ will increase with time as the integrated helium release below the sediment surface increases with increasing sediment thickness.

Jaeger (1965) lists values of $F(p)$ for given values of p from which a graph of p versus $F(p)$ can be constructed (in this chapter, $F(p) = -\psi(p)$ of Jaeger). Table 9-2 lists values of $F(p)$ for combinations of U and t relevant to the marine sedimentary environment. Since table 9-2 assumes a constant D'_{He} of 1.75×10^{-5} cm^2/s, the table would have to be recalculated for other values of D'. However, values of $F(p)$ are not very sensitive to the small changes in D' that would occur due to changes in porosity and temperature in a typical marine sediment. For practical purposes, those variations can be ignored.

This model also ignores vertical advective water movements within the sediment. That is not a serious omission if we assume that the sediment reaches a relatively stable porosity within a few tens of meters of the sediment/water interface. The model is not very sensitive to conditions near the sediment/water interface.

From table 9-2 we can see that $F(p)$ is quite sensitive to variations in sedimentation rate U. Sensitivity to time variations is a factor of 10 less.

Although some nearshore organic-rich sediments exhibit high concentrations of uranium, published and unpublished measurements of U and Th in surface sediments of basins of the Southern California borderland show values within the range of normal pelagic sediments (Koide et al., 1973, 1976; Veeh, 1967; Koide, personal communication). Based on those measurements, the best estimates for U and Th concentrations at our profile locations are

	U (ppm)	Th (ppm)
San Clemente Basin (SCB)	4	8
Santa Monica and San Pedro Basins (SMB and SPB)	4	10

The U and Th concentrations at the pelagic station (figure 9-1a) are probably lower but not by more than a factor of 2.

Using the U and Th concentrations just listed for SMB-SPB sediments, a ϕ of 70 percent, a grain density of 2.6 g/cm^3, and helium production rates for U and Th of 1.0×10^5 and 2.5×10^4 atoms/g·sec), respectively, a helium production of 6×10^{-13} ml STP/(cm^3·yr) is calculated. A large fraction of

Inert Gas Gradients

the helium will be immediately or eventually released to pore waters. I will therefore assume that $R = 5 \times 10^{-13}$ ml STP/(cm^3·yr).

Let us assume an initial gradient [$g_0 = 0.2 \times 10^{-10}$ ml STP/(cm^3·cm)] similar to that in figure 9–1a, where the sedimentation-rate effect is minimal. Such a gradient would be produced by helium production in about 200 m of sediment with the preceding value of R.

For SMB-SPB, with an assumed $U = 0.3$ cm/yr, $R/Ug_0 = 0.08$. Therefore, substitution into a rearranged equation 9.1 gives

$$\frac{g_t(z=0)}{g_0} = 1 - F(p)(1 - 0.08) \qquad (9.2)$$

$$= 1 - 0.92\, F(p)$$

From equation 9.2 and table 9–2, g_0 will be reduced to 25 percent of its original value in only 10^4 years and after only 30 m of sediment have been deposited. Such an excess helium gradient would not be distinguishable from a zero gradient given the accuracy of our procedures and the measured variability of noble gas contents in pore waters. In 10^5 years, the surface gradient would reach a stable value equal to 9 percent of g_0.

If we assume $U = 0.001$ cm/yr for the pelagic station (figure 9–1a) and the same R as in SMB-SPB, then $R/Ug_0 = 8$ and

$$\frac{g_t(z=0)}{g_0} = 1 + 7F(p) \qquad (9.3)$$

Table 9–2
Values of $F(p)$ (See Text) for Various Combinations of Elapsed Time (t) and Sedimentation Rate (U)

t (yr)	U (cm/yr)							
	1.0	0.3	0.1	0.03	0.01	0.003	0.001	0.0003
1×10^3	0.84	0.39	0.16	0.06	0.02	<0.01	—	—
3×10^3	0.96	0.58	0.26	0.09	0.04	0.01	—	—
1×10^4	0.99	0.82	0.41	0.16	0.06	0.02	<0.01	—
3×10^4	0.99+	0.95	0.60	0.25	0.10	0.04	0.01	—
1×10^5	—	0.99	0.84	0.39	0.16	0.06	0.02	<0.01
3×10^5	—	0.99+	0.96	0.58	0.26	0.09	0.04	0.01
1×10^6	—	—	0.99	0.82	0.41	0.16	0.06	0.02
3×10^6	—	—	0.99+	0.95	0.60	0.25	0.10	0.04
1×10^7	—	—	—	0.99	0.84	0.39	0.16	0.06
3×10^7	—	—	—	0.99+	0.96	0.58	0.26	0.09
1×10^8	—	—	—	—	0.99	0.82	0.41	0.16

Here the sedimentation-rate effect is minimal (< 10 percent) for periods up to 3×10^6 yr. The surface gradient increases with time, reflecting the gradual increase of integrated helium production as the sediment thickness increases.

Between the two extremes of high and low sedimentation rate, $g_i(z=0)$ is very dependent on the actual sedimentation history and helium release rate within the sediment column. In general, those factors are not known with sufficient certainty to make accurate predictions from the model. However, San Clemente Basin, with a sedimentation rate intermediate between the two cases just presented, exhibits inert gas concentration profiles intermediate in character between those of figure 9–1a and those of figure 9–d and e. About 50 percent of the total He concentration gradient there can be attributed to excess radiogenic helium (Barnes and Bieri, 1976).

Gas Concentration Anomalies

The large pore-water concentration changes for the three heavier inert gases were an unexpected and surprising result of my measurements. I do not at present have a satisfactory explanation of those concentration anomalies. However, one can narrow the field of possible explanations by a process of elimination.

Barnes (1973b) discussed in detail a number of possible mechanisms. A condensed and updated version of that discussion follows. The preceding sedimentation model would suggest that, in areas of high sedimentation rate where the large concentration anomalies are found, the source of the observed anomalies most probably lies within the upper tens of meters of the sediment column since gas fluxes from deep in the sediment are greatly attenuated at the sediment surface. Any reasonable changes in paleotemperature or paleoatmospheric composition could not produce anomalies of the magnitude or temporal wavelength observed. Nor has any consistent profile pattern been observed that would suggest the necessary global changes. The presence of sedimentary thermal gradients cannot explain the observed patterns because equilibrium gas solubility (and chemical potential) changes with temperature follow the order He < Ne < Ar < Kr, which is opposite to the magnitude of the observed concentration changes.

The results of duplicate pore-water sampling and collection and analysis of bottom waters show that our measurements are accurate to within a few percent (see Table 1, Barnes, 1973b; Barnes and Bieri, 1976; Barnes et al., 1975). Furthermore, my profiles show systematic trends, not the random variations that one would expect from experimental difficulties.

The low permeability of clay-rich marine sediments suggests that free thermal convection of pore fluid is not significant in most types of marine sediments. However, such a process is being invoked as a possible explana-

Inert Gas Gradients 183

tion of anomalous temperature gradients in certain east equatorial Pacific calcareous oozes (Von Herzen, personal communication). There are sandy turbidite layers in SMB-SPB (and probably with less frequency in SCB) sediments that could act as channels for horizontal movements of pore water. Such movements could maintain a vertically localized concentration anomaly against diffusive dissipation, but one is still left with the problem of generating the anomalous pore water concentrations, whether the waters originate locally or are advected from elsewhere.

Figure 9-1c, a profile from the Samoan Passage area in the South Pacific, is most interesting and possibly instructive (Barnes, 1973b). The sample at 4 m is from a layer of carbonate ooze apparently deposited as a turbidite bed from some nearby topographic high above the carbonate compensation depth. All other samples analyzed to date have come from predominantly clay-silt sediments. Some of the samples have suffered postcollection diffusive loss of He and Ne, as illustrated by the low surface concentrations for those two gases, but a correction for the loss would only magnify the apparent anomalies. The profile is unique because He and Ne show large concentration increases, while Ar and Kr concentrations show large decreases at the 4 m depth. The concentration anomaly in the Samoan Passage core with a vertical extent of 2 m would have to be actively supported or it would be removed by diffusion in less than 700 years (Barnes, 1973b).

There are several possible mechanisms of gas exchange between pore waters and mineral phases: (1) gas exchange between cooling pillow basalts and lavas and the surrounding pore water or seawater (Dymond and Hogan, 1973, 1974; Lupton and Craig, 1975); (2) low-temperature weathering of unstable volcanic minerals and glasses; (3) mineral dissolution and crystallization; and (4) surface adsorption and desorption due to changing mineral surface characteristics.

Table 9-3 lists the average ratios of the gas content of the glassy and holocrystalline basalt samples analyzed by Dymond and Hogan (1973) to the gas content of an equal weight of deep-ocean water. The difference between the numbers for glassy and holocrystalline rocks is the result of gas exchange during cooling of the hot rock. He and Ne are lost from the rock, and Ar and Kr are introduced into the rock. If such a process were to affect pore-water gas concentrations, the change in the helium would be about 100 times more than the change in the heavier gases.

The concentrations of Ar, Kr, and Xe in deep-sea basalts measured by Fisher (1970) and He and Ne measured by Lupton and Craig (1975) are similar to those reported by Dymond and Hogan. In addition, a review of the He and Ar concentrations of basaltic rocks obtained as a by-product of K-Ar age studies shows concentrations of a similar magnitude (Dalrymple and Moore, 1968; Funkhouser et al., 1968; Noble and Naughton, 1968;

Table 9-3
The Average Abundance of He, Ne, Ar, and Kr in Holocrystalline Mid-ocean Ridge Tholeiites and Glassy Margins of Pillow Basalts Relative to the Abundance of These Gases in the Same Weight of Deep Ocean Water

Gas	Glassy	Holocrystalline
He	10^2	2.5×10^{-3}
Ne	3×10^{-1}	5×10^{-3}
Ar^{36}	2×10^{-4}	10^{-2}
Kr	2×10^{-4}	10^{-2}

Source: The hard rock concentrations are from Dymond and Hogan (1973).
Note: The numbers are round numbers.

Dymond, 1970). Unpublished data on the He and Ar concentrations of a variety of marine sedimentary minerals (apatite, glauconite, barite) and phillipsite (Bernat et. al., 1970) obtained in my laboratory confirm the preceding conclusion that the content of the heavier gases in the mineral phases is too small to significantly affect pore-fluid concentrations without much larger effects for helium.

The one remaining possibility for gas exchange between pore water and solid phases is surface adsorption-desorption. However, there is no experimental data on gas adsorption on solid surfaces immersed in liquid media.

More interstitial gas concentration measurements are needed to further elucidate the relationships between various types of inert gas concentration profiles and sedimentary characteristics. Such work may suggest which of the possible explanations are worth pursuing.

Acknowledgments

M. Prisbylla provided necessary assistance in the mass-spectrometric analysis of the new profiles reported here. I am also grateful for the support of the scientific staff and crew of the R/V *Melville* and R/V *E.B. Scripps* during the collection of the samples at sea. Supported in part by ONR contract N00014-69-200-6049-Goldberg, NSF grant DES75-19383, and G.S.A. grant 1749-73.

References

Barnes, R.O. 1973*a*. An *in situ* interstitial water sampler for use in unconsolidated sediments. *Deep-Sea Res.* 20:1125–1128.

Barnes, R.O. 1973b. Noble gas concentrations in the pore fluids of marine sediments and the construction of an *in situ* pore water sampler. Ph.D. dissertation, University of California, San Diego.

Barnes, R.O., Bertine, K.K., and Goldberg, E.D., 1975. N_2: Ar, nitrification and denitrification in Southern California borderland basin sediments. *Limnol. Oceanogr.* 20:962–970.

Barnes, R.O., and Bieri, R.H., 1976. Helium flux through marine sediments of the northeast Pacific Ocean. *Earth Planet. Sci. Lett.* 28:331–336.

Bernat, M., Bieri, R.H., Koide, M., Griffin, J.J., and Goldberg, E.D., 1970. Uranium, thorium, potassium, and argon in marine phillipsites. *Geochim. Cosmochim. Acta* 34:1053–1071.

Bieri, R.H., Koide, M., and Goldberg, E.D., 1968. Noble gas contents of marine waters. *Earth Planet. Sci. Lett.* 4:329–340.

Carslaw, H.S., and Jaeger, J.C., 1959. *Conduction of Heat in Solids*, 2d. ed. London: Oxford Univ. Press.

Dalrymple, G.B., and Moore, J.G., 1968. Argon-40: Excess in submarine pillow basalts from Kilauea Volcano, Hawaii. *Science* 161:1132–1135.

Dymond, J. 1970. Excess argon in submarine basalt pillows. *Bull. Geo. Soc. Am.* 81:1229–1232.

Dymond, J., and Hogan, L. 1973. Noble gas abundance patterns in deep-sea basalts-primordial gases from the mantle. *Earth Planet. Sci. Lett.* 20:131–139.

Dymond, J., and Hogan, L. 1974. The effects of deep-sea volcanism on the noble gas concentration of ocean water. *J. Geophys. Res.* 79:877–879.

Emery, K.O. 1960. *The Sea Off Southern California*. New York: Wiley.

Fisher, D.E. 1970. Heavy rare gases in a Pacific seamount. *Earth Planet. Sci. Lett.* 9:331–335.

Funkhouser, J.G., Fisher, D.E., and Bonatti, E., 1968. Excess argon in deep-sea rocks. *Earth Planet. Sci. Lett.* 5:95–100.

Jaeger, J.C. 1965. Application of the theory of heat conduction to geothermal measurements. In W.H.K. Lee and S. Uyeda (Eds.), *Terrestrial Heat Flow*. American Geophysical Union, Monograph 8, pp. 13–14.

Koide, M., Bruland, K.W., and Goldberg, E.D. 1973. Th-228/Th-232 and Pb-210 geochronologies in marine and lake sediments. *Geochim. Cosmochim. Acta* 37:1171–1187.

Koide, M., Bruland, K., and Goldberg, E.D. 1976. ^{226}Ra chronology of a coastal marine sediment. *Earth Planet. Sci. Lett.* 31:31–36.

Koide, M., Soutar, A., and Goldberg, E.D. 1972. Marine geochronology with ^{210}Pb. *Earth Planet. Sci. Lett.* 14:442–446.

Lupton, J.E., and Craig, H. 1975. Excess ^{3}He in oceanic basalts: Evidence for terrestrial primordial helium. *Earth Planet. Sci. Let.* 26:133–139.

Nobel, C.S., and Naughton, J.J. 1968. Deep-ocean basalts: Inert gas content and uncertainties in age dating. *Science* 162:265–266.

Veeh, H.H. 1967. Deposition of uranium from the ocean. *Earth Planet. Sci. Lett.* 3:145–150.

10 Methane Production, Consumption, and Transport in the Interstitial Waters of Coastal Marine Sediments

Christopher S. Martens

Abstract

Methane distribution in the interstitial waters of anoxic coastal marine sediments is controlled by a combination of chemical, biological, and physical processes. Methane concentrations remain below approximately 0.1 mM until about 90 percent of seawater sulfate is removed by sulfate-reducing bacteria. As sulfate concentrations approach zero, saturation concentrations of methane occur. Incubation studies of laboratory sediment using labeled and unlabeled substrates suggest that competition for acetate and H_2 by sulfate-reducing bacteria may inhibit methanogenesis in the presence of dissolved sulfate. Concave-upward depth profiles of methane concentration are consistent with methane production at greater depth in the sediments because the methane diffuses upward and is consumed in the zone of sulfate reduction.

Methane production in coastal sediments can be indirectly influenced by burrowing macrofauna, which irrigate sediments with dissolved sulfate. In nonbioturbated organic-rich coastal sediments, transport by both bubbles and diffusion across the sediment/water interface can be important sources of methane for overlying waters.

Introduction

The distribution of dissolved methane in the interstitial waters of anoxic marine sediments is directly influenced by a number of processes. Bacteria mediate its production and oxidation, and it can be transported by diffusion and advection, which includes the migration of bubbles. Through control of dissolved interstitial sulfate, macrofaunal irrigation can indirectly affect methane production and distribution.

The objective of this chapter is to assess the relative significance of those processes in coastal environments. Hypotheses and models developed through consideration of combined laboratory and field studies will be tested utilizing representative methane distributions from the interstitial waters of organic-rich coastal sediments.

Methane Distribution in Coastal Sediments

Methane concentrations in the interstitial waters of anoxic coastal sediments remain below approximately 0.1 mM until about 90 percent of seawater sulfate is removed by sulfate-reducing bacteria. As sulfate concentrations approach zero, saturated methane concentrations occur (Martens and Berner, 1974).

Distributions of dissolved methane and sulfate observed at three nearshore Long Island Sound stations (Martens and Berner, 1977) are illustrated in figure 10–1. Station TH cores were collected approximately 2 km offshore near Guilford, Connecticut, while station SC and BS cores were collected in two small harbors approximately 4 km west of Guilford. Reeburgh and Heggie (1977) have summarized available data which suggest that concave-upward methane distributions such as that found in core TH-51 are representative of anoxic marine sediments.

Methane Production and Consumption

Methanogenic bacteria isolated in the laboratory have been shown to utilize H_2 plus CO_2, formate and ethanol, and acetate as substrates for methanogenesis (Bryant et al., 1967; Torien and Hattingh, 1969; Wolfe, 1971). Methane production by a symbiotic community of microbes can be viewed as the result of stepwise degradation of simple organic compounds followed by CO_2 reduction and acetate conversion to methane (Torien and Hattingh, 1969). The exact mechanism for conversion of the substrates to methane remains undetermined (see review by Zeikus, 1977); however, acetate and CO_2 plus H_2 appear to be the major substrates in sediments. Studies of anaerobic lake sediments (e.g., Koyama, 1963; Cappenberg and Prins, 1974) suggest that acetate accounts for approximately 60 to 70 percent of methane formed, with CO_2 reduction accounting for the balance (e.g., Winfrey and Zeikus, 1977).

Influence of Dissolved Sulfate

A number of studies have demonstrated that dissolved sulfate in the interstitial waters of marine sediments affects methanogenesis. The distributions of dissolved sulfate and methane in the interstitial waters of anoxic

Methane Production/Consumption/Transport 189

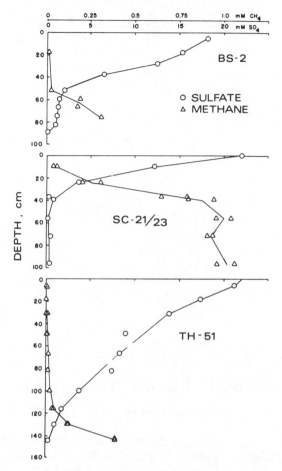

Figure 10-1. Dissolved methane and sulfate concentrations in the interstitial waters at three nearshore Long Island Sound stations.

coastal sediments (figure 10-1) can be explained by several hypotheses, including two of Martens and Berner (1974):

1. Methane production at approximately the same rate throughout the sediment column with bacterial consumption in the zone of sulfate reduction.
2. Significant methane production only in the absence of bacterial sulfate reduction, followed by upward diffusion and bacterial consumption in the zone of sulfate reduction.

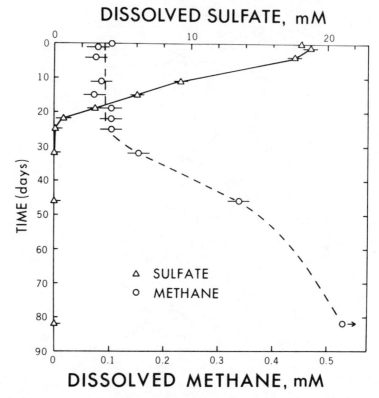

Source: Martens and Berner, *Limnol. Oceanogr.* 22: 10–25, 1977. Reprinted with permission.

Figure 10–2. Changes in dissolved methane and sulfate observed in natural sediments homogenized and sealed in the laboratory and opened sequentially over an 82-day period.

Direct evidence indicating no methane production until sulfate concentration approaches zero (figure 10–2) has been obtained by monitoring changes in dissolved methane and sulfate in natural sediments sealed in glass jars and opened sequentially over an 82-day period in the laboratory (Martens and Berner, 1974). Those data, along with field results indicating build-up of methane concentrations in the absence of sulfate (e.g., Claypool and Kaplan, 1974), have been taken to support the hypothesis that significant methane production occurs only in the absence of sulfate reduction. One explanation is that sulfate reducers compete favorably for available H_2 (Nissenbaum et al., 1972; Goldhaber and Kaplan, 1974; Martens and

Berner, 1974), utilizing their hydrogenase system to couple sulfate reduction with H_2. Accumulation of H_2 is inhibitory to hydrogen-producing bacteria (Bryant et al., 1967; Mechalas, 1974). Large populations of H_2 producers may thus be initially linked with sulfate reducers and subsequently support CO_2 reduction by methanogenic bacteria following exhaustion of sulfate.

Cappenberg (1975) has alternatively suggested that methanogenesis is inhibited by sulfide produced during sulfate reduction. Recent laboratory studies (e.g., Winfrey and Zeikus, 1977; Oremland and Taylor, 1977) do not appear to support his hypothesis; however, most workers have reported total sulfide concentrations rather than pH-dependent pS^{2-} values, as specified by Cappenberg (1975). Experiments in which inhibition of methane production is reversed by addition of either H_2 or acetate in the presence of high concentrations of sulfate and sulfide (see subsequent discussion) disagree with the sulfide-inhibition hypothesis.

Laboratory studies using labeled substrates added to natural lake sediments (Winfrey et al., 1977; Winfrey and Zeikus, 1977) have resulted in findings which may be significant for understanding the microbially mediated processes controlling methanogenesis in marine sediments. Winfrey et al. (1977) found that the rate of methane production in Lake Mendota sediments was greatly stimulated by H_2 and was, in fact, proportional to the amount of H_2 present. Addition of 50-μM concentrations of organic substrates also stimulated methanogenesis in the decreasing order glucose, formate, ethanol, and acetate. The authors suggested that the amount of stimulation by organic substrates may be related to the number of reducing equivalents (H_2) and carbon equivalents (acetate) derived from the degradation of the substrate.

In an associated series of experiments utilizing labeled acetate, they found that H_2 greatly stimulated $^{14}CH_4$ production from [1-^{14}C] acetate through reduction of $^{14}CO_2$ released from the carboxyl group of acetate. Under an N_2 atmosphere $^{14}CO_2$ was evolved rapidly, and only small amounts of $^{14}CH_4$ were produced. Experiments with [2-^{14}C] acetate resulted in rapid release of $^{14}CH_4$ and $^{14}CO_2$ under an N_2 atmosphere, with the addition of H_2 resulting in only a slight increase in $^{14}CH_4$ production and decrease in $^{14}CO_2$.

Winfrey and Zeikus (1977) continued those experiments to determine the effects of dissolved sulfate on methane production. Sulfate additions as low as 0.2 mM to incubated lake sediments inhibited methanogenesis. The inhibition was reversed by the addition of *either H_2 or acetate*. [2-^{14}C] acetate was converted to $^{14}CH_4$ and $^{14}CO_2$ only in the absence of sulfate, whereas only $^{14}CO_2$ was produced in the presence of sulfate. The $^{14}CO_2$ was not the result of $^{14}CH_4$ oxidation, since no conversion of $^{14}CH_4$ to $^{14}CO_2$ was observed in similar experiments with or without sulfate present.

When H_2 was added to sediment that contained both 10 mM sulfate and [2-^{14}C] acetate, a rapid evolution of $^{14}CO_2$ was observed from the methyl position of acetate. The small amount of $^{14}CH_4$ produced was attributed to reduction of some of the $^{14}CO_2$. Winfrey and Zeikus (1977) proposed that the acetate was being utilized by *nonmethanogenic organisms* which were stimulated by sulfate. Their arguments were supported by thermodynamic considerations which showed that utilization of acetate in the presence of sulfate was energetically favored over conversion of acetate to CH_4 and CO_2. They postulated that competition for both acetate and H_2 was responsible for inhibition of methane production by sulfate in the Lake Mendota sediment samples.

Widdel and Pfennig (1977) have described a new anaerobic bacterium *Desulfotomaculum acetoxidans* which respires acetate. Sorokin (1966) has previously reported a *Desulfovibrio* species that can grow on H_2, CO_2, and acetate, with the latter utilized for cell carbon synthesis. The utilization of acetate by sulfate-reducing organisms would have important implications for competition between sulfate reducers and methanogens, as pointed out by Winfrey and Zeikus (1977) and Zeikus (1977).

Laboratory experiments with anoxic marine sediments by Oremland and Taylor (1977) appear to agree in most respects with the lake studies previously cited. Methane production was stimulated in sulfate-containing sediments incubated in an H_2 atmosphere. Inhibition of sulfate reduction by addition of β-fluorolactate or sodium molybdate resulted in greater total methane production, although initial production rates between experiments were the same. Acetate additions had no immediate effect on methane production, leading to the hypothesis that carbon was not limiting to methanogenic activity during initial stages of incubation. Incubations under an N_2 atmosphere in the presence of sulfate as well as during inhibition of sulfate reduction indicated an initial period of rapid methane production followed by little production. Those results agree with the results of Winfrey and Zeikus (1977) in that it is likely that carbon substrates initially present in the marine sediments represented an acetate source for methanogens, allowing low rates of methane production during sulfate reduction.

It should be pointed out that in the experiments of Oremland and Taylor, H_2 uptake was dominated by sulfate reducers, indicating that the sediments incubated came from a depth zone dominated by sulfate reducers. Cappenberg (1974) has previously found that the highest numbers of methanogenic bacteria are found beneath sulfate reducers in lake sediments.

The studies of lake sediment suggest that when available H_2 and acetate are both limiting, methane production should occur only in the absence of sulfate reduction in marine sediments. Oremland and Taylor (1977) and Winfrey and Zeikus (1977) have pointed out that low rates of methanogenesis may occur in sulfate-containing marine sediments extremely rich in organic matter, such as sea-grass beds (Oremland, 1975), as a result of lessened competition for acetate and H_2. The effect of varying sulfate concentrations under such circumstances is unknown. The lake-sediment experiments of Winfrey and Zeikus (1977) indicate that concentrations of less than 0.2 mM may effectively inhibit methanogenesis when H_2 and acetate are limiting.

Further laboratory and field work are needed to identify and quantify the relative importance of microbially mediated processes controlling methane production in marine sediments. Sediment-incubation experiments with labeled substrates and identification of acetate-respiring, sulfate-reducing organisms in marine sediments will be required to substantiate the hypothesis that competition for both acetate and H_2 can be responsible for inhibition of methanogenesis by sulfate in the marine environment. Field studies linking sites and in situ rates of methanogenesis to available substrate (i.e., acetate) concentrations may then be interpreted in terms of specific microbial interactions.

Methane Consumption during Sulfate Reduction

Distributions of methane in marine sediments are also affected by consumption (oxidation) and transport processes. Evidence for methane consumption during sulfate reduction comes mainly from the methane-sediment distributions discussed earlier. Barnes and Goldberg (1976) and Reeburgh (1976) have demonstrated that the concave-upward profiles in anoxic sediments could only be explained by consumption in the zone of sulfate reduction through the reaction

$$CH_4 + SO_4^{2-} \rightarrow HS^- + HCO_3^- + H_2O$$

That reaction is thermodynamically favored under conditions found in sediments of Long Island Sound where concentrations of dissolved methane, sulfate, bisulfide, and bicarbonate are approximately 0.1, 10, 0.5, and 30 mM respectively (Martens and Berner, 1977).

The concave-upward methane concentration profile observed in Long Island Sound core TH-51 (figure 10-3) can be fitted utilizing a steady-state kinetic model (Martens and Berner, 1977) incorporating the influences of diffusion, compaction, and first-order methane consumption:

$$D_s \frac{\partial^2 c}{\partial z^2} - w \frac{\partial c}{\partial z} - k_1 C = 0$$

where z = depth below the sediment/water interface measured positively downward; Z = methane saturation depth

C = concentration of dissolved methane in interstitial water
w = rate of deposition
D_s = whole sediment diffusion coefficient for methane
k_1 = first-order rate constant

Solution for the boundary conditions

$$z = 0, C = 0$$
$$z = Z, C = C_Z$$

yields the solution:

$$C = \frac{\exp \alpha z - \exp \gamma z}{\exp \alpha Z - \exp \gamma Z} C_Z$$

where

$$\alpha = w + \frac{(w^2 + 4k_1 D)^{1/2}}{2D}$$

$$\gamma = w - \frac{(w^2 + 4k_1 D)^{1/2}}{2D}$$

Goldhaber et al. (1977) have determined that w at the core TH-51 area is 10^{-8} cm/s. Previous work of Hammond et al. (1975) and our recent studies in Cape Lookout Bight (Martens et al., unpublished data) suggest that 1×10^{-5} cm^2/s is a reasonable D_s value for methane. A saturation methane concentration C_Z of 2.6 mM at 175 cm is calculated for station TH based on temperature and pressure conditions and the sulfate distribution observed in core TH-51 and a sister core collected beside it.

The core TH-51 data are well fitted using a k_1 value of 8×10^{-8} s^{-1} (figure 10-3). The inferred slow rate of methane consumption would not be observed over the 82-day jar experiment described earlier (figure 10-2). Use of an alternative D_s value of 2×10^{-6} cm^2/s for methane results in a good fit to core TH-51 data with a k_1 value of 8×10^{-9} s^{-1} (Martens and Berner, 1977). Recently, in a 420-day laboratory jar experiment with Cape Lookout Bight sediments, Martens et al. (unpublished data) have observed decreases in methane concentrations during sulfate concentration best fitted with a first-order model with a k_1 of approximately 10^{-7} s^{-1}.

Combined field and laboratory results from Long Island Sound support the hypothesis that methane is produced mainly after sulfate is depleted but is consumed in the sulfate-reduction zone when diffusing upward.

Methane Production/Consumption/Transport

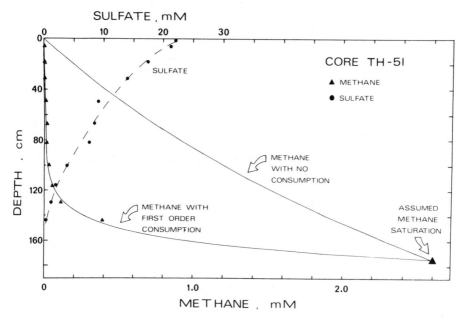

Source: Martens and Berner, *Limnol. Oceanogr.* 22:10–25, 1977. Reprinted with permission.

Figure 10–3. Long Island Sound core TH-51 methane distribution fitted with a kinetic model incorporating the influences of diffusion, compaction, and first-order consumption ($D_s = 1 \times 10^{-5}$ cm²/s; $k_1 = 8 \times 10^{-8}$ s^{-1} for the case with first-order consumption).

Methane Transport

Transport of methane across the sediment/water interface occurs as a result of both molecular diffusion and advective processes, including mixing and bubble ebullition (e.g., Hammond et al., 1975; Martens, 1976). Recent studies in Cape Lookout Bight, North Carolina (figure 10–4) have demonstrated the important indirect influence of macrofaunal irrigation on bubble transport (Martens, 1976). A typical estuarine community of burrowing macrofauna present at Station 3 near the bight entrance (figure 10–4) irrigates the highly reducing sediments, maintaining vertical linear sulfate concentration profiles in the upper 20 to 30 cm at that location (figure 10–5). The continuous resupply of sulfate prevents saturation concentrations of methane and thus bubble formation.

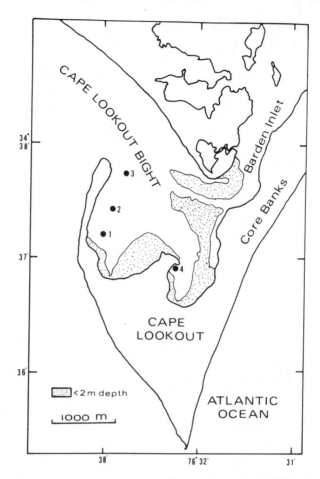

Figure 10–4. Cape Lookout Bight study site. Locations of Stations 1, 2, and 3 at approximately 8.5 m are shown.

At Stations 1 and 2, where macrofauna are absent and sulfate reduction is complete within the upper 10 to 25 cm (figure 10–5), saturation methane concentrations occur within 20 cm of the sediment/water interface (figure 10–6), and bubbles can be observed leaving the sediments (see figure 10–7). Lack of upward concavity in the methane profile observed at Station 1 may be the result of methane redistribution in interstitial waters caused by upward migrating bubbles.

Diving surveys conducted near Station 1 revealed that the bubbles were leaving the sediments through large numbers of randomly distributed cylindrical tubes approximately 0.3 to 2 cm in diameter, for which the term *bubble tubes* was coined. Diver-inserted, Plexiglas, 1-in-thick box cores were

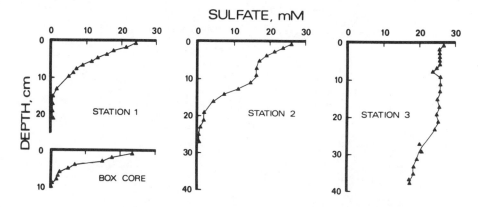

Figure 10-5. Concentrations of dissolved sulfate in interstitial waters at Stations 1, 2, and 3.

utilized to obtain x-rays (figure 10-8) of the upper 20 cm at Station 1. Several bubble tubes are visible in the x-ray, one clearly connecting pockets of bubbles (appearing as light-colored, irregularly shaped areas) with the surface waters. The bubble tubes appear to be of abiotic origin, probably created and maintained by bubbles pushing their way up through the soft, easily resuspended mud. The depth of the layer of bubbles seen in the x-ray corresponds well with the level of zero sulfate observed in gravity and box cores collected at Station 1 (see figure 10-5).

The significance of the diffusive flux of methane from Cape Lookout Bight sediments is currently under investigation. Initial flux measurements obtained by measuring in situ concentration increases in benthic chambers placed on the bottom by divers range from 30 to greater than 100 $\mu M/(m^2 \cdot h)$ (Martens et al., unpublished data).

Methane transport across the sediment/water interface by bubbles is probably not significant in marine environments where normal communities of irrigating macrofauna occur and where sulfate reduction is not complete at very shallow depths (i.e., much less than 1 m). Molecular diffusion and transport by mixing processes may also be limited in such environments by the concave-upward-shaped methane concentration profile and by aerobic methane oxidation. The latter sink is probably especially important where low oxygen concentrations are maintained by physical mixing and/or irrigation and relatively high concentrations of dissolved inorganic nitrogen occur. Facultative microaerophilic methane oxidizers are known to function optimally under such conditions in marine waters (Sansone and Martens, 1978). Undetectable methane concentrations observed in the upper 10 to 30 cm of heavily irrigated Long Island Sound sediments where low oxygen and

198   Dynamic Environment of the Ocean Floor

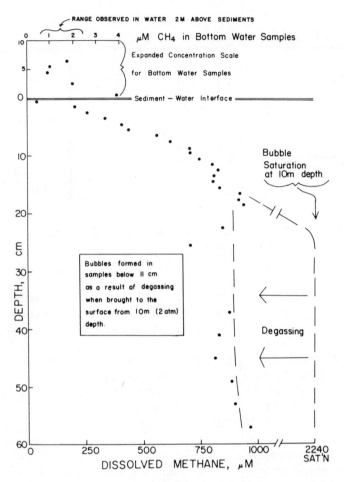

Figure 10–6. Concentrations of dissolved methane in interstitial waters at Station 1 as determined by dialysis equilibration sampler.

high ammonia concentrations occur (Martens and Berner, 1977) appear to have resulted from such aerobic oxidation.

Acknowledgments

Constructive critical remarks by N.R. Andersen led to significant improvements in this chapter. Financial support was provided by NSF Oceanography Section Grant OCE 75-06199.

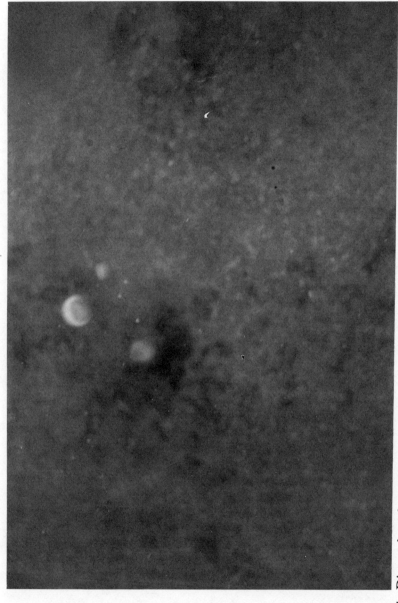

Figure 10–7. Diver-taken photo showing bubble salvo exiting bubble tube opening at sediment/water interface. Bubbles shown range from approximately 0.1 to 1.5 cm in diameter.

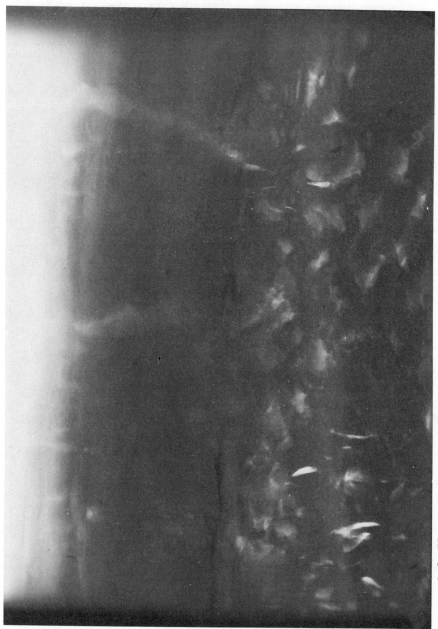

Figure 10-8. X-ray of a 1-in-thick box core collected by a diver at Station 1. Bubbles appear as irregularly shaped light-colored areas below approximately 10 cm.

References

Barnes, R.O., and Goldberg, E.D. 1976. Methane production and consumption in anoxic marine sediments. *Geology* 4:297–300.

Bryant, M.P., Wolin, E.A., Wolin, M.J., and Wolfe, R.S. 1967. *Methanobacillus omelianskii*, a symbiotic association of two species of bacteria. *Arch. Mikrobiol.* 59:20–31.

Cappenberg, Th. E. 1974. Interrelations between sulfate-reducing and methane-producing bacteria in bottom deposits of a freshwater lake. I. Field observations. *Antonie van Leeuwenhoek* 40:285–295.

Cappenberg, Th. E. 1975. A study of mixed continuous cultures of sulfate reducing and methane producing bacteria. *Microb. Ecol.* 2:60–72.

Cappenberg, Th. E., and Prins, R.A. 1974. Interrelations between sulfate-reducing and methane-producing bacteria in bottom deposits of a freshwater lake. III. Experiments with ^{14}C-labeled substrates. *Antonie van Leeuwenhoek* 40:457–469.

Claypool, G., and Kaplan, I.R. 1974. The origin and distribution of methane in marine sediments. In I.R. Kaplan (Ed.), *Natural Gases in Marine Sediments*. New York: Plenum, pp. 99–139.

Goldhaber, M.B., Aller, R.C., Cochran, K., Rosenfeld, J.K., Martens, C.S., and Berner, R.A. 1977. Sulfate reduction, diffusion and bioturbation in Long Island Sound sediments: Report of the FOAM group. *Am. J. Sci.* 277:193–237.

Goldhaber, M.B. and Kaplan, I.R. 1974. The sulfur cycle. In E.D. Goldberg (Ed.), *The Sea*, Vol. 5. New York: Wiley, pp. 569–655.

Hammond, D.E., Simpson, H.J., and Mathieu, G. 1975. Methane and radon-222 as tracers for mechanisms of exchange across the sediment-water interface in the Hudson River estuary. In T. Church (Ed.), *Chemistry in the Coastal Marine Environment*. Washington: Am. Chem. Soc. Symposium Series 18, pp. 119–132.

Koyama, T. 1963. Gaseous metabolism in lake sediments and paddy soils and the production of atmospheric methane and hydrogen. *J. Geophys. Res.* 68:3971–3973.

Martens, C.S. 1976. Control of methane sediment-water transport by macroinfaunal irrigation in Cape Lookout Bight, North Carolina. *Science* 192:998–1000.

Martens, C.S., and Berner, R.A. 1974. Methane production in the interstitial waters of sulfate-depleted marine sediments. *Science* 185:1167–1169.

Martens, C.S., Berner, R.A. 1977. Interstitial water chemistry of anoxic Long Island Sound sediments. 1. Dissolved gases. *Limnol. Oceanogr.* 22:10–25.

Mechalas, B.J. 1974. Pathways and environmental requirements for biogenic gas production in the ocean. In I.R. Kaplan (Ed.), *Natural Gases in Marine Sediments*. New York: Plenum, pp. 11–25.

Nissenbaum, A.,Presley, B.J., and Kaplan, I.R. 1972. Early diagenesis in a reducing fjord, Saanich Inlet, British Columbia. I. Chemical and isotopic changes in major components of interstitial water. *Geochim. Cosmochim. Acta* 36:1007–1027.

Oremland, R.S. 1975. Methane production in shallow water, tropical marine sediments. *Appl. Microbiol.* 30:602–608.

Oremland, R.S., and Taylor, B.F. 1977. Sulfate reduction and methanogenesis in tropical, marine sediments. Unpublished manuscript.

Reeburgh, W.S. 1976. Methane consumption in Cariaco Trench waters and sediments. *Earth Planet. Sci. Lett.* 28:337–344.

Reeburgh, W.S., and Heggie, D.T. 1977. Microbial methane consumption reactions and their effect on methane distributions in freshwater and marine environments. *Limnol. Oceanogr.* 22:1–9.

Sansone, F.J., and Martens, C.S. 1978. Methane oxidation in Cape Lookout Bight, North Carolina. *Limnol. Oceanogr.* 23:349–355.

Sorokin, Yu. I. 1966. Sources of energy and carbon for biosynthesis in sulfate-reducing bacteria. *Microbiol. (USSR)* 35:643–647.

Torien, D.F., and Hattingh, W.H.J. 1969. Anaerobic digestion. 1. The microbiology of anaerobic digestion. *Water Res.* 3:385–416.

Widdel, F., and Pfennig, N. 1977. A new anaerobic sporing bacterium *Desulfotomaculum* (emerd.) *acetoxidans*. *Arch. Microbiol.* 112:119–122.

Winfrey, M.R., Nelson, D.R., Klevickis, S.C., and Zeikus, J.G. 1977. Association of hydrogen metabolism with methanogenesis in Lake Mendota sediments. *Appl. Environ. Microbiol.* 33:312–318.

Winfrey, M.R., and Zeikus, J.G. 1977. Effect of sulfate on carbon and electron flow during microbial methanogenesis in freshwater sediments. *Appl. Environ. Microbiol.* 33:275–281.

Wolfe, R.S. 1971. Microbial formation of methane. *Adv. Microbial. Physiol.* 6:107–146.

Zeikus, J.G. 1977. The biology of methanogenic bacteria. *Bacteriolog. Rev.* 41:514–541.

11 A Major Sink and Flux Control for Methane in Marine Sediments: Anaerobic Consumption

William S. Reeburgh

Abstract

A major portion of the upward methane flux from marine sediments appears to be consumed before it reaches the sediment/water interface. Slope changes in depth distributions of methane, total carbon dioxide, and sulfate as well as a minimum in $\delta^{13}CO_2$ at the depths of the slope changes are consistent with the presence of a thin subsurface anaerobic zone of methane consumption. The zone coincides with a depth where the product (CH_4) (SO_4^{2-}) is maximum, suggesting kinetic control of the zone's location. Differences in methane distributions from marine and freshwater sediments suggest that sulfate reducers are responsible for anaerobic methane consumption in marine systems.

Introduction

When practical methods for sampling interstitial water were devised just prior to 1970, a great deal of research activity was focused on the chemical composition of interstitial waters. Because of the high solid-solution ratio and the resulting restricted circulation, the effects of biological and chemical reactions were expected to be intensified and retained in interstitial waters, providing a key to many problems in early diagenesis. Those initial expectations have been blunted somewhat by the complicating factors of air oxidation, bioturbation, and pressure and temperature effects on solid-solution equilibria. We appreciate now that the depth distribution of an interstitial solute is a composite of the effects of physical, chemical, and biological processes, all of which may operate with different time scales and intensities. Very few of the reactions occurring in interstitial waters lend

---

Contribution number 360 from the Institute of Marine Science, University of Alaska. Contribution number 233 from the Marine Sciences Research Center, SUNY, Stony Brook. This work was supported in part by NSF Grants GA-41209, GA-19380, and OCE 76-08891.

themselves to direct study. As a result, most of our information about reactions in interstitial waters has been gleaned from field measurements in different environments coupled with thermodynamic and microbiological insights, laboratory studies, and mathematical models. A reasonable data base exists for gases in sediments, but the data have not been reviewed as a whole. This chapter attempts to review and extend the available measurements of CH_4, ΣCO_2, SO_4^{2-}, and $\delta^{13}CO_2$ in marine sediments, emphasizing anaerobic methane oxidation and its importance in methane geochemistry.

Methane production is currently a poorly understood subject. Evidence for two production reactions—carbon dioxide reduction and acetate fermentation—has been advanced, but the question of which is most important in nature remains unanswered. There is, however, no compelling evidence for large-scale methane production in any environment other than interstitial waters, and work by Reeburgh and Heggie (1977) shows that consideration of consumption reactions and the conditions under which they occur leads to consistent explanations for methane distributions in a wide variety of environments.

This chapter considers the net reaction

$$CH_4 + SO_4^{2-} \rightarrow HS^- + HCO_3^- + H_2O \qquad (11.1)$$

and presents depth-distribution data from a variety of environments, suggesting that the reaction occurs in a zone below the sediment/water interface. Possible reasons for the zone's subsurface location are considered, and estimates of reaction rates are used to show that it serves as a major sink for methane in marine systems.

Data

Figure 11–1 shows schematic depth distributions of methane, total carbon dioxide, sulfate, and $\delta^{13}CO_2$[a] in marine sediments. The distributions have the dubious distinction of never having been determined in one environment as part of a single investigation involving methane. There are, however, cross-links, where pairs or trios of these parameters have been measured in single investigations or where other data for the same environment are available. Table 11–1 summarizes the data sources. Combining data in such a manner is clearly not desirable, but the lack of complete data for any one environment makes this approach necessary. Since the depth distributions of the

---

[a] $\delta^{13}C = \dfrac{(^{13}C/^{12}C) \text{ sample} - (^{13}C/^{12}C) \text{ standard}}{(^{13}C/^{12}C) \text{ standard}} \times 1{,}000$

A Major Sink and Flux Control

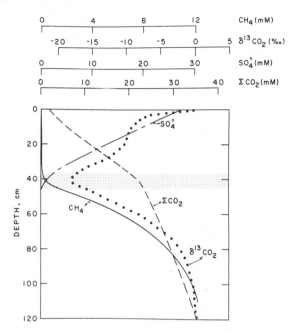

Figure 11-1. Schematic diagram showing depth distribution of methane, sulfate, total carbon dioxide, and the carbon isotope ratio of carbon dioxide in interstitial waters of marine sediments. All the distributions show breaks or slope changes in the stippled area, which represents the zone of anaerobic methane oxidation.

constituents common to several data sets are similar in different environments and all show slope or concentration changes at the same depths relative to other constituents, a general explanation may be suspected, and presentation of the combined results in a schematic diagram appears reasonable. Figure 11-1 has been drawn with an arbitrary depth scale; all distributions are relative to the bottom of the low-methane-concentration surface zone. Ancillary data were chosen from adjacent stations where possible.

The interstitial methane distribution in figure 11-1 is characteristic of marine sediments (Reeburgh and Heggie, 1977). The distributions are concave upward and have low methane concentrations (< 0.05 M) in a surface zone located between the sediment/water interface and depths that range from 20 cm (Long Island Sound) to 1 m (Santa Barbara Basin). Methane gradients in the surface zone are either non-existent or 100-fold

Table 11-1
Interstitial Water Data Sources

Location	CH_4	CO_2	SO_4	$^{13}CO_2$
Chesapeake Bay	Reeburgh (1969)	Reeburgh (1969)	Matisoff, et al. (1975)	—
Long Island Sound	Martens and Berner (1977)	—	Martens and Berner (1974)	—
Cariaco Trench	Reeburgh (1976)	Reeburgh (1976)	Presley (1974)	—
Saanich Inlet[a]	—	Nissenbaum et al. (1972)	Nissenbaum et al. (1972)	Nissenbaum et al. (1972)
Gulf of California				
South Guymas Basin	—	Goldhaber (1974)	Goldhaber (1974)	Goldhaber (1974)
California Basins				
Santa Barbara	Emery and Hoggan (1958)	Sholkovitz (1973)	—	
	Barnes and Goldberg (1976)			
San Pedro	—	Presley and Kaplan (1968)	Presley and Kaplan (1968)	Presley and Kaplan (1968)
Santa Catalina	—	Presley and Kaplan (1968)	Presley and Kaplan (1968)	Presley and Kaplan (1968)

[a]Core 1 data only. Other cores showed almost complete reduction of SO_4^{2-} at surface.

A Major Sink and Flux Control

smaller than those in the zone below it. Methane accumulates to much higher concentrations (5 to 15 mM) below the surface zone, sometimes reaching concentrations where bubble formation is possible.

The sulfate distributions show more variability between environments than do methane distributions. The actual distributions vary between linear and exponential decreases with depth; the key feature of the sulfate distributions are their approach to zero concentration at the base of the low-methane-concentration surface zone.

The total carbon-dioxide distributions increase with depth, following either an exponential distribution or one approximated by two straight lines. As with the sulfate distributions, the slopes of the total carbon dioxide distributions also increase above the base of the surface zone of concentration.

Continuous sets of fine-scale carbon-isotope-ratio measurements in interstitial waters are rare. The measurements shown schematically in figure 11-1 are from Goldhaber (1974) and were summarized in Claypool and Kaplan (1974). The important feature of the distribution is the $\delta^{13}CO_2$ minimum, which also occurs at the base of the low-methane-concentration surface zone. Claypool and Kaplan (1974) also commented on the general nature of the $\delta^{13}CO_2$ minimum in sediments.

Discussion

Depth Distributions

The processes controlling depth distributions of methane in marine sediments have been considered by Reeburgh (1969, 1976), Martens and Berner (1974), and Claypool and Kaplan (1974). Reeburgh (1969) interpreted the surface zone of low methane concentration as being due to admixture of overlying surface water and oxidation by molecular oxygen. Further work by Claypool and Kaplan (1974) and Martens and Berner (1974) showed that sulfate was nearly depleted in sediments before methane accumulated in quantity. Claypool and Kaplan suggested that sulfate reduction and methane production were mutually exclusive processes, while Martens and Berner presented four alternate hypotheses to account for their field and laboratory observations, favoring either mutual exclusion or limited methane production in the presence of sulfate without methane consumption.

It should be noted that no process which inhibits methane production can by itself produce the observed concave-upward distributions and the low-methane-concentration surface zone; an active consumption process is required to dispose of the methane flux from deep in the sediments. Martens and Berner (1977) presented a discussion and calculation showing that

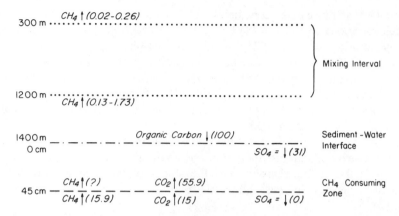

Source: Reeburgh, *Earth Planet. Sci. Lett.* 28:337–344, 1976. Reprinted with permission.

Figure 11–2. Summary of methane fluxes in the Cariaco Trench water column and methane, carbon dioxide, sulfate, and organic carbon fluxes in the sediments [in μmole/(cm$^2 \cdot$yr)].

diffusion, burial, and production at depth can produce a slightly concave-upward distribution. They concluded that there must be consumption to produce the large degree of concavity observed.

When methane distributions similar to those obtained from other marine environments were obtained in sediments from the anoxic Cariaco Trench (Reeburgh, 1976) and the nearly anoxic Santa Barbara Basin (Barnes and Goldberg, 1976), it became clear that the earlier methane measurements of Emery and Hoggan (1958) were reliable and that bioturbation and oxidation by molecular oxygen could not account for the distributions in those environments. Model calculations indicating methane consumption in the anoxic Cariaco Trench water column (Reeburgh, 1976) suggested that anaerobic methane oxidation was possible in sediments.

The low interstitial methane concentration of the surface zone indicates that methane in sediments is almost completely consumed at the base of the zone. As pointed out earlier and shown in figure 11–1, the carbon dioxide distributions show a slope change indicating a flux increase in the surface zone, while sulfate is completely consumed there. Figure 11–2 shows fluxes calculated from concentration gradients in Cariaco Trench sediments (Reeburgh, 1976). Methane and total carbon dioxide fluxes are from Reeburgh's (1976) measurements; the organic carbon flux to the sediment surface is from Presley (1974); and the sulfate flux is from Presley (1974) and Fanning and Pilson (1972). All fluxes are expressed in μmoles cm^{-2} yr^{-1}.

A Major Sink and Flux Control

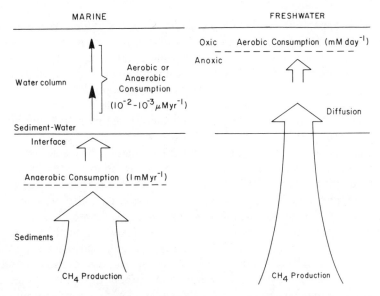

Figure 11-3. Schematic diagram showing how the locations of methane consumption reactions in marine and freshwater environments affect methane distributions. Methane concentrations are usually quite low in marine water columns because the major sink for methane is in the sediments; in freshwater systems, methane accumulates to high concentrations before rapid consumption in a hypolimnetic zone by aerobic methane oxidizers. Typical measured and calculated consumption rates are shown adjacent to the reaction zones.

If the reaction responsible for anaerobic methane consumption is as shown in reaction 11.1, anaerobic methane oxidation in the Cariaco Trench (figure 11-2) accounts for about half of the downward sulfate flux and for nearly 25 percent of the upward carbon dioxide flux in the surface zone. About 40 percent of the organic carbon flux is preserved in the sediments. The free-energy change for the reaction at typical reactant and product concentrations is -6 kcal mole^{-1}, indicating that reaction 11.1 proceeds as written. There is little question that the reaction is mediated biologically.

Microbial Ecology

Work by Cappenberg (1974, 1975) clarified the situation regarding mutual exclusion advanced by Claypool and Kaplan (1974) and Martens and Berner

(1974). Cappenberg (1974) used specific inhibitors to identify the principal substrates and products of sulfate reducers and methane producers. His later work (1975) demonstrated a commensal relationship between the two bacterial groups using continuous culture techniques. Those studies showed that methane producers require acetate for methane production as well as the low redox potentials supplied by the sulfate reducers. The commensal relationship between the two groups is complicated by sulfide inhibition of the methane producers at pS^{2-} values below 10.5.

Laboratory studies by Davis and Yarbrough (1966) have shown that sulfate reducers were able to oxidize methane at low rates in a lactate medium, but Sorokin (1957) found no evidence of the reaction using sulfate reducers and methane as the sole carbon source. Also, Quale (1972) indicated that anaerobic organisms capable of using methane as the sole carbon source are unknown. The fact that anaerobic methane oxidation occurs in nature suggests that the studies of Davis and Yarbrough and of Sorokin represent end members of a range of possible reaction conditions. Sulfate reducers may derive energy and carbon from the co-metabolic oxidation of recalcitrant molecules (Mechalas, 1974), a process that significantly broadens the spectrum of substrates available to them. Since sulfate is the only likely oxidant available, co-metabolism by sulfate reducers must be the consumption process.

Rates

The rate of anaerobic methane oxidation may be estimated from the upward diffusive flux of methane into the anaerobic zone where the oxidation occurs; (Reeburgh, 1976; and Barnes and Goldberg, 1976). The estimated methane fluxes to the zone are about 10 μmole cm^{-2} yr^{-1}. The nearly complete consumption of the methane produces the low methane concentrations observed in the surface zone. The consumption rate (expressed in μmole l^{-1} yr^{-1}) is sensitive to the choice of zone thickness. Based on sample spacing and the observed methane distributions, Reeburgh (1976) estimated the thickness of the zone to be 10 cm; Barnes and Goldberg (1976) estimated a thickness of 46 cm using the same criteria. The methane consumption rates range between 200 and 2,000 μM yr^{-1}, which are 10^6 times faster than the rates calculated for anaerobic methane oxidation in the water column of the Cariaco Trench (Reeburgh, 1976).

Methane production rates were estimated using an advection-diffusion model (Berner, 1971) and yielded rates of 3 to 22 μmole l^{-1} yr^{-1}. The previously calculated consumption rates are 10^2 times larger than the production rate, showing that the methane is rapidly oxidized as it diffuses into the zone where anaerobic consumption takes place.

Carbon Isotopes

A further suggestion that anaerobic methane oxidation occurs and is located in a subsurface zone is provided by the carbon isotope distributions of Goldhaber (1974) and Claypool and Kaplan (1974). The latter reference also summarizes measurements by Presley and Kaplan (1968) and Nissenbaum et al. (1972). The data were collected from several environments (see table 11-1) and were interpreted prior to reports of anaerobic methane oxidation. With some data, particularly in DSDP measurements, sample intervals were too large for good depth resolution. Claypool and Kaplan (1974) commented on the general nature of the $\delta^{13}CO_2$ minimum and interpreted it as being due to addition of isotopically light carbon dioxide at a rate faster than carbon dioxide is fractionated during methane production. The process was not specified.

The principal difficulty in using $\delta^{13}CO_2$ to elucidate processes occurring in sediments lies in the fact that there are several as yet unresolvable sources for carbon dioxide in sediments, each contributing carbon dioxide with a reasonably consistent carbon isotope ratio. In sediments, such carbon dioxide may result from admixture of seawater biocarbonate ($\delta^{13}C = \sim -1‰$), dissolution of or exchange on calcareous material ($\delta^{13}C = \sim 0‰$), oxidation of organic matter ($\delta^{13}C = -5$ to $-17‰$) during sulfate reduction, oxidation (aerobic or anaerobic) of methane ($\delta^{13}C = -40$ to $-80‰$), and fractionation of residual carbon dioxide due to methane production under closed or semi-closed conditions, resulting in $\delta^{13}C$ values of 0 to $+5‰$. Those processes may be expected to vary with time (depth), so there may be a means of assessing the various contributions. The presence of negative $\delta^{13}CO_2$ in contact with a large (100 to 1000-fold) excess of calcium carbonate ($\delta^{13}C = 1‰$) (Presley and Kaplan, 1968, table 2) suggests that isotopic equilibration of dissolved carbon dioxide with mineral surfaces is slow compared to the rate of injection of isotopically light carbon dioxide.

Presley and Kaplan (1968) reported a decrease in $\delta^{13}CO_2$ with depth from the sediment surface in the San Pedro Basin (~ -1 to $21‰$) and the Santa Catalina Basin (~ -1 to $-19‰$) and concluded that the isotopically light carbon was of metabolic origin. Sulfate was present at all depths in the cores, except for the bottom-most samples from the San Pedro Basin. Nissenbaum et al. (1972) suggested that the decreases in $\delta^{13}CO_2$ indicated release during sulfate reduction of biogenic carbon dioxide with the same isotope ratio as the parent organic matter. The Saanich Inlet cores studied by Nissenbaum et al. (1972) contained virtually no sulfate and had positive (up to $+17‰$) $\delta^{13}CO_2$ values, which decreased with depth. Core 1, however, contained sulfate to depths of 50 cm and had negative ($-37‰$) $\delta^{13}CO_2$ values in that interval.

Nissenbaum et al. (1972) and Claypool and Kaplan (1974) have amplified the finding of positive $\delta^{13}CO_2$ values using a Rayleigh distillation model to suggest that the positive $\delta^{13}CO_2$ values result from isotopic fractionation of carbon dioxide during carbon dioxide reduction to form methane, which has a characteristic low $\delta^{13}C$ value.

Kaplan and his co-workers have offered two different explanations for $\delta^{13}CO_2$ distributions above and below the $\delta^{13}CO_2$ minimum. Above the minimum, the trend toward light $\delta^{13}CO_2$ values is thought to result from the addition of carbon dioxide with an isotopic composition identical to the parent material. Below the minimum, the same isotopically light carbon dioxide was interpreted as being fractionated during reduction to methane, yielding isotopically heavy carbon dioxide.

Those interpretations on unconstrained systems were advanced prior to reports of anaerobic methane oxidation in sediments. In the following, the $\delta^{13}CO_2$ minimum itself is taken as being due to a continuous injection of isotopically light carbon dioxide from anaerobic methane oxidation.

If we consider that the situation deep in sediments is as shown in Claypool and Kaplan (1974, figures 7 to 10) and is thus similar to a sewage digester operating at steady state producing methane ($\delta^{13}C = -47\textperthousand$) and carbon dioxide ($\delta^{13}C = +5\textperthousand$) from organic matter ($\delta^{13}C = -23\textperthousand$) (Nissenbaum et al., 1972), we can estimate the fraction of methane-derived carbon dioxide in the $\delta^{13}CO_2$ minimum if we know the carbon dioxide and methane fluxes to the anaerobic methane consuming zone. That seems reasonable in view of the slowness of surface equilibration reactions mentioned earlier and the abundance of organic carbon buried in sediments, where 40 percent of the organic matter flux is preserved (figure 11–2). Unfortunately, no methane data exist for environments where carbon isotope data are available. We can, however, estimate how much of the carbon dioxide in the $\delta^{13}CO_2$ minimum is methane-derived and compare that fraction with fluxes of methane and carbon dioxide from other environments. The appropriate relationship (from Shultz and Calder, 1976) is

$$\%CO_{2(CH_4)} = \frac{\delta^{13}C_{min} - \delta^{13}C_{bkg}}{\delta^{13}C_{CH_4} - \delta^{13}C_{bkg}} \times 100$$

where $\%CO_{2(CH_4)}$ = percent methane-derived CO_2

$\delta^{13}C_{min} = \delta^{13}C$ of the minimum ($-17‰$)

$\delta^{13}C_{bkg} = \delta^{13}C$ of the background CO_2 (0 to $+4‰$)

$\delta^{13}C_{CH_4} = \delta^{13}C$ of the methane (-40 to $-70‰$)

Substituting known $\delta^{13}C$ values, the fraction of methane-derived carbon dioxide in the $\delta^{13}CO_2$ minimum ranges between 30 and 50 percent, depending on the value chosen for $\delta^{13}CH_4$. Those proportions are in very reasonable agreement with the carbon dioxide and methane fluxes shown in figure 11-2.

The concentration distributions of methane, total carbon dioxide, and sulfate (figure 11-1) and the flux compilation for Cariaco Trench sediments (figure 11-2) show that anaerobic methane oxidation is occurring in marine sediments and suggest the process is located in a narrow depth interval. The distribution of $\delta^{13}CO_2$ seems to confirm this view. Anaerobic methane oxidation is a process that will inject characteristically low $\delta^{13}CO_2$, and restriction of the process to a particular depth interval will produce a $\delta^{13}CO_2$ minimum with the magnitude and location that have been observed. The $\delta^{13}CO_2$ distribution above the minimum reflects upward diffusion of $\delta^{13}CO_2$; below the minimum, burial, exchange on carbonate surfaces, and downward diffusion account for the distribution.

Zonation of Reactions

All work done so far on anaerobic methane oxidation in marine systems has considered the distribution of measured concentrations, advection-diffusion models, and fluxes. The notion of anaerobic methane oxidation is derived from loss of methane accompanied by compensating changes in the total carbon dioxide and sulfate distributions. No organisms have been isolated or cultured. Outside of what we can gather from figure 11-1, the substrate requirements of the anaerobic consumers are unknown.

Puzzling aspects of the anaerobic consumption process are its completeness (the methane flux is almost entirely consumed) and localization into such a narrow zone, as indicated by the sudden concentration changes and the $\delta^{13}CO_2$ minimum. Another puzzling aspect is the scale of the process; anaerobic methane oxidation accounts for almost half of the downward sulfate flux in Cariaco Trench sediments (figure 11-2). If anaerobic methane oxidation were due to co-metabolism, as suggested by Reeburgh (1976), we might reasonably expect the process to be smeared over a greater depth interval in the sediments and be far less effective in removing methane.

The localization of the reaction into a zone, shown in the stippled area in figure 11-1, suggests that the organism responsible for the anaerobic oxidation has fairly restricted substrate and/or environmental requirements; methane and sulfate must be present in proper proportions for the reaction to proceed. We tend to think of absolute concentrations as most important in driving reactions. Actually, the *product* of substrate concentrations is most important to organisms mediating chemical reactions. Using data from figure 11-1, the product $[CH_4] \times [SO_4^{2-}]$ is zero or nearly so above and below the stippled zone because of the near absence of methane above and sulfate below the zone. The product is a maximum within the zone. An organism anaerobically oxidizing methane according to reaction 11.1 would either localize its numbers or maximize its activities in such a zone because of the mass-action advantage.

The location of the anaerobic methane-consuming zone in sediments appears to be dynamically conditioned, much like the activities of the aerobic methane oxidizers studied by Rudd et al. (1974, 1976). Above the zone, too little methane is present, and other substrates are oxidized by sulfate; below the zone, no sulfate is available as an oxidant, preventing consumption of methane. The sediment methane profiles are maintained by diffusion of sulfate and methane into a zone where methane is anaerobically oxidized at a rate 100-fold higher than the methane production rate. The sedimentation rate and flux of organic carbon to the sediments appear to control the depth of the anaerobic methane consumption zone.

Martens and Berner (1977) fitted Long Island Sound methane data to a model similar to those reported by Berner (1971, 1974), incorporating a term for methane consumption. They were unable to obtain a fit using a consumption term following zeroth order kinetics ($-k_0$), but succeeded with a fit considering the consumption followed first-order kinetics ($-k_1 C$). This approach explains the methane distribution and provides rate constants consistent with those reported for other species by Berner (1974). Consumption following first-order kinetics varies directly with methane concentration, so maximum concentration should be at the highest methane concentration. However, if the model's lower boundary is improperly assigned, little or none of the only available oxidant, sulfate, will be included. Reaction 11.1 may be correctly considered to be first-order with respect to methane, as shown by Martens and Berner (1977), first-order with respect to sulfate, and second-order overall. To fit the distributions and describe the location of the consumption reaction, it is necessary to consider consumption following higher-order kinetics. A consumption term of the form $-k[CH_4][SO_4^{2-}]$ would properly reflect zonal consumption without arbitrarily fixing boundaries.

Comparison of Freshwater and Marine Systems

Methane distributions in marine and freshwater systems were reviewed by Reeburgh and Heggie (1977), who concluded that the distributions differ between environments because the controlling consumption reactions are located in different places. Two processes are important: aerobic methane oxidation in the water columns of freshwater systems and anaerobic methane oxidation located in marine sediments.

Rudd et al. (1974) showed that aerobic methane oxidation takes place in a thin meta-limnetic zone in stratified lakes where both methane and oxygen are present. The organisms responsible were regarded as micro-aerophiles because they appeared to function only in low oxygen concentrations, but further work by Rudd et al. (1976) showed that the organisms were able to function in high-oxygen environments provided sufficient inorganic nitrogen was available. These workers concluded that the restricted distribution of aerobic methane oxidation in the meta-limnion was a result of nitrogen limitation (and lack of methane) in the epi-limnion and the inability of the organisms to function in the anoxic hypo-limnion. Most of the work on aerobic methane oxidizers has been limited to freshwater environments, but Sansone and Martens (1978) have observed similar activity in marine systems. Methane distributions in freshwater sediments show no surface zone of low methane concentration, indicating little, if any, consumption in the sediments. The principal methane sink in freshwater systems is aerobic methane oxidation located in the water column.

Figure 11-3 schematically compares the locations of methane consumption reactions in freshwater and marine systems, showing typical consumption rates. Sulfate concentrations in freshwater systems range between 1 and 10 μM; in marine systems they usually exceed 10 mM. In view of the 1000-fold concentration difference, inhibition of methane production and anaerobic methane oxidation will not operate when the small quantities of sulfate present in freshwater systems are exhausted. Accumulation of methane to concentrations larger than those observed in marine environments is possible in the absence of other consumption reactions. Sulfate, then, is required for anaerobic methane oxidation; the reaction cannot proceed in the absence of sulfate (Reeburgh and Heggie, 1977).

Conclusions and Future Work

Anaerobic methane oxidation takes place in marine sediments and emerges as a major control on methane in marine systems. The reaction appears to be located in a narrow subsurface zone where its rate and location are controlled by the product of methane and sulfate concentrations. The consumption reaction proceeds at a rate some 10^2 times faster than methane production and appears to be responsible for the $\delta^{13}CO_2$ minimum found in marine sediments.

Several suggestions for the future emerge from this work. It is clear that many more species should be measured in a single environment as part of a single investigation, but it is also clear from published work that individuals are working at near their limit in the number of parameters they are able to measure. Cooperative efforts on carefully chosen type environments should permit collection of consistent sets of data. The work should include measurements of low-molecular-weight organic species known to be important in sulfate reduction and methane production and should fully exploit the potential offered by carbon and sulfur isotopes. By constraining natural systems with concentration and isotope ratio measurements on active as well as passive fractions, a great deal can be learned about sources and transformations.

Workers should also be aware of how different controlling mechanisms affect fluxes to the overlying water. That can be illustrated by comparison of the silica (Kent Fanning, personal communication) and methane (Reeburgh, 1976) distributions in Cariaco Trench sediments. Variations in silica were confined to the upper 10 cm of these sediments, while changes in methane extended to depths of 60 cm.

References

Barnes, R.O., and Goldberg, E.D. 1976. Methane production and consumption in anoxic marine sediments. *Geology* 4:297–300.

Berner, R.A. 1971. *Principles of Chemical Sedimentology*. New York: McGraw-Hill.

Berner, R.A. 1974. Kinetic models for the early diagenesis of nitrogen, sulfur, phosphorus and silicon in anoxic marine sediments. In E.D. Goldberg, (Ed.), *The Sea*, Vol. 5. New York: Wiley-Interscience. pp. 427–449.

Cappenberg, Th. E. 1974. Interrelations between sulfate-reducing and methane-producing bacteria in bottom deposits of a freshwater lake. II. Inhibition experiments. *Antonie van Leeuwenhoek* 40:297–306.

Cappenberg, Th. E. 1975. A study of mixed continuous cultures of sulfate-reducing and methane-producing bacteria. *Microb. Ecol.* 2:60–72.

Claypool, G.E., and Kaplan, I.R. 1974. The origin and distribution of methane in marine sediments. In I.R. Kaplan, (Ed.), *Natural Gases in Marine Sediments*. New York: Plenum, pp. 99–140.

Davis, J.B., and Yarbrough, H.F. 1966. Anaerobic oxidation of hydrocarbons by *Desulfovibrio desulfuricans*. *Chem. Geol.* 1:137–144.

Emery, K.O., and Hoggan, D., 1958. Gases in marine sediments. *Bull. Am. Assoc. Petrol. Geol.* 42:2174–2188.

Fanning, K.A., and Pilson, M.E.Q. 1972. A model for the anoxic zone of the Cariaco Trench. *Deep-Sea Res.* 19:847–863.

Goldhaber, M.B. 1974. Equilibrium and dynamic aspects of the marine geochemistry of sulfur. Ph.D. Thesis, University of California, Los Angeles.

Martens, C.S., and Berner, R.A. 1974. Methane production in the interstitial waters of sulfate-depleted marine sediments. *Science* 185:1167–1169.

Martens, C.S., and Berner, R.A. 1977. Interstitial water chemistry of anoxic

Long Island Sound sediments. I. Dissolved gases. *Limnol. Oceanogr.* 22:10-25.
Nissenbaum, A., Presley, B.J., and Kaplan, I.R. 1972. Early diagenesis in a reducing fjord, Saanich Inlet, British Columbia. I. Chemical and isotopic changes in major components of interstitial water. *Geochim. Cosmochim. Acta* 36:1007-1027.
Matisoff, G., Bricker, O.P., III, Holdren, G.R., Jr., and Kaerk, P. 1975. Spatial and temporal variations in the interstitial water chemistry of Chesapeake Bay sediments. In T.M. Church (Ed.), *Marine Chemistry in the Coastal Environment*. Washington: Am. Chem. Soc. Symposium Series 18, pp. 343-363.
Mechalas, B.J. 1974. Pathways and environmental requirements for biogenic gas production in the ocean. In I.R. Kaplan (Ed.), *Natural Gases in Marine Sediments*. New York: Plenum, pp. 11-25.
Presley, B.J., and Kaplan, I.R. 1968. Changes in dissolved sulfate, calcium and carbonate from interstitial water of near-shore sediments. *Geochim. Cosmochim. Acta* 32:1037-1048.
Presley, B.J. 1974. Rates of sulfate reduction and organic carbon oxidation in the Cariaco Trench (Abstract). *EOS-Trans. Am. Geophys. Union* 55:319.
Quale, J.R. 1972. The metabolism of one-carbon compounds by microorganisms. *Adv. Microb. Physiol.* 7:119-203.
Reeburgh, W.S. 1969. Observations of gases in Chesapeake Bay sediments. *Limnol. Oceanogr.* 14:368-375.
Reeburgh, W.S. 1976. Methane consumption in Cariaco Trench waters and sediments. *Earth Planet. Sci. Lett.* 28:337-344.
Reeburgh, W.S., and Heggie, D.T. 1977. Microbial methane consumption reactions and their effect on methane distributions in freshwater and marine environments. *Limnol. Oceanogr.* 22:1-9.
Rudd, J.W.M., Hamilton, R.D., and Campbell, N.E.R. 1974. Measurement of microbial oxidation of methane in lake water. *Limnol. Oceanogr.* 19:519-524.
Rudd, J.W.M., Furutani, A., Flett, R.J., and Hamilton, R.D. 1976. Factors controlling methane oxidation in shield lakes: The role of nitrogen fixation and oxygen concentration. *Limnol. Oceanogr.* 21:357-364.
Sansone, F.J., and Martens, C.S. 1978. Methane oxidation in Cape Lookout Bight, North Carolina. *Limnol. Oceanogr.* 23:349-355.
Sholkovitz, E.R. 1973. Interstitial water of the Santa Barbara Basin sediments. *Geochim. Cosmochim. Acta* 37:2043-2073.
Shultz, D.J., and Calder, J.A. 1976. Organic carbon $^{13}C/^{12}C$ variations in estuarine sediments. *Geochim. Cosmochim. Acta* 40:331-335.
Sorokin, Y.I. 1957. On the ability of sulfate-reducing bacteria to utilize methane for the reduction of sulfate to hydrogen sulfide. *Mikrobiol.* 115:713-715.

12 The Presence of Methane Bubbles in the Acoustically Turbid Sediments of Eckernförder Bay, Baltic Sea

Michael J. Whiticar

Abstract

Interstitial concentrations of methane and sulfate were determined in the sediments of Eckernförder Bay. A comparison of the methane profiles with the acoustic properties of the sediments showed that the zones of sediment which were acoustically turbid were also zones in which the measured concentration of methane in interstitial solution exceeded the calculated solubility of methane under in situ conditions. The mechanism suggested for the phenomenon is formation of bubbles of methane gas in the sediments because the presence of such bubbles in the water column is known to interfere with the penetration of sound pulses. Sediments that were undersaturated with respect to interstitial methane were acoustically transparent. The methane profiles yielded estimates for the production, transport, and consumption of the gas that were consistent with values from elsewhere in oceanic sediments. The acoustically turbid sediments and highest interstitial methane values appeared in the most rapidly accumulating sediments in the Bay.

Introduction

Interpretation of results from seismic, sediment echographic, and other geophysical remote-sensing techniques has relied, to a large extent, on their correlation with laboratory-based experiments or on direct determinations, either by sampling or by in situ measurement. Such is the case with acoustically turbid sediments. In reducing, organically rich unconsolidated sediments such as lakes (Levin, 1962), channels (Van Weering, 1975, Van Weering et al., 1973), and bays (Hinz et al., 1969, 1971; Schubel, 1974), there are extensive acoustically diffuse zones which are seemingly impenetrable to acoustic sediment profilers and prevent the mapping of underlying

---

This abstract was prepared by the editors.

reflectors. Although those zones may have resulted from the presence of gases (Schüler, 1952; Edgerton et al., 1966; Schubel, 1974), the acoustical effect of gas bubbles in recent sediments has been examined only to a limited extent (Brandt, 1960; Jones et al., 1964; Hampton and Anderson, 1974; Schubel, 1974).

In addition to influencing acoustic properties of the sediments, the production of large amounts of methane through bacterial CO_2 reduction (Claypool and Kaplan, 1974) or acetate fermentation (Cappenberg, 1974) has been shown to affect other dissolved gas concentrations such as Ar and N_2 (Reeburgh, 1969, 1972; Martens and Berner, 1977) through equilibrium stripping processes. Sediment stabilities also may be directly influenced by gas-bubble entrapment (Whelan and Roberts, 1976) or indirectly by carbonate cementation (Garrison et al., 1969; Allen et al., 1969; Hathaway and Degens, 1969; Whelan and Roberts, 1973).

Background

Eckernförder Bay is a 17-km-long, 3-km-wide inlet extending southwesterly from Kiel Bay of the western Baltic Sea. Overlying Tertiary bedrock, glacial sediments determine the major morphological features of the Eckernförder Bay, although shaped by erosion and infilling since the Littorina Transgression (Seibold et al., 1971). Fine-grained, organically rich muds have sedimented over pre-existent glacial channels covering the deeper ($>$ 20 m water depth) central basin of the Eckernförder Bay. The channels divide around Mittelgrund, an isolated morainic ridge that rises to 6-m water depth at the Bay's mouth (Werner, 1967) (see figure 12-1).

Several workers (Schüler, 1952; Edgerton et al., 1966; Werner, 1968; Hinz et al., 1969, 1971; Deutsches Hydrographisches Institut, Hamburg, personal communication) have reported that acoustically turbid sediments were encountered during seismic and sediment echographic investigations performed in the Eckernförder and Kiel Bays (see figures 12-2 and 12-3). Those zones, referred to locally as "basin effect" areas, appear in the recent fine-grained mud sediments normally deposited in the channels and basins of Kiel Bay with thicknesses greater than 7 m.

Penetration of these turbid zones was unsuccessfully attempted using different sediment echographs with frequencies of 3, 7, 12, 18, and 30 kHz, as well as with variable emission energies from a Pneuflux (air gun) seismic reflection profiler (Hinz et al., 1969).

Accurate mapping of the acoustically turbid sediments in the Eckernförder Bay was made in March 1975 (figure 12-1). The turbid zone parallels the mud/sand facies boundary (Werner, 1968) and is separated by a

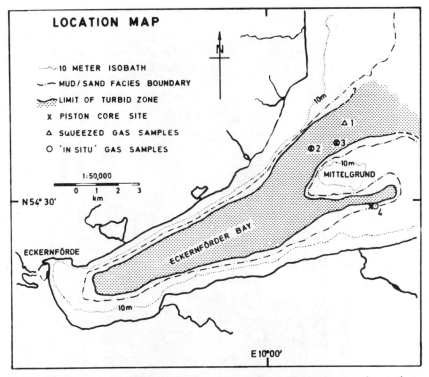

Figure 12–1. Map of Eckernförder Bay showing the four sample stations, the acoustically turbid areas (shaded), and the sand/mud boundary.

horizontal distance necessary for the required 7-m mud thickness to overlay the steeply sloping glacial sediment.

In an initial attempt, a 7.6-m-long piston core was taken at station 1 (table 12–1) in the acoustically turbid zone north of Mittelgrund (figure 12–1), sectioned into 20-cm intervals, and kept at 4°C. Interstitial waters were expressed the same day using a sediment filter press (Hartmann, 1965), and sulfate (gravimetric $BaSO_4$ method) and alkalinity (titration) were determined. From the same core, seven samples were placed in a helium-flushed Plexiglas glovebox, similar to that used by Reeburgh (1968) and Martens (1974), and the interstitial waters were expressed directly into glass collection cylinders. The cylinders were connected to an in-line stripping system similar to Barnes et al. (1975) and modified from Swinnerton et al (1962). The extracted gases were analysed by gas chromotography (Beckman model 419 with T.C. detector, M.S. 5A at 60 and $-30°C$ to separate O_2 from Ar).

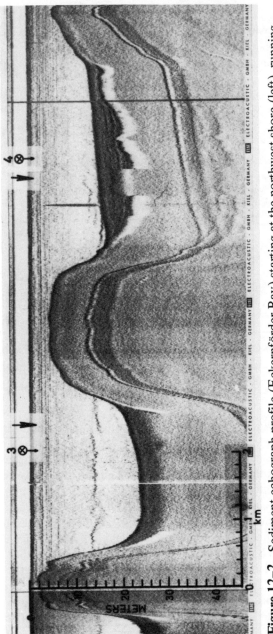

Figure 12-2. Sediment echograph profile (Eckernförder Bay) starting at the northwest shore (left), running southeast close to Station 3 (marked) over Mittlegrund, past Station 4 (marked), and ending on the south shore. Arrows indicate acoustically turbid sediments.

The Presence of Methane Bubbles

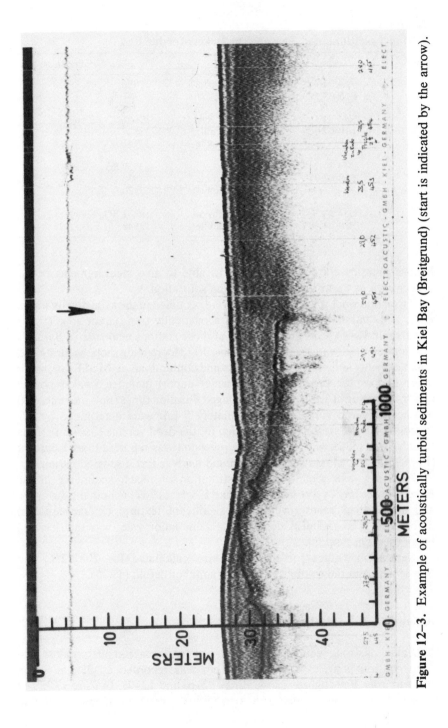

Figure 12–3. Example of acoustically turbid sediments in Kiel Bay (Breitgrund) (start is indicated by the arrow).

Table 12–1
Station Locations and Protocol in Eckernförder Bay

Station	Position	Type	Ref. No.	Water Depth (m)
1	N54°31'42" E10°03'21"	Piston core	12896	28
2	N54°31'15" E10°02'06"	Piston core Gas profile Gas profile	12897 SO_4^{2-} only 13827 13941	28
3	N54°31'21" E10°03'07"	Piston core Gas profile Gas profile	13939 SO_4^{2-} only 13938 13826	27
4	N54°29'57" E10°04'40"	Piston core Gas profile	12891 SO_4^{2-} only 13948	21

Such a pressure-extraction method was able to give measurements of gas concentrations up to values approaching saturation.

Two additional stations (2 and 3) were occupied in the acoustically turbid zone, and station 4 was placed in an acoustically transparent zone (figure 12–1, table 12–1). Gas profiles taken at those stations made use of Barnes's in situ interstitial water sampler (Barnes, 1973), enabling more accurate and reliable results and minimizing gas loss and contamination. Modifications for lower hydrostatic pressures (28 m water depth) included weaker poppet springs (opening at 1.5 atm) and prolonged filtration times (approximately 30 min.). Recovered volumes (approximately 5 ml) were determined by differential cylinder weighings. Correction for the dead volume (0.34 ml) of the sampler was made, and the sampling precision was reported to be 1 percent (Barnes, 1973). Gases were stripped and analysed as discussed earlier.

Analyses, classification, and distribution of recent sediments in the Eckernförder Bay by Werner (1968) and Kögler (1967) in both the disturbed and undisturbed zones revealed no significant textural or compositional differences. The sediment depth, i.e. accumulation rate, was the deciding factor (see Discussion).

Saturation values for methane were calculated for $T = 12°C$ and $S = 16‰$ from the solubility data of Yamamoto et al. (1976).

Results

Indicative of strong reducing conditions, a sharp decease in interstitial sulfate concentration between 150 and 360 cm was recorded at all stations in Eckernförder Bay (figures 12–4, 12–5, 12–6, and 12–7). In Core 13939 at station 3, Eh varied from −168 to −55 mV over 6 m. Complete sulfate

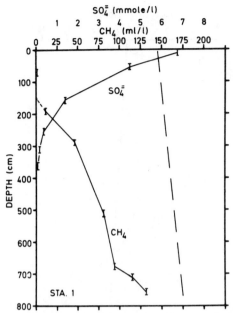

Figure 12–4. Dissolved sulfate and methane profiles for station 1 (acoustically turbid zone) of pressure-extracted interstitial waters from piston core. Dashed vertical line is the methane saturation limit.

depletion at station 1 occurred at 360 cm (figure 12–4). Expansion of sediments due to degassing upon core retrieval was quite strong, with end caps being pushed off and sediment being extruded. The smell of H_2S from the surface sediments ceased quickly with depth in the core. In agreement with results obtained from other areas (Nissenbaum et al., 1972; Claypool and Kaplan, 1974; Martens, 1974; Cappenberg, 1974; Martens and Berner, 1977), methane concentrations in Eckernförder Bay were generally lower than 0.4 mM where dissolved sulfate concentrations were above 0.5 mM (see figures 12–4 through 12–7). Those low methane values were in acoustically transparent sediment layers. Then, at greater depths in the sediments, there were sharp increases in interstitial methane concentration. At stations 2 and 3 (figures 12–5 and 12–6), the deeper methane levels exceeded the methane saturation concentration in sediment layers which appear to be acoustically turbid. Methane values in deeper sediments at station 1 (figure 12–4), although in the acoustically turbid zone, failed to reach saturation. It is quite possible that methane gas was lost from these pore fluids because of degassing during squeezing. Unfortunately, no in situ sampling was performed at station 1. At station 4 (figure 12–7), low methane values in the presence of sulfate were also found in the acoustically

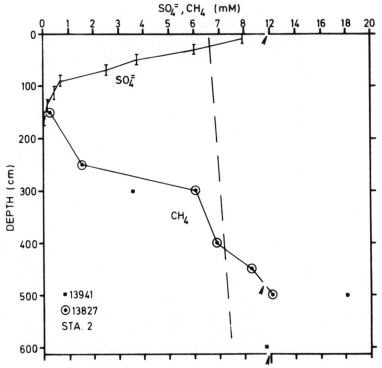

Figure 12–5. Dissolved sulfate and methane profile for station 2 (acoustically turbid zone) showing the oversaturation of interstitial methane and the presence of interstitial bubbles. Note: scale break (arrows) at 9 mM. Dashed line is methane saturation limit.

transparent zone. However, unlike stations 2 and 3, interstitial methane values at greater depths remained below saturation levels.

Discussion

Perhaps the most striking feature is the correlation between the appearance of the acoustically turbid zone and the occurrence of the in situ methane saturation. The possibility of in situ gas-bubble formation as a result of oversaturation has been proposed by several workers (Koyama, 1954; Reeburgh, 1969, 1976; Claypool and Kaplan, 1974; Hammond, 1974; Martens, 1974, 1976; Whelan, 1974a; Barnes and Goldberg, 1976; Martens and Berner, 1977). Dissolved methane partial pressures at stations 2 and 3 do exceed the hydrostatic pressure at sediment depths greater than 4.5 m, permitting bubble formation and entrapment. It is known that high

The Presence of Methane Bubbles

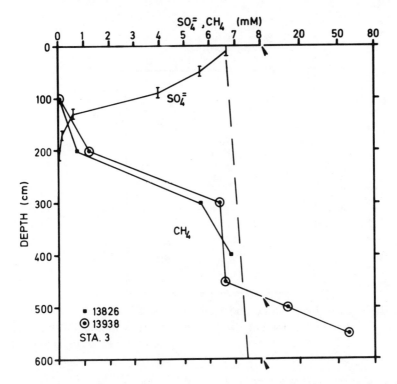

Figure 12-6. Dissolved sulfate and methane profiles for station 3 (acoustically turbid zone) showing bubble formation (< 4.5 m). Note scale break at 8 mM. Dashed line is methane saturation limit.

interstitial methane concentrations exist in deep-water sediment, where the total pressure prevents bubble formation or where hydrates are stable (Emery and Hoggan, 1958; Hammond et al., 1973; Barnes and Goldberg, 1976; Reeburgh, 1976). Reeburgh (1969), Martens (1974), Whelan (1974a), and Martens and Berner (1977) have shown for shallow-water marine sediments that methane concentrations approach saturation and have suggested vertical ebullition and migration as mechanisms preventing oversaturation. Methane concentrations in Eckernförder Bay sediment actually exceed the saturation limit, producing a zone with oversaturated (gas-bubble) conditions.

Vertical bubble migration at stations 2 and 3 may be present, starting from a zone of bubbles below 6 m, ascending to the level of undersaturation (approximately 4.5 m), and redissolving into the interstitial waters. Such migration would account for the values close to the saturation limit between 3 and 4 m of sediment depth.

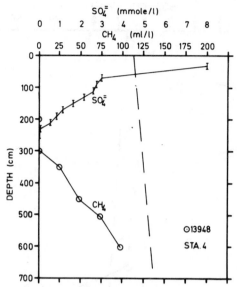

Figure 12-7. Dissolved sulfate and methane profiles for station 4 (acoustically transparent zone). Dashed line is methane saturation limit. Note that the underlying sand layer is encountered below 6 m.

Extrapolation of the methane-distribution curve at station 4 (figure 12-7) taken in the acoustically transparent zone would yield saturation concentrations at around 7 m of sediment depth provided the sediment characteristics remained constant. Thus one might ask why the methane profile for station 4 does not increase as rapidly with depth as do those for stations 2 and 3, which cross from unsaturation to oversaturation at only 4.5 m of depth. Since the postglacial sediments at all stations have been depositing for approximately the same period of time, those with a greater thickness (e.g., basins and channels such as stations 2 and 3) have a higher accumulation rate (Werner, personal communication). A higher accumulation rate means a higher organic carbon content (Suess, 1976), leading to an increased decomposition rate (Berner, 1974; Goldhaber and Kaplan, 1974, 1975). My estimates show that the rate of sulfate reduction is 10.9 μM/yr at stations 2 and 3 and is 4.89 μM/yr at station 4, due to the higher sedimentation rate. Thus stations 2 and 3 could have a greater methane production rate than station 4.

Methane concentrations at station 1, although in the acoustically turbid zone (figure 12-1), do not exceed saturation. As mentioned, it appears that gas loss through sample preparation and squeezing is responsible. That and a shallow depth of sampling may explain why other workers in shallow-water sediment using similar techniques have failed to measure oversaturated (bubble-forming) concentrations.

Attenuation and reflection of sound by a screen of bubbles in water was shown by Carstensen and Foldy (1947) to be high for near-resonance frequencies and to decrease for off-resonance frequencies. A variation in bubble size would widen the resonance and attenuated bands. Lowered sound velocities and increased attenuation and scattering have been reported for sediments containing gas bubbles (see, for example, Hampton and Anderson, 1974; Brandt, 1960). The increases of both sound attenuation and reflectivity in gas-rich sediments combine to produce the acoustically turbid zone, obscure the sediment profile, and cause diffuse shadow traces on sediment echographs (figures 12-2 and 12-3). The fact that methane concentrations exceed the saturation point, and produce bubbles in the turbid zone and not in the clear area, confirms the phenomenon. Penetration did not improve over a frequency range of 3 to 30 kHz, indicating varying bubble radii.

In situ sound velocity and attenuation in sediments of Eckernförder Bay (Ulonska, 1968; Schirmer, 1970) were, unfortunately, recorded only to a sediment depth of 2 m and not from depths where oversaturation existed (> 4.5 m). However, plans are underway to continue the measurements to a greater sediment depth. In situ sound attenuation profiles (up to 3 m of sediment depth) in Chesapeake Bay (Schubel, 1974) indicated higher values for acoustically turbid sediments than the acoustically clear ones, with the explanation for the turbidity being gas bubbles. Cores taken by Reeburgh (1969) just south of the Chesapeake Bay Bridge are in the same area as Schubel's (1974) seismic and attenuation work. The latter registered the greatest attenuations just below 100 cm of sediment depth, the same depth where Reeburgh's methane profiles, close to saturation, ended. Assuming conditions similar to Eckernförder Bay, an extension of those profiles would probably reveal methane-oversaturated interstitial waters producing gas bubbles, corresponding to the depth of the turbid sediment expression.

The vertical methane flux was calculated by adapting Berner's (1976) flux equation (5):

$$J_{CH_4} = C_{z_1} \frac{J_s}{\rho_s} \left(\frac{\phi_{z_2}}{-\phi_{z_1} + \phi_{z_2}} \right) - \phi_{z_1} D_{z_1} \frac{dc}{dz}$$

where J_{CH_4} is the vertical methane flux in $\mu mole/cm^2/year$; C_{z_1} is the methane concentration at the sediment depth z_1; J_s is the flux of solid particles to the sediment depth z_1 (accumulation rate 1.4 mm/yr, Erlenkeuser et al., 1974), assumed constant and corrected for burial compaction to depth z_1; ρ_s is the density of solids (2.5 g/cm^3, Kögler, 1967); ϕ_{z_1} is the porosity at sediment depth z_1 (found to be 0.85, Kögler, 1967); ϕ_{z_2} is the measured porosity at a sediment depth z_2, below which porosity did not change (0.74, as in the case of stations 2 and 3 or 0.78 at station 4 where an underlying sand layer was encountered); D_{z_1} is the interstitial diffusion coefficient of methane (3×10^{-6} cm^2/s after Reeburgh, 1976, in agreement with Martens and Berner, 1977); and dc/dz is the concentration gradient for stations 2 and 3 combined. The gradient was calculated in the linear region between 2 and 3 m, the methane values in the older (> 3 m) sediments being influenced by bubble migration and those in younger sediments (< 2 m) by methane consumption (Reeburgh, 1976; Barnes and Goldberg, 1976; Martens and Berner, 1977; Oremland and Taylor, 1978; Reeburgh and Heggie, 1977). It is assumed for calculation that the methane profile remains stationary with time, seasonal variations not having as great an effect as in other areas (Reeburgh, 1969; Martens, 1976).

The vertical methane flux to the consuming zone at stations 2 and 3 was found to be 4.76 μmole/(cm$^2 \cdot$ yr). Comparison with values from the Cariaco Trench (Reeburgh, 1976) and the Santa Barbara Basin (Barnes and Goldberg, 1976) is given in table 12-2. The values of the latter were recalculated using a diffusion coefficient of 3×10^{-6} cm^2/s (Reeburgh, 1976) for comparison purposes. The strong decrease of the methane flux between 150 and 200 cm depth is suggested to be the result of anaerobic methane consumption (see preceding references), preventing the diffusion or ebullition of methane to the overlying waters. A consumption rate of 95.2 μmole/(L$\cdot$yr) was calculated for stations 2 and 3 using a consuming-zone thickness of 50 cm. Comparison rates are listed in table 12-2 and vary according to the consuming-zone thickness.

The vertical methane flux to the methane-consuming zone at station 4 where bubble migration does not affect the concentration gradient was 1.16 μmole/(cm$^2 \cdot$yr) (see table 12-2). A nearly linear concentration profile below 300 cm shows that methane production should be low; however, a selection of a 25-cm consumption zone between 275 and 300 cm yields a consumption rate of 46.4 μmole/(L$\cdot$yr).

Methane production rates for stations 2 and 3 are probably incorrect, since they were calculated from the concentration gradients which were influenced by bubble migration. A production rate of 1.78 μmole/(L$\cdot$yr) is estimated for station 4 (see table 12-2), which is over 25 times smaller than the consumption rate precluding the escape of methane into the water column.

Conclusions

The ability of interstitial methane concentrations to become so large that bubbles form in sediments has a significant influence on acoustic properties.

Table 12–2
A Comparison of Methane Consumption and Production in Sediments of Eckernförder Bay, the Cariaco Trench, and the Santa Barbara Basin

Location	CH_4 [μmole/(cm^2·yr)]	Consumption [μmole/(L·yr)]	Consumption Zone Thickness (cm)	Production [μmole/(L·yr)]
Cariaco Trench (Reeburgh, 1976)				
Eastern basin	15.9	1.59×10^3	10	8.45
Western basin	5.48	0.55×10^3	10	2.92
Santa Barbara Basin (Barnes and Goldberg, 1976)	3.22	70.0	46	20.5
Eckernförder Bay (this chapter)				
Stations 2 and 3	4.76	95.2	50	8.5
Station 4	1.16	46.4	25	1.78

The presence of these bubbles in a layer of reducing sediment seems to attenuate sound, affect the sound velocity, and obscure underlying reflectors. The calculated rates of production, transport, and consumption of methane in sediments of Eckernförder Bay are consistent with estimates from other reducing sediments.

The calculation of fluxes and production rates of methane are dependent on the concentration gradient, which, in turn, is affected by ascending methane bubbles. Provided the rate of methane consumption exceeds the flux to the consuming zone (due to methane diffusion, advection, and bubble migration), then there should be no release of methane to the overlying waters.

The in situ sampler has proven to be reliable, minimizing both atmospheric contamination and gas loss normally experienced through core retrieval and pressure-extraction methods.

Acknowledgments

The author is indebted to M. Hartmann, P.J. Müller, and E. Suess for critically reviewing the manuscript and for providing advice and assistance throughout its preparation. The in situ samplers were generously supplied by R.O. Barnes, to whom warm thanks is extended. The help given by the crew of R/V *Littorina* and M. Arthur is also appreciated. This chapter is part of the Joint Research Program 95 at Kiel University (Sonderforschungsbereich 95 der Deutschen Forschungsgemeinschaft, contribution number 167).

References

Allen, R.C., Garvish, E., Friedman, G.M., and Sanders, J.E. 1969. Aragonite cemented sandstone from the outer continental shelf off Delaware Bay: Submarine lithification mechanism yields product resembling beachrock. *J. Sed. Petrol.* 39:136–149.

Barnes, R.O. 1973. An "in situ" interstitial water sampler for use in unconsolidated sediments. *Deep-Sea Res.* 20:1125–1128.

Barnes, R.O., Bertine, K.K., and Goldberg, E.D. 1975. N_2: Ar, nitrification and denitrification in southern California borderland basin sediments. *Limnol. Oceanogr.* 20:962–970.

Barnes, R.O., and Goldberg, E.D. 1976. Methane production and consumption in anoxic marine sediments. *Geology* 4:297–300.

Berner, R.A. 1974. Kinetic models for the early diagenesis of nitrogen, sulphur, phosphorous, and silicon in anoxic marine sediments. In E.D. Goldberg (Ed.), *The Sea*, Vol. 5: *Marine Chemistry*. New York: Wiley-Interscience, pp. 427–450.

Berner, R.A. 1976. The benthic boundary layer from the viewpoint of a geochemist. In I.N. McCave (Ed.), *The Benthic Boundary Layer*. New York: Plenum, pp. 33–55.

Brandt, H. 1960. Factors affecting compressional wave velocity in unconsolidated marine sand sediments. *J. Acoust. Soc. Am.* 32:171–179.

Cappenberg, Th.E. 1974. Interrelations between sulphate-reducing and methane-producing bacteria in bottom deposits of a fresh-water lake. II. Inhibition experiments. *Antonie van Leeuwenhoek* 40:297–306.

Carstensen, E.L., and Foldy, L.L. 1947. Propagation of sand through a liquid containing bubbles. *J. Acoust. Soc. Am.* 19:481.

Claypool, G.E., and Kaplan, I.R. 1974. The origin and distribution of methane in marine sediments. In I.R. Kaplan (Ed.), *Natural Gases in Marine Sediments*. New York: Plenum, pp. 99–139.

Edgerton, H.E., Seibold, E., Vollbrecht, K., and Werner, F. 1966. Morphologische Untersuchungen am Mittelgrund (Eckernförder Bucht, westliche Ostsee). *Meyniana* 16:37–50.

Emery, K.O., and Hoggan, D. 1958. Gases in marine sediments. *Bull. Am. Soc. Petrol. Geol.* 42:2174–2188.

Erlenkeuser, H., Suess, E., and Willkomm, H. 1974. Industrialization affects heavy metal and carbon isotope concentrations in recent Baltic Sea sediments. *Geochim. Cosmochim. Acta* 38:823–842.

Garrison, R.E., Luternaur, L.J., Grill, E.V., McDonald, R.D., and Murray, J.W. 1969. Early diagenetic cementation of recent sands, Fraser River delta, British Columbia. *Sedimentol.* 12:27–46.

Goldhaber, M.B., and Kaplan, I.R. 1974. The sulphur cycle. In E.D. Goldberg (Ed.), *The Sea*, Vol. 5: *Marine Chemistry*. New York: Wiley-Interscience, pp. 569–655.

Goldhaber, M.B., and Kaplan, I.R. 1975. Controls and consequences of sulphate reduction rates in recent marine sediments. *Soil Science*, 119:42–55.

Hammond, D.E., Horowitz, R.M., and Broecker, W.S. 1973. Interstitial water studies. Leg 15. Dissolved gases at site 147. *Initial Reports DSDP* 20:765–771.

Hammond, D.E. 1974. Dissolved gases in Cariaco Trench sediments: Anaerobic diagenesis. In I.R. Kaplan (Ed.), *Natural Gases in Marine Sediments*. New York: Plenum, pp. 71–89.

Hampton, L.D., and Anderson, A.L. 1974. Acoustics and gas in sediments: Applied Research Laboratories (ARL) experience. In I.R. Kaplan (Ed.), *Natural Gases in Marine Sediments*. New York: Plenum, pp. 249–273.

Hartmann, M. 1965. An apparatus for the recovery of interstitial water from recent sediments. *Deep-Sea Res.* 12:225–226.

Hathaway, J.C., and Degens, E.T. 1969. Methane-derived marine carbonates of Pleistocene age. *Science* 165:690–692.

Hinz, K., Kögler, F.C., and Seibold, E. 1969. Reflexions-seismische Untersuchungen mit einer pneumatischen Schallquelle und einem Sedimentecholot in der westlichen Ostsee. *Meyniana* 19:91–102.

Hinz, K., Kögler, F.C., Richter, I., and Seibold, E. 1971. Reflexions-seismische Untersuchungen mit einer pneumatischen Schallquelle und einem Sedimentecholot in der westlichen Ostsee. *Meyniana* 21:17–24.

Jones, J.L., Leslie, C.B., and Barton, L.E. 1964. Acoustic characteristics of underwater bottoms. *J. Acoust. Soc. Am.* 36:154–157.

Kögler, F.C. 1967. Geotechnical properties of recent marine sediments from the Arabian Sea and the Baltic Sea. *Marine Geotechnique*, U. of Illinois Press, 170–176.

Koyama, T. 1954. Distribution of carbon and nitrogen in lake muds. *J. Earth Sci., Nagoya Univ.* 2:5–14.

Levin, F.K. 1962. The seismic properties of Lake Maracaibo. *Geophysics* 27:35–47.

Martens, C.S. 1974. A method for measuring dissolved gases in pore waters. *Limnol. Oceanogr.* 19:525–530.

Martens, C.S. 1976. Control of methane sediment-water bubble transport by macroinfaunal irrigation in Cape Lookout Bight, North Carolina. *Science* 192:998–999.

Martens, C.S., and Berner, R.A. 1977. Interstital water chemistry of anoxic Long Island Sound sediments. I. Dissolved gases. *Limnol. Oceanogr.* 22:10–25.

Nissenbaum, A., Presley, B.J., and Kaplan, I.R. 1972. Early diagenesis in a reducing fjord, Saanich Inlet, British Columbia. I. Chemical and isotopic changes in major components of interstitial water. *Geochim. Cosmochim. Acta* 36:1007–1027.

Oremland, R.S., and Taylor, B.F. 1978. Sulfate reduction and methanogenesis in marine sediments. *Geochim. Cosmochim. Acta.* 42:209–214.

Reeburgh, W.S. 1968. Determination of gases in sediments. *Environ. Sci. Technol.* 2:140–141.

Reeburgh, W.S. 1969. Observations of gases in Chesapeake Bay sediments. *Limnol. Oceanogr.* 14:368–375.

Reeburgh, W.S. 1972. Processes affecting gas distributions in estuarine sediments. *Mem. Geol. Soc. Am.* 133:383–389.

Reeburgh, W.S. 1976. Methane consumption in Cariaco Trench waters and sediments. *Earth Planet. Sci. Lett.* 28:337–344.

Reeburgh, W.S., and Heggie, D.T. 1977. Microbial methane consumption reactions and their effect on methane distributions in freshwater and marine environments. *Limnol. Oceanogr.* 22:1–9.

Schirmer, F. 1970. Schallausbreitung im Schlick. *Deutsche Hydrographische Zeitschrift* 1:24–30.

Schubel, J.R. 1974. Gas bubbles and the acoustically impenetrable or turbid character of some estuarine sediments. In I.R. Kaplan (Ed.), *Natural Gases in Marine Sediments.* New York: Plenum, pp. 275–297.

Schüler, F. 1952. Untersuchungen über die Mächtigkeit von Schlickschichten mit Hilfe des Echograms. *Deutsche Hydrographische Zeitschrift* 5:220–231.

Seibold, E., Exon, N., Hartmann, M., Kögler, F.C., Krumm, H., Lutze, G.F., Newton, R.S., and Werner, F. 1971. Marine geology of Kiel Bay. In *Sedimentology of Parts of Central Europe.* Guidebook VIII, pp. 209–235.

Suess, E. 1976. *Porenlösungen Mariner Sedimente—Ihre chemische Zusammensetzung als Ausdruck frühdiagenetischer Vorgänge.* Habilitationsschrift, Christian-Albrechts-Universität, Kiel.

Swinnerton, J.W., Linnenbom, V.J., and Cheek, C.H. 1962. Determination of dissolved gases in aqueous solutions by gas chromatography. *Anal. Chem.* 34:483–485.

Ulonska, A. 1968. Versuche zur Messung der Schallgeschwindigkeit und Schalldämpfung in Sediment in situ. *Deutsche Hydrographische Zeitschrift* 21:49–58.

Van Weering, T., Jansen, J.H.F., and Eisma, D. 1973. Acoustic reflection

profiles of the Norwegian channel between Oslo and Bergen. *Netherlands J. Sea Res.* 6:241–263.
Van Weering, T. 1975. Late Quarternary history of the Skagerrak: An interpretation of acoustical profiles. *Geologie en Mijnbouw* 54:130–145.
Werner, F. 1967. Sedimentation und Abrasion am Mittelgrund (Eckernförder Bucht, westliche Ostsee). *Meyniana* 17:101–110.
Werner, F. 1968. Gefügeanalyse feingeschichteter Schlicksedimente der Eckernförder Bucht (westliche Ostsee). *Meyniana* 18:79–105.
Whelan, T. 1974a. Methane and carbon dioxide in coastal marsh sediments. In I.R. Kaplan (Ed.), *Natural Gases in Marine Sediments*. New York: Plenum, pp. 47–61.
Whelan, T. 1974b. Methane, carbon dioxide and dissolved sulphate from interstitial waters of coastal marsh sediments. *Estuar. Coast. Mar. Sci.* 2:407–415.
Whelan, T., and Roberts, H.H. 1973. Carbon isotope composition of carbonate nodules from a fresh water swamp. *J. Sed. Petrol.* 43:54–58.
Whelan, T., and Roberts, H.H. 1976. The occurrence of methane in recent deltaic sediments and its effect on soil stability. Marine Slope Stability Conference, L.S.U. Baton Rouge, La., October 14–15, 1976.
Yamamoto, S., Alcauskas, J.B., and Crozier, T.E. 1976. Solubility of methane in distilled water and seawater. *J. Chem. Eng. Data* 21:78–80.

Part V
Transition Metals in Deep-Sea Sediments

13 Pacific Sediments from Japan to Mexico: Some Redox Characteristics

A.G. Rozanov

Abstract

Data on Eh, pH, iron species (Fe^{2+}, Fe^{3+}, $Fe_{S^{2-}}$, Fe_{tot}), Mn species, sulfur species, C_{org}, and general petrographic observations are reported for a transect across the Pacific Ocean.

Eh values were low in borderland areas and reached -350 mV at station 670 off Mexico in the presence of free H_2S, high organic carbon, and high ΣS_{H_2S} concentration. Sediment columns in such areas have thin oxidized surface layers or none at all. Approximately 500 to 1,000 nautical miles offshore from Mexico, and even further offshore from Japan, sediment cores were oxidized over their entire lengths (4 to 5 m), had low C_{org} (< 0.5 percent), and negligible reduced sulfur species (0.01 to 0.1 percent or undetectable).

The redox "barrier" between oxidizing and reducing sediments is an important zone of transition between forms of iron and is also a zone of Mn enrichment. Within some sediments that were partly or wholly reducing, some layers were found far below the redox barrier but still had high contents of Mn. Their existence is attributed to a rather rapid burial of the layer after its accumulation above an ancient redox barrier.

Introduction

Geochemical investigations in the Pacific Ocean in 1969 (R/V *Vityaz*) and 1973 (R/V *Dmitri Mendeleev*) allowed us to complete a study of recent bottom sediments on a latitudinal profile from the shores of Japan to the shores of Mexico (figure 13–1). The sediments studied are characterized by diverse redox conditions, some of which are discussed in this chapter.

According to Lisitsyna and Dvoretskaya (1972), those sediments belong to lithological and facial types that change consecutively from the coast seaward.

This abstract was prepared by the editors.

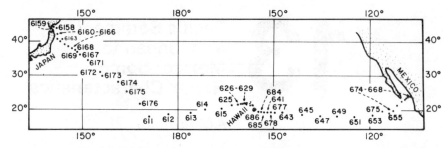

Figure 13-1. Locations of sampling stations (expedition 46, R/V *Vityaz*; expedition 9, R/V *Dmitri Mendeleev*).

1. Terrigenous sediments in the coastal zone are represented by clayey silt with admixtures and streaks of continental fragments and siliceous and calcareous organisms (stations 6158–6161 and stations 668–670).
2. Hemipelagic sediments are represented by siliceous-clayey or slightly calcareous materials, wherein the sandy silt content is less and the pelite fraction (< 0.01 mm) may increase to as much as 94 percent (stations 6162–6166 and stations 671–672).
3. Intermediate-type pelagic clays are represented by a dominant pelite fraction (over 80 percent), with slight admixtures of terrigenous fragments and siliceous organisms (stations 6168–6171 and stations 655, 673–674).
4. Pelagic red clays are represented by the finest material (pelite fraction, 92 to 99 percent; station 6172 and further east; station 675 and further west). With the transition from coast to ocean and lower sedimentation rates, the red clays decompose to form authigenic zeolites (station 6175 and further east; station 651 and further west).

Sediments in the western Pacific are affected by andesitic vulcanism, which provides the admixture of ash material observed far to the east (station 6174). In contrast, sediments bordering the volcanic center of the Hawaiian Islands show signs of basaltic vulcanism. The Pacific profile in figure 13–1 also crosses large zones of foraminiferal and other calcareous sediments (stations 611–613), and East Pacific sediments bordering the Mexican coast display effects of juvenile exhalations characteristic of the East Pacific Rise zone.

Redox processes in marine sediments involve a large number of reactive substances, some of which (O_2, C_{org}, and S^{2-}) control redox processes, while others (Mn, Fe, and reactive forms of other elements) reflect the changing environmental conditions. In time, newly formed sediment changes its original appearance under the influence of diagenetic processes. Such transformations are characteristic of all ocean sediments but are especially pronounced in coastal sediments. Here, an increased supply of organic matter leads to extensive recombination of reactive substances during diagenesis. Redox processes are accompanied by diffusive exchange, whereby additional amounts of oxidants (like O_2 and SO_4^{2-}) are transported to the sediments, and the resulting decomposition products and mineral forms of carbon, nitrogen, and phosphorus are transferred back to the bottom waters.

Many diagenetic processes are considerably affected by microorganisms, such as aerobic and anaerobic bacteria. Redox processes involving those organisms include decomposition of organic matter, processes of reduction (sulfate reduction, denitrification, etc.), and processes of oxidation of substances like manganese, iron, and sulfide sulfur.

Methods

Samples of bottom sediment were obtained with a piston corer (diameter 190 mm). Simultaneously, the top 25 cm of the sediment were sampled with a grab. Lengths of piston cores ranged from 4 to 8 m.

Eh was estimated using platinum-needle electrodes with saturated silver–silver chloride reference electrodes. Critiques (e.g., Stumm, 1966; Whitfield, 1969) regarding the use of Eh for determining the redox level of marine sediments have pointed out that the absence of a uniform measurement procedure (choice and preparation of electrodes, retention time in sediment, etc.) leads to different results. The development of a standard thermodynamically valid procedure for measuring Eh would promote wider use of Eh for characterizing redox processes.

In my measurements, electrodes were inserted into a fresh sediment core immediately after separation. Platinum electrodes were vibrated to shorten considerably the time needed for establishing the equilibrium Eh value (Rozanov, 1975).

pH was estimated by inserting glass combination electrodes (ESKL-03, Instrumentation Mfg. Works, Gomel, USSR) into the core.

Total iron (Fe_{tot}) and manganese (Mn_{tot}) content were estimated by volumetric and photometric methods (Ponomarev, 1961) after decomposition of samples with hydrofluoric and sulfuric acids and final fusion with potassium pyrosulfate.

In a sample of fresh sediment, the iron contained in dilute acid-soluble fractions of carbonates, silicates, sulfides, and hydroxides was solubilized by dilute H_2SO_4. Several forms of iron in sediment were determined. Bivalent and trivalent iron (Fe^{2+} and Fe^{3+}) in that solution were determined titrimetrically with potassium dichromate (Zalmanzon, 1957). Such acid-soluble iron comprises the hydrogenous portion of Fe_{tot} and is the fraction of Fe_{tot} which enters into diagenetic reactions to the greatest extent. Along with pyrite iron, that fraction is generally authigenic in marine sediments, and the three types were included in "reactive iron": $Fe_{react} = Fe^{2+} + Fe^{3+} + Fe_{S^{2-}}$. Sulfide iron—$FE_{S^{2-}}$—was estimated by determining the total amount of sulfide sulfur (hydrotroilite) and pyrite sulfur present: $Fe_{S^{2-}} = Fe_{FeS} + Fe_{FeS_2}$. Another portion of the iron was represented by detrital forms ($Fe_{det} = Fe_{tot} - Fe_{react}$) and is the fraction not readily subject to diagenetic transformations.

Tetravalent manganese (Mn^{4+}) was assayed after reacting a fresh sediment sample in 7 N H_2SO_4 with Mohr's salt ("available oxygen" determination procedure). The method is applicable to oxidized sediments where bivalent iron and forms of sulfide are absent. The method determines the amount and degree of oxidation (Mn^{4+}/Mn_{tot}) of the manganese in a sample.

Sulfur compound forms were determined by the procedure of Ostroumov (1953) and Volkov (1959) as slightly modified by Volkov and Zhabina (1975). The total of the derivatives of H_2S (ΣS_{H_2S}) was used to characterize the extent of sulfate bacterial reduction. Thus the total sum includes elemental sulfur, sulfide (hydrotroilite), pyrite, and organic forms of sulfur in the sediments; it does not include free H_2S (see table 13-1). In sediments, specific methods permit separate determination of sulfur as hydrotroilite (FeS), as pyrite (FeS_2), and as elemental and organic sulfur. There are also methods for free hydrogen sulfide, sulfates, sulfides, and thiosulfates in pore water.

Results

The deep waters of the Pacific are abundantly aerated. The oxygen content in the bottom waters over the entire area from Japan to Mexico does not decrease below 3 ml/L (Chernyakova, 1966). Sedimentation conditions in ocean water promote partial or complete oxidation of mineral species and particulate matter settling to the bottom. The finest fractions of oceanic suspended matter transported into the pelagic area undergo the fullest oxidation.

Reduction processes begin after sediments are buried. Active reduction of sedimentary material involving organic matter and microorganisms begins at the sediment-water interface and in the upper sediment layers. Our data show that microbiological reduction of sulfate to hydrogen sulfide is most intense in the coastal zone, and the degree of reduction in sediments depends in part on the transport of organic matter to the sediments. The low content of organic matter in pelagic sediments is evidently insufficient to produce appreciable reduction of those deposits.

The redox potential (Eh) characterizes the extent of reduction within the sediments. Eh values exceeding +400 mV characterize oxidized sediments such as red clays. Where sulfide forms (usually iron sulfides) appear in the sediments, the Eh declines to negative values. The intermediate range (from 0 to +400 mV) is transitional, and under such conditions reactive forms of Mn^{4+} and Fe^{3+} are reduced to bivalent forms. Strong reduction conditions in the sediments generally involve the appearance of free hydrogen sulfide and Eh values less than −200 mV.

The transect from Japan to Mexico illustrated the variations in Eh within a major ocean basin (Figure 13-2). Values of Eh in sediments bordering the coast of Mexico were quite low, especially in locations where free hydrogen sulfide was detected (stations 668–670). The lowest Eh values (−350 mV) were observed at station 670. In coastal regions, sediment surfaces have a more oxidized layer, and the thickness and Eh of that layer increase seaward. In the most reduced sediments (stations 668–670), an oxidized layer was virtually absent; only a slight elevation of Eh was observed in the 0- to 2-cm horizon (up to +110 mV at station 670). Seaward from station 672, the Eh value in the oxidized layer reached +580 mV (0 to 3 cm), and beginning from station 674, it did not fall below +440 mV throughout the entire sampled sediment column. Thus we see the transition from coastal reduced sediments to oxidized red clays. A comparable picture was observed in sediments in the West Pacific, where the oxidized stratum appeared at station 6164. The stratum persisted and thickened seaward. At station 6171, it was 50 cm thick, and at station 6172 the sediments were oxidized as far down as we

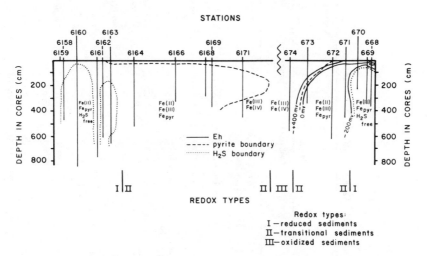

Figure 13-2. Redox characteristics of Pacific sediments (Eh, Fe^{2+}, Fe^{3+}, Fe_{pyr}, Mn^{4+}, H_2S). I = reduced sediments; II = transitional sediments; and III = oxidized sediments.

Table 13-1
Redox Characteristics of Pacific Sediments

	Sedimentation Redox Types						
	West				East		
	Reduced I	Transitory II	Oxidized III	Hawaiian	Oxidized III	Transitory II	Reduced I
Stations (Fig. 13-1)	6158–6163	6164–6171	6172–625	626–677	643–674	671–673	668–670
Distance from coast (km)	Up to 600	600–1,500	Over 1,500	300–700	Over 700	150–700	Up to 150
Depths (km)	0.49–7.43	5.3–5.99	5.99	1.72–5.41	3.47	2.65–3.28	0.14–1.45
Sedimentation rate (mm/1,000 yrs)[a]	100–30	30–3	3–1–<1	1–30	3–1–<1	30–3	120 (sta. 669)
C_{org} (percent)	1–2	0.8–0.3	0.3–0.05	0.03–0.93	0.3–0.05	1.4–0.6	2–8
Eh (mV)		None	+600	+400–+600	+600	−150–+600	Up to −350
Free H_2S max (mg/ml)	153 (sta. 6160)	None	None	None	None	None	26.0 (sta. 668)
ΣS_{H_2S} max (percent)	0.04 (sta. 6160)	0.04 (sta. 6171)	None	0.25	None	0.05 (sta. 673)	1.22 (sta. 669)
Fe_{tot} (percent)	4.1 (sta. 6160)	4.5 (sta. 6171)	4.5–9.0	7–10	5.0–6.9 (Up to 10 at sta. 675)	5.0 (sta. 673)	3.2 (sta. 669)

Some Redox Characteristics

Fe_{react} (percent) $= Fe(II) + Fe(III) + Fe_{S^{2-}}$	1.8 (sta. 6160)	1.2 (sta. 6171)	1.2–0.5	0.26–4.9	1.2–0.4	1.2 (sta. 673)	1.4 (sta. 669)
Fe(II) (percent)	0.9 (sta. 6160)	0.9 (sta. 6171)	None	0–1.46	None	0.4 (sta. 673)	0.3 (sta. 669)
Fe(III) (percent)	0.1 (sta. 6160)	0.3 (sta. 6171)	1.2–0.5	0.26–4.1	1.2–0.4	0.8 (sta. 673)	0.1 (sta. 669)
Fe_{S^2} = $Fe_{hydr} + Fe_{pyr}$	0.8 (sta. 6160)	0.01 (sta. 6171)	None	0–0.13	None	0.3 (sta. 673)	1.0 (sta. 669)
Mn_{tot} (percent)	0.04 (sta. 6160)	0.25 (sta. 6171)	0.3–1.5 (sta. 6176)–2.0 (sta. 613)	0.07–1.26	0.5–1.5 (sta. 645) (Up to 3.0 at sta. 675)	0.46 (sta. 673)	0.03 (sta. 669)
Specific features	Terrigenous and volcanic particles; oxidized layer absent	Oxidized layer 0–65 (sta. 6171)	Ferro-manganese nodules	Ferro-manganese nodules	Ferro-manganese nodules	Oxidized layer 0–120 cm (sta. 673)	Terrigenous particles; oxidized layer absent
		Layers (30–50, 240–250, 305–308 cm) enriched by Mn				Layers (0–120, 200–300, 400–500 cm at sta. 674) enriched by Mn	

[a] See Lysitsin, 1974; and Van Andel, 1964.

could detect. In pelagic sediments (red clays and foraminiferal silts), Eh values were high and stable at +600 mV.

The distribution of Eh values, organic matter content (C_{org}), and free hydrogen sulfide and sulfides, as well as the position and boundaries of oxidized and reduced layers, permitted division of the sediments into the following three groups according to redox characteristics (figure 13-2): reduced (I), transitional (II), and oxidized sediments (III) (Volkov et al., 1975; Rozanov et al., 1976).

Reduced sediments (I) of the continental margins (stations 668–670 and 6158–6163) are characterized by rather low Eh values (−200 mV) and the highest C_{org} (over 1 percent) and usually contain free hydrogen sulfide and abundant pyrite (1 percent or more). As a rule, an oxidized layer is absent in those regions which also include some of the pelagic sediments adjoining the continental slope.

Oxidized sediments (III) occupy virtually the entire pelagic part of the central Pacific Ocean (east of station 6171 and west of station 673) and have high Eh values (+600 mv) and low C_{org} content (below 0.5 percent). Also, they are characterized by an almost complete absence of reduction processes.

Transitional sediments (II) occupy an intermediate position between oxidized and reduced layers both with respect to intensity of redox processes and to location in the ocean (stations 6164–6171 and 671–673). The chief characteristics of transitional sediments are a superficial oxidized layer (Eh +600 mV) and an underlying reduced stratum (Eh −200 mV). The C_{org} content amounts to 0.5 to 1.5 percent, free hydrogen sulfide is absent, and the pyrite content in the reduced layer does not exceed several tenths of 1 percent.

The pH values of oxidized oceanic sediments (III) rarely deviate from 7.5, although we did note a minimum pH of 7.1. Bruevich (1973) explained the decrease in pH values of oxidized pelagic sediments below the pH values of supernatant water (pH 8.0 according to Ivanenkov, 1966) as due to the acidic properties of manganese dioxide, a characteristic component of red clays. In contrast, the maximum pH value in reduced sediments was 8.1. This value confirms the idea that pelagic pore waters can be of the chloride-alkaline type (Bruevich, 1973; Shishkina, 1972). The higher interstitial pH values in reduced sediments are due to the formation of increased amounts of bicarbonate ions in the course of the mineralization of organic matter. Accumulation of bicarbonates in pore water has still greater impact on alkalinity, which, according to Shishkina (1972), may rise to 16.3 mEq/L (station 668) in deeper sediments, while alkalinity in the surface layer is 3.9 mEq/L. Bruevich (1973) reports that alkalinity in near-bottom Pacific waters averages 2.46 mEq/L.

Sediments bordering the Hawaiian Islands consist of a specific redox type owing to the interactions of basaltic ash and volcanic-terrigenous matter within the pelagic matrix of red clays (Lisitsyna et al., 1975). A slight decline of Eh values (+400 mV) in these sediments is caused by increased organic matter (C_{org} up to 0.9 percent). Sulfate reduction, however, occurs to a minor degree and in isolated pockets. Pyrite content does not exceed 0.1 to 0.2 percent. Another reason for the higher reducing character of the Hawaiian sediments is the increased transport of bivalent iron by local pyroclastic matter.

Table 13-1 shows the redox characteristics of the most typical Pacific stations in terms of averages at each station for the types of sediments discussed. Some of those characteristics deserve closer attention.

Sulfur Compounds

In ocean water and in the pore waters of oxidized sediments, sulfur occurs only in the form of sulfates. It is an important characteristic of sediments on Pacific continental margins that sufficient organic matter is often present for sulfate-reducing organisms to convert interstitial sulfate to free hydrogen sulfide (Chebotarev and Ivanov, 1976). Formation of free hydrogen sulfide in sediments causes, on the one hand, a chain of transformations in the system of sulfur compounds and, on the other, the reduction and binding of other substances as sulfides. The problem regarding the mechanism and scope of the process of sulfate reduction in Pacific sediments was discussed previously (Rozanov et al., 1971; Volkov et al., 1972, 1976*a*).

In this chapter, redox conditions in sediments are characterized by total sulfur content in hydrogen sulfide derivatives (ΣS_{H_2S}) comprising sulfide, elemental sulfur, pyrite, and organic forms of sulfur. ΣS_{H_2S} reflects the total magnitude of sulfate reduction. In coastal sediments (I), ΣS_{H_2S} values are maximal (over 1 percent; see table 13-1 and figure 13-3). The farther away the sediment is from land, the less sulfate reduction. In transitional sediments (II), ΣS_{H_2S} is characterized by tenths or hundredths of 1 percent; in oxidized sediments (III), reduced sulfur forms are undetectable.

Because of the fact that the supply of available sulfates to the sediments from near-bottom water is virtually inexhaustible, organic-matter content may be the chief factor limiting the process of sulfate reduction. Indeed, the maximal ΣS_{H_2S} and C_{org} values in the sediments coincide, and decreased C_{org} in the sediments corresponds to a decrease in ΣS_{H_2S} (figure 13-4). However, with C_{org} over 3 percent, the correlation between the two values breaks down, and ΣS_{H_2S} no longer increases (see sediment type I near Mexico). That development, observed in strongly reduced sediments, is due to saturation of forms of reactive iron with hydrogen sulfide (Volkov et al., 1976*b*). The appearance of free hydrogen sulfide in the sediments is an indicator showing

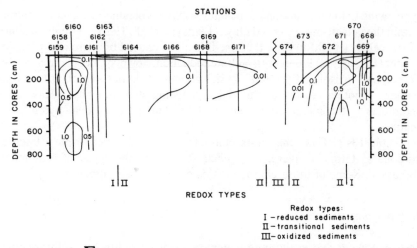

Figure 13-3. ΣS_{H_2S} (sulfur of hydrogen sulfide derivatives) content (%) in Pacific sediments. Symbols I, II, and III as in figure 13-2.

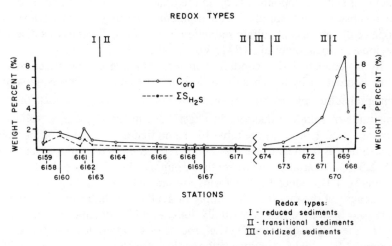

Figure 13-4. Organic carbon content (C_{org}) in superficial layer and mean values of ΣS_{H_2S} in Pacific sediments. Symbols I, II, and III as in figure 13-2.

that the transformations of forms of reactive iron in the sediment are virtually complete.

Forms of Iron

The total iron content (Fe_{tot}) in coastal sediments and red clays is 3 to 4 percent and 5 to 6 percent, respectively. The tendency for Fe_{tot} to increase seaward is apparently due to its transport by streams in dilute concentration as part of the finest fractions of suspended matter. Indeed, direct measurement on the less than 0.001-mm size fraction of Pacific sediments showed that its iron content increased seaward from 3.73 (station 6158) to 9.1 percent (station 6174) (Lisitsyna and Dvoretskaya, 1972).

In addition to zones where uniform increase of iron content is observed, some stations or individual sediment horizons show locally high values of Fe_{tot}. Such anomalies are related to changing conditions of sediment formation, supply of terrigenous material (coastal sediments), influence of volcanic centers (West Pacific, Hawaiian sediments), and influence of exhalations (East Pacific). Some other irregularities in Fe_{tot} content may be caused by diagenetic redistribution.

The portion of Fe_{tot} subject to reduction, transfer to pore solution, and migration is more important than Fe_{tot} for characterizing redox processes. That reactive mobile portion is, to a considerable extent, represented by Fe_{react} as defined earlier.

Iron sulfides in sediments are chiefly represented by pyrite. The abundance of amorphous sulfides soluble in dilute HCl, i.e., hydrotroilites, is negligible (usually $<$ 0.01 percent), whereas pyrite may exceed 1 percent. The zone where sulfide forms are present was restricted to coastal sediments (figure 13-2, west of station 6171 and east of station 673). In reduced sediments, sulfides or pyrite maxima are detected even in near-surface horizons. For example, maximum S_{H_2S} occurred at approximately 30 to 50 cm of depth at stations 6160 and 669 (figure 13-2). Also, the same sediments had free H_2S concentrations up to 152 and 26 mg/ml, respectively, and were from the only area in the Pacific that seemed to have any (table 13-1).

In transitional sediments, iron sulfides were detected only below the oxidized layers, which are 0.5 to 2 cm thick near the coast (stations 6164 and 671), thicken to 70 to 190 cm seaward (stations 6171 and 673), and subsequently force out the reduced layer altogether. Sulfide forms were not detected seaward of the preceding stations.

Trivalent iron (Fe^{3+}), which I have included as part of Fe_{react}, is a characteristic feature of oxidized sediments. However, it is present virtually everywhere, even in strongly reduced sediments, although its concentration there declines drastically. In these sediments, Fe^{3+} is much less abundant than iron in sulfides or silicates. In the sediments of the oxidized zone (III), e.g., station 6172, all the Fe_{react} was present in the form of Fe^{3+} to the maximum depth sampled. By contrast, the sediments in the transitional zone (II) had Fe^{3+} only in an upper layer, e.g., the upper 65 cm at station 6171 (figure 13–5). Deeper, we found Fe^{2+}, which was partially in the sulfide form. A similar situation existed at station 672 on the eastern edge of the Pacific. The trends of the transitional zone are even more pronounced in the reducing zone (I); see stations 6162 and 668 in figure 12–5. At these stations, the preceding relationships are sharply displaced toward Fe^{2+}, even quite close to the uppermost sediment layer. At depths of 15 to 25 cm, where forms of H_2S become quite abundant and remain so to greater depths (figure 13–3), Fe_{S2^-} is dominant in the Fe_{react} component. Deeper in the sediments, the relationship between Fe_{react} forms remains constant, unlike the depth-associated changes in zone II (stations 6171 and 672).

At the threshhold of the $Fe^{3+} - Fe^{2+}$ transitions, Fe_{react} content was higher (up to 1.90 percent), the increase being chiefly caused by higher Fe^{3+} values. The reason is that accumulation of Fe^{3+} is characteristically found at the interface between oxidized and reduced sediments. Dissolved constituents in pore waters of reduced sediments become oxidized at contact with the oxidized zone and precipitate because of the lower solubility of oxidized forms. That is true for iron, manganese, and some other hydroxide-forming elements. Enrichment of the hydroxide layer in micro-elements is also possible owing to co-precipitation and adsorption.

Forms of iron differ in coastal and pelagic sediments. Whereas in surface pelagic sediments the Fe_{tot} content showed a slight increase in red clays

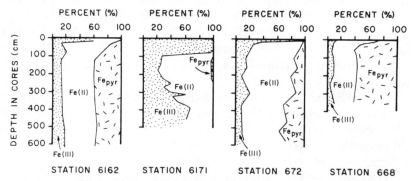

Figure 13–5. Forms of iron (percent, Fe^{2+}, Fe^{3+}, and Fe_{pyr}) contained in Fe_{react} in reduced (stations 6162 and 668) and transitional (stations 6171 and 672) Pacific sediments.

compared with coastal sediments, the values of Fe_{react} decreased seaward (table 13-1). Such a distribution probably occurs because the principal carriers of Fe_{react} are chiefly iron hydroxides, which are transported in the finest suspended sediment and also in dissolved form. In the course of sediment deposition and subsequent diagenesis, those iron forms age, crystallize, and become less susceptible to acid leaching. Therefore, to seaward there may be even more iron present, but less of it is reactive. The high values of Fe_{react} in coastal reduced sediments are frequently associated with processes of sulfate reduction and formation of hydrogen sulfide (Carroll, 1958; Drever, 1971). Quite probably iron is extracted from clay minerals to form iron sulfide, and a resultant increase of Fe_{react} occurs in the surface layer, with increases twofold to threefold higher in deeper reduced sediments (Rozanov et al., 1972, 1976).

Manganese

Manganese in oceanic sediments is particularly subject to the effect of redox processes. Unlike iron, most of the manganese in those sediments is in hydrogenous reactive form, which is readily mobilized in the course of reduction: $MnO(OH)_2$ (sediment) $\rightleftharpoons Mn^{2+}$ (solution). That in turn leads to diagenetic redistribution of manganese to form nodules and layers with increased manganese content and to the formation of authigenic manganese minerals. Under reducing conditions (pH 8.0, Eh below $+500$ mV), divalent manganese can form and diffuse in pore solutions and precipitate as $MnCO_3$ or, alternatively, oxidize at the boundary of the redox barrier to oxidized sediments (Calvert and Price, 1972).

To seaward, manganese content increased on an average from 0.03 to 0.04 percent in coastal reduced sediments to more than 1 percent in red clays (table 13-1). At some horizons and stations, manganese content was found to increase to several percent. If we ignore sediments where the preceding increases were associated with additional Mn supply due to exhalation, for example, at station 655 (Butusova et al., 1975), these changes are largely due to postsedimentary (diagenetic) changes, and only broad-scale average Mn_{tot} values can be reliably used to characterize regional Mn fluxes.

When the redox barrier passes across the sediment/water interface, manganese can be transported directly from sediment to supernatant water, as in reduced sediments lacking oxidized layers altogether (stations 6158–6163 near Japan and 668–670 near Mexico). In such sediments, manganese is chiefly represented by detrital residues.

In transitional sediments (stations 6164–6171 and 671–673) a reduced layer under the oxidized layer allows migration of Mn^{2+} upward followed by oxidation and accumulation in the redox barrier. Thus the superficial oxidized layer is enriched to as much as several percent Mn (11 percent at station 6171). However, irregular layers of increased manganese content are observed deep in reduced sediment. The layers may be due to buried Mn-enriched layers that previously occuped the position of the redox barrier.

Similar irregularities in manganese content were likewise observed throughout the oxidized layer of transitional sediments and also in oxidized sediments where there was no reduced layer, but local increases in C_{org} content may have been sufficient to cause reduction and migration of manganese (stations 6172 and 674). Stronger reducing conditions are essential for iron to be reduced and migrate. Hence diagenetic layers with increased iron content are less pronounced than those with enriched manganese. A large portion of iron (80 to 90 percent) in the sediments is represented by detrital (non-reactive) material, whereas a major portion of manganese is reactive. Thus, in oxidized sediments, the Mn^{4+} value amounts to 90 percent of the total manganese content.

A characteristic feature of oxidized sediments in the profiles studied is the presence of ferromanganese concretions in the superficial layer (Glagoleva, 1972; Volkov et al., 1976*b*). However, they form only at high Eh values (+600 mV) and are suggested by Volkov (chapter 15) to form through colloidal migration and aggregation of hydrated manganese and iron oxides rather than by redox processes.

Conclusion

Geochemical studies of the upper 4 to 5 m of recent Pacific sediments from the coast of Japan to the Mexican margin indicate a successive change in redox conditions from each coast seaward. Coastal reduced sediments, characterized by high content of organic matter (C_{org} over 1 percent) and active sulfate reduction, are replaced by oxidized pelagic sediments, wherein reduction processes virtually cease due to almost complete absence of organic material. As a result, we have a symmetric structure of the redox profile of the Pacific sediments. A major portion of it (4/5) is dominated by oxidized sediments.

The widths of the areas of reduced sediments localized in the coastal zone are primarily determined by hydrodynamic factors. In the east, the area is cut off from the ocean by the California Current and is limited to 700 km. In the west, the zone of reduced sediment are almost twice as wide because highly productive coastal waters with a high content of suspended C_{org} are abundantly transported by the Japan Current into the open sea.

The main features of coastal sediments are intensive reduction processes in mineral matter and destruction of organic matter. The processes provide active exchange of chemical elements with near-bottom water and result in formation of authigenic minerals. Diagenetic changes in oxidized sediments also lead to an authigenic mineral formation, which is virtually independent of redox processes.

The sediments located between oxidized and reduced layers are characterized by a rather variable picture of diagenetic alterations associated with the presence of an oxidized superficial sediment layer, whose thickness increases seaward.

Acknowledgments

The present study is part of the lithologic and geochemical investigations conducted on a profile across the North Pacific. The work was preformed under the leadership of N.M. Strakhov (Geological Institute, USSR Academy of Sciences) and I.I. Volkov (Shirshov Institute of Oceanology, USSR Academy of Sciences).

Data on this profile were partly discussed in papers written by the author in collaboration with I.I. Volkov, V.S. Sokolov, L.S. Fomina, N.N. Zhabina, T.A. Yagodinskaya, and M.F. Pilipchuk, who also participated in chemical analysis of the samples. The author expresses his gratitude for their collaboration.

References

Bruevich, S.V. 1973. Alkaline reserve of waters and ground solutions in seas and oceans. In *Issledovaniya po Khimiyi Morya*. Moscow: Izdat. Nauka (in Russian).

Butusova, G.Yu., Lysitsyna, N.A., Volkov, I.I., and Lubchenko, I.Yu. 1975. Signs of exhalant activity in bottom sediments of the Pacific Ocean south of Gulf of California. *Litologiya i Polezniye Iskopayemye* 6 (in Russian).

Calvert, S.E., and Price, N.B. 1972. Diffusion and reaction profiles of dissolved manganese in the pore wates of marine sediments. *Earth Planet. Sci. Lett.* 16:245–249.

Carroll, D. 1958. Role of clay minerals in the transportation of iron. *Geochim. Cosmochim. Acta* 14:1–27.

Chebotarev, E.N., and Ivanov, M.V. 1976. Distribution and activity of sulphate-reducing bacteria in bottom sediments of the Pacific Ocean and Gulf of California. In I.I. Volkov (Ed.), *Biogeokhimiya Diageneza Osadkov Okeana*. Moscow: Izdat. Nauka (in Russian), pp. 68–74.

Chernyakova, A.M. 1966. Dissolved oxygen. In S.V. Bruevich (Ed.), *Tikhii Okean. Khimiya*. Moscow: Izdat. Nauka (in Russian), pp. 82–116.

Drever, J.J. 1971. Magnesium-iron replacement in clay minerals in anoxic marine sediments. *Science* 172:1334–1336.
Glagoleva, M.A. 1972. Regularities in changes of chemical composition of iron-manganese concretions in northwest Pacific sediments. *Litologiya i Polezniye Iskopaemye* 4 (in Russian):40–49.
Ivanenkov, V.N. 1966. Carbonate system. In S.V. Bruevich (Ed.), *Tikhii Okean, Khimiya*. Moscow: Izdat. Nauka (in Russian), pp. 57–81.
Lisitsyna, N.A., Butusova, G.Yu., Volkov, I.I., Glagoleva, M.A., and Sokolov, V.S. 1975. Influence of Hawaiian volcanism on sediment accumulation. In *Problemy Litologii i Geokhimii Osadochnykh Porod i Rud*. Moscow: Izdat. Nauka (in Russian), pp. 130–149.
Lisitsyna, N.A., and Dvoretskaya, O.A. 1972. Lithologic profile across the northwestern hollow in the Pacific Ocean. *Litologiya i Polezniye Iskopaemye* 4 (in Russian):3–25.
Lysitsin, A.P. 1974. *Sedimentation in the Oceans*. Moscow: Izdat. Nauka (in Russian).
Ostroumov, E.A. 1953. Method for determining sulphur compound forms in Black Sea sediments. *Trudy Inst. Okaenol. USSR Acad. Sci.* 7 (in Russian):57–69.
Ponomarev, A.I. 1961. *Methods for Chemical Analysis of Silicate and Carbonaceous Rocks*. Moscow: Izdat. Akad. Nauk, SSSR, (in Russian), pp. 19, 111, and 216.
Rozanov, A.G., Volkov, I.I., Zhabina, N.N., and Yagodinskaya, T.A. 1971. Hydrogen sulphide in coastal slope sediments of the northwest Pacific. *Geokhimiya* 5 (in Russian):543–552.
Rozanov, A.G., Sokolov, V.S., and Volkov, I.I. 1972. Iron and manganese forms in sediments of the northwest Pacific. *Litologiya i Polezniye Iskopaemye* 4 (in Russian):26–39.
Rozanov, A.G. 1975. On determination of redox potential value in marine sediments. In E.A. Ostroumov (Ed.), *Khimicheskii Analysiz Morskikh Osadkov*. Moscow: Izdat. Nauka (in Russian), pp. 5–16.
Rozanov, A.G., Volkov, I.I., Sokolov, V.A., Pushkina, Z.V., and Pilipchuk, M.F. 1976. Redox processes in the sediments of Gulf of California and neighboring Pacific area (iron and manganese compounds). In I.I. Volkov (Ed.), *Biogeokhimiya Diageneza Osadkov Okeana*. Moscow: Izdat. Nauka (in Russian), pp. 96–135.
Shishkina, O.V. 1972. *Geochemistry of Sea and Ocean Silt Waters*. Moscow: Izdat. Nauka (in Russian).
Stumm, W. 1966. Redox potential as an environmental parameter: Conceptual significance and operational limitation. *3rd International Conf. Water Pollution Research*, Sec. 1, Paper No. 13 (Munich), pp. 1–16.

Van Andel, T.H. 1964. Recent marine sediments of the Gulf of California. In *Symposium Marine Geology of the Gulf of California*. American Assoc. Petrol. Geol. Memoir 3.

Volkov, I.I. 1959. Determination of various sulphur compound forms in marine sediments. *Trudy Inst. Okeanologii, USSR Akad. Sci.* 33 (in Russian):194–208.

Volkov, I.I., Rozanov, A.G., Zhabina, N.N., and Yagodinskaya, T.A. 1972. Sulphur in Pacific sediments east of Japan. *Litologiya i Polezniye Iskopaemiye* 4 (in Russian):50–64.

Volkov, I.I., and Zhabina, N.N. 1975. Determination of free sulphur by reduction to H_2S with bivalent chromium chloride solution. *Zhurnal Analyticheskoi Khimiyi* 30(8) (in Russian):1572–1575.

Volkov, I.I., Rozanov, A.G., and Sokolov, V.S. 1975. Redox processes in diagenesis of sediments in the northwest Pacific Ocean. *Soil Science* 119(1):28–35.

Volkov, I.I., Rozanov, A.G., Zhabina, N.N., and Fomina, L.S. 1976a. Sulphur compounds in the sediments of Gulf of California and neighboring Pacific area. In I.I. Volkov (Ed.), *Biogeokhimiva Diageneza Osadkov Okeana*. Moscow: Izdat. Nauka (in Russian), pp. 136–170.

Volkov, I.I., Fomina, L.S., and Yagodinskaya, T.A. 1976b. Chemical composition of iron-manganese concretions in the Pacific along the profile Wake Island-coast of Mexico. In I.I. Volkov (Ed.), *Biogeokhimiya Diageneza Osadkov Okeana*. Moscow: Izdat. Nauka (in Russian), pp. 186–204.

Volkov, I.I. 1981. On the mechanisms of the formation of ferromanganese concretions in recent sediments (chapter 15 of this book).

Whitfield, M. 1969. Eh as an operational parameter in estuarine studies. *Limnol. Oceanogr.* 14:547–558.

Zalmanzon, E.S. 1957. Determination of forms of certain elements and analysis of clay colloid fraction. In *Metody Izucheniya Osadochnykh Porod*. Moscow: Gosgeoltekhizdat (in Russian), pp. 59–63.

14 Migration of Manganese in the Deep-Sea Sediments

Shizuo Tsunogai
and *Masashi Kusakabe*

Abstract

To study the early diagenesis of manganese in sediments, we sampled various kinds of core sediments such as red clay, calcareous ooze, siliceous ooze, and volcanic turbidite from the Pacific Ocean. Interstitial water and the metals in reducible weak-acid soluble form were separated from the sediments. The fractions were analyzed for manganese and iron, and the total concentrations of metals in the sediments were determined as well. The results suggest that manganese in the deep-sea sediments is diffusing from the sediments to bottom water and that manganese is geochemically balanced in the ocean because the deeper sediment layers contain considerably lower manganese than the surface does. By applying a three-layer model to the vertical distribution of manganese in the red clay obtained in the western North Pacific, the following values are calculated: 1.5×10^{-5} cm^2/s for the apparent diffusivity of interstitial manganese in the surface oxidizing layer, 5.4×10^{-7} cm^2/s for that in the deeper reducing layer, and 530 and 5.2×10^4 years for half-lives of manganese in the rapidly and slowly soluble parts of the reducing solid phase, respectively.

Introduction

There are many uncertainties concerning manganese in the ocean. In addition to the question of the origin of manganese nodules on the deep sea floor, we must investigate the reasons why the residence time of manganese in the ocean is larger by one order of magnitude than that of iron (Goldberg and Arrhenius, 1958: Tsunogai and Kusakabe, 1973; Brewer, 1975), even though both manganese and iron have thermodynamically insoluble solid oxides in seawater. Also of interest is the problem of making a geochemical balance for manganese in seawater, as discussed by Horn and Adams (1966). Those questions can be easily answered if manganese in the surface layer of sediments is recycled between seawater and sediments because the cycling lengthens the residence time of elements in a particulate form in seawater and because previous geochemical-balance calculations

257

have usually been based on the manganese concentrations in the surface sediments. The migration of manganese in coastal and other sediments is rather well known e.g., Bender, 1971, and Tsunogai and Uematsu, 1978). However, the coastal area is too small to explain quantitatively the total amount of manganese recycling in the ocean expected for the geochemical balance.

Some studies (e.g., Lynn and Bonatti, 1965; Li et al., 1969) have indicated that manganese migrates through interstitial water in the deep-sea sediments when manganese is reduced from $Mn(IV)$ to $Mn(II)$. The reducing environment arises from the reaction of oxygen with organic matter. However, the amounts of manganese reduced and escaping to the bottom water have not been estimated precisely because of the scarcity of data. We sampled various kinds of core sediments such as coastal sediments, red clay, calcareous ooze, siliceous ooze, and volcanic turbidites from the deeper Pacific Ocean. The sediments were analyzed for manganese and some other elements in various states in the sediments. The detailed results will be reported elsewhere. Here we report studies on the migration of manganese in the deep-sea sediments, including red clay, as well as calcareous and siliceous oozes that contain comparatively large amounts of organic matter.

Methods and Observed Results

The sampling sites from several cruises are shown in figure 14–1. The 9.6-m-long core sample of red clay which is discussed here in detail was taken from the bottom under 6,052 m of water at 35°23′N, 149°34′E in the western North Pacific (KH-74-4, station 1). The detailed description of the core is given by Nasu (1975), in which several patches of brown clay surrounded by light brown clay were found in the core sediments between 25 and 35 cm from the top. The cores were cut into 1- to 5-cm sections. Interstitial water was squeezed out of each slice of sediment soon after sampling, and the soluble fraction of the associated solids was extracted with a weakly acidic reducing solution. Those fractions, as well as total concentration in the sediments, were analyzed for manganese, iron, and other constituents. The experimental procedure is schematically illustrated in figure 14–2, where the extraction procedure for solid sediments is principally the same as that of Chester and Hughes (1967).

As an example of the general trends in the distribution of manganese in the sediments, the results for red clay at station 1 of KH-74-4 are shown in figure 14–3 and summarized for station 1 and other sites in table 14–1. The trends are as follows:

1. Total manganese is more concentrated in the upper sediments than at greater depths (i.e., a few meters). That confirms some of the results of Li et

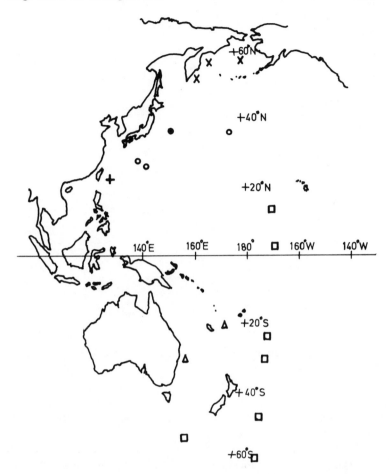

Figure 14-1. Sampling sites. Samples were collected during the cruises, KH-68-4 (□), KH-71-2 (+), KH-74-4 (○ and ●; the latter is the station KH-74-4, station 1), KH-75-4 (×), and *Vitya*-7010 (△).

al. (1969), Bonatti et al. (1971), Van der Weijden et al. (1970), etc., but we have not found the thick high manganese layer observed by Bender (1971) and others (e.g., Landergren, 1964).

2. One large maximum of total manganese content is present in the upper manganese-rich layer, although the maximum is sometimes accompanied by a few smaller peaks. The magnitude and depth of the maximum is highly dependent on the type of sediments. For instance, the maximum layer in the

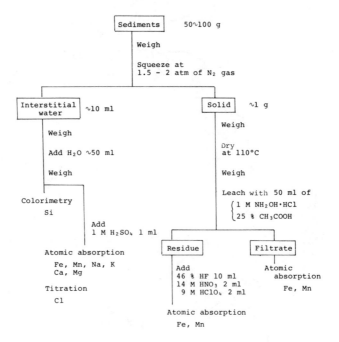

Figure 14-2. Experimental procedure.

red clay is more striking and deeper than those of the calcareous ooze and the siliceous ooze.

3. The concentration of refractory manganese, probably manganese in silicate phases, is nearly constant at 500 to 1,000 ppm on a carbonate-free basis for all the samples and comprises the major fraction of the total manganese in coastal sediments and in the deeper parts of pelagic sediments. Thus the manganese-rich layer results from the accumulation of reducible weak-acid soluble manganese.

4. The concentration of manganese in the interstitial water increases with depth and achieves maximum values at depths of several meters. The maximal values of interstitial manganese concentration are usually higher for red clay than for calcareous or siliceous oozes.

5. In contrast to manganese, the vertical distributions of various states of iron are fairly constant with depth, and the concentrations of reducible weak-acid soluble iron in the solids are extremely low. Concentrations of interstitial iron are also low.

6. Compared to most pelagic sediments, coastal sediments are depleted in all forms of manganese. The concentrations of manganese in the reducible weak-acid soluble fraction and in interstitial water are slightly larger in the near-surface sediments than in the deeper sediments.

Migration of Manganese

Table 14-1
Manganese and Iron in Pelagic and Coastal Sediments. Total Mn concentration (Σ) and Mn concentration in the weak-acid soluble fraction (Oxide) are given for dry sediments, and concentration in intersitital water (Water) is given for water. Symbols s, m, and c are marked for the layers, respectively, the surface, the maximum layer for total manganese, and the layer of nearly constant concentration for total manganese in long piston cores.

Station and Depth	Depth in core (cm)	Mn Σ (ppm)	Mn Oxide (ppm)	Mn Water (ppm)	Fe Σ (percent)	Fe Oxide (percent)	Fe Water (ppm)
KH–74–4, Station 1 (red clay), 6,052 m							
35°23'N	s: 0–5	4,240	3,790	0.01	4.28	1.18	0.01
149°34'E	m: 30–35	41,500	41,200	0.9	4.44	1.12	0.36
	c: 150–960	930	320	16	4.37	0.55	0.28
KH–74–4, Station 12 (calcareous ooze), 4,397 m							
34°51'N	s, m: 0–5	7,520	7,200	0.28	3.09	1.08	
175°03'E	c: 265–417	390	260	1.07	0.96	0.09	
KH–74–4, Station 21 (turbidite), 3,850 m							
26°46'N	s: 0–5	2,290	1,440	0.10	4.48	0.82	
139°42'E	m: 44–49	6,540	5,850	0.11	4.31	0.74	
	c: 74–195	1,510	380	3.0[a]	5.88	0.44	
KH–74–4, Station 24 (red clay with Mn sheets), 5,050 m							
37°26'N	s: 0	3,960	3,530	< 0.01	5.07	1.21	
137°36'E	m: 58–61	56,400	55,300	<0.01	6.99	2.31	
KH–75–4, Station 5 (siliceous ooze), 5,805 m							
52°00'N	s, m: 0–5	2,670	2,250	0.2	4.62	1.42	
161°58'E	c: 512–1007	669	254	4.5	3.85	0.63	
KH–71–2, Station 9 (red clay), 5,700 m							
22°00'N	s: 0–2	2,760	2,470		4.76	2.27	
125°00'E	m: 24–26	4,470	4,090		4.69	1.76	
Vitya–7010, Station 4 (calcareous ooze), 3,674 m							
22°38'S	s, m: 0–2	2,490	2,380		3.38	0.84	
170°01'E							
Vitya–7010, Station 11 (calcareous ooze), 4,090 m							
31°22'S	s: 0–2	1,010	836		2.57	0.14	
156°58'E	m: 22–24	9,890	9,010		2.58	0.20	
Funka Bay–7, Station 1 (blue silty clay), 102 m							
42°24'N	s: 0–3	447	92	0.62	3.47	0.83	
140°31'E	m: 18–23	602	64	0.23	3.85	0.54	
Funka Bay–8, Station 1 (blue silty clay), 102 m							
42°24'N	s: 0–5	401	76		2.95	0.55	
140°32'E	m: 40–45	771	48		4.26	0.40	
	c: 50–96	347	75		3.48	0.66	

[a]Concentration at the lowest depth.

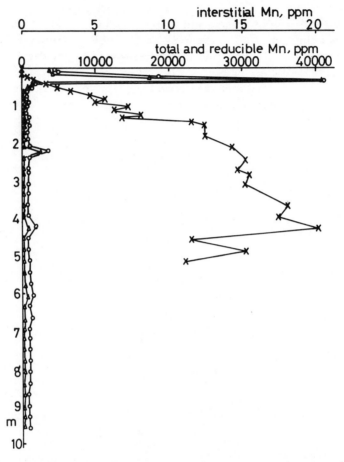

Figure 14–3. The vertical distribution of total manganese (○) and reducible manganese (△) in units of parts per million of dry sediments and interstitial manganese (×) in units of parts per million of pore water at the station KH-74-4, station 1. Vertical scale is in meters of sediment.

Our results suggest that manganese in pelagic sediments is migrating from the deeper layer, depositing in the near surface layer, and partly escaping into the deep seawater across the sediment/water interface. That idea can be presented more quantitatively in a mathematical model.

Model of Manganese Migration

The observed distribution of manganese in the red clay (KH-74-4, station 1, figure 14–3) occurs because the oxidation potential of the sediments de-

Migration of Manganese

creases with time and thus with depth. Reducing conditions in sediments are produced by the oxidation of sedimentary organic matter by interstitial electron acceptors such as oxygen. Some of the acceptors may diffuse downward from bottom water before reacting. Therefore, for convenience, the sediments can be divided vertically into three layers (figure 14–4): (1) an

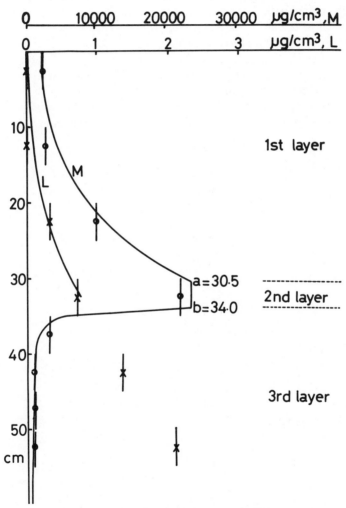

Figure 14–4. The three-layer model used and calculated curves for manganese in the liquid (L, the equation 14.25′) and solid (M, the equation 14.26′), which are compared to observed concentrations of manganese in the liquid ($\times$) and solid ($\bigcirc$) phases. a and b are the depths to the base of the first and second sediment layers, respectively.

oxidizing layer at the top ($0 \leq z \leq a$), where manganous ion introduced through interstitial water is oxidized to manganese oxide and then precipitates; (2) an intermediate layer ($a \leq z \leq b$), where neither manganous ion nor manganese oxide changes its chemical form owing to the lack of activation energy; and (3) a reducing layer ($z \geq b$), where manganese oxide is reduced to manganous ion and diffuses through interstitial water. As a whole, steady-state conditions will be assumed, and thus the boundaries a and b are moving upward with the same rate as the sedimentation rate v, although their depths below the sediment-water interface are always fixed.

This model is similar to that of Michard (1971), except for the following. Michard (1971) considered the deposition of manganese carbonate in the third layer. We, however, have not found much acid-soluble manganese in the third layer of the red clay but instead have found that manganese converges to a definite concentration in this layer and is chiefly refractory.

A diffusion-advection model with three layers is given as follows:

First layer (oxidizing layer, $0 \leq z \leq a$):

$$\frac{\partial L}{\partial t} = D\frac{\partial^2 L}{\partial z^2} - w\frac{\partial L}{\partial z} - k_1 L \tag{14.1}$$

$$\frac{\partial M}{\partial t} = -v\frac{\partial M}{\partial z} + k_1 L \tag{14.2}$$

Second layer (intermediate layer, $a \leq z \leq b$):

$$\frac{\partial L}{\partial t} = D\frac{\partial^2 L}{\partial z^2} - w\frac{\partial L}{\partial z} \tag{14.3}$$

$$\frac{\partial M}{\partial t} = -v\frac{\partial M}{\partial z} \tag{14.4}$$

Third layer (reducing layer, $z \geq b$)

$$\frac{\partial L}{\partial t} = D\frac{\partial^2 L}{\partial z^2} - w\frac{\partial L}{\partial z} + k_{21}M_1 + k_{22}M_2 \tag{14.5}$$

$$\frac{\partial M}{\partial t} = -v\frac{\partial M}{\partial z} - k_{21}M_1 - k_{22}M_2 \tag{14.6}$$

where L and M are the concentrations of manganese in the liquid and solid phases, respectively; D and w are the vertical diffusivity and advection velocity of interstitial water; v is the sedimentation rate; k is the first-order reaction rate for the oxidation of manganous ion and the deposition of manganic oxide; k_{21} and k_{22} are the first-order reaction rates for rapid and slow dissolution rate of manganese in the solid phase, respectively; M_1 and M_2 are the rapidly and slowly soluble fractions of manganese in the third layer at the lower boundary, depth b, respectively; M_∞ is the nonreactive fraction of manganese in the third layer; z is the depth; t is the time.

In our equations, first-order reactions are assumed for all the chemical reactions. As shown in figure 14–4, the curve showing the decrease of solid manganese with depth in the sediment is not a simple exponential curve in the third layer. Therefore the mechanism for the dissolution of manganese from the solid phase is not simple. We therefore, arbitrarily assume the two components for the soluble fractions in the solid phase as the second simplest case. Thus the concentration of manganese M below the boundary b is the sum of $M_1 + M_2 + M_\infty$.

To obtain solutions for the preceding differential equations, we assume that the upward velocity of interstitial water is balanced by the sedimentation rate, which means w is equal to zero, because total water content is constant with time in the infinitely thick sediments. That assumption was introduced by Anikouchine (1967), who considered the effect of the compaction of sediments. Although we do not know whether his assumption is the case or not in this region, the assumption does not seriously affect the result of the calculation.

Another assumption is that in such abyssal depths, bioturbation does not play a large role in the transport of either solids or solutes in sediments.

The following solutions are obtained at a steady state:

$0 \le z \le a$:

$$L = A \exp\left(\sqrt{\frac{k_1}{D}}\, z\right) + B \exp\left(-\sqrt{\frac{k_1}{D}}\, z\right) \qquad (14.7)$$

$$M = \sqrt{\frac{k_1 D}{v}}\, A \exp\left(\sqrt{\frac{k_1}{D}}\, z\right) - B \exp\left(-\sqrt{\frac{k_1}{D}}\, z\right) + C \qquad (14.8)$$

$a \le z \le b$:

$$L = Ez + F \qquad (14.9)$$

$$M = G \qquad (14.10)$$

$z \ge b$:

$$L = Hz + I - \frac{v^2}{Dk_{21}} J_1 \exp\left(-\frac{k_{21}}{v}z\right) - \frac{v^2}{Dk_{22}} J_2 \exp\left(-\frac{k_{22}}{v}z\right) \quad (14.11)$$

$$M = J_1 \exp\left(-\frac{k_{21}}{v}z\right) + J_2 \exp\left(-\frac{k_{22}}{v}z\right) + J_3 \quad (14.12)$$

where, $A, B, C, E, F, G, H, I, J_1, J_2,$ and J_3 are integral constants. Those constants can be eliminated by introducing the boundary conditions and continuous conditions for concentration and flux at the boundaries.

1. The boundary conditions at the surface $z = 0$ are $M = M_0$ and $L = 0$. The latter condition is assumed because the concentration of manganous ion in seawater is negligibly small. Thus we obtain

$$B = -A \quad (14.13)$$

$$C = M_0 - 2\frac{\sqrt{k_1 D}}{v} A \quad (14.14)$$

2. At the infinite depth $z \to \infty$, the boundary conditions are $L = L_\infty$, $M = M_\infty$. We get

$$H = 0 \quad (14.15)$$

$$I = L_\infty \quad (14.16)$$

$$J_3 = M_\infty \quad (14.17)$$

3. At a steady state, the flux out of the sediments should be balanced by the difference between input to the surface and output to a sediment layer at infinite depth. Then we get

$$D(\partial L/\partial z)_{z=0} = 2A\sqrt{k_1 D} = v(M_0 - M_\infty) \quad (14.18\text{a})$$

$$\therefore A = v(M_0 - M_\infty)/2\sqrt{k_1 D} \quad (14.18\text{b})$$

4. The continuous steady-state conditions at the boundary $z = a$ are written for the flux, for the concentration in the liquid phase, and for that in the solid phase as follows:

$$2A \sqrt{\frac{k_1}{D}} \cosh\left(\sqrt{\frac{k_1}{D}} a\right) = E \quad \text{Flux} \quad (14.19)$$

Migration of Manganese

$$2A \sinh\left(\sqrt{\frac{k_1}{D}}\,a\right) = Ea + F \qquad \text{Liquid} \qquad (14.20a)$$

$$F = -2A\sqrt{\frac{k_1}{D}}\,a \cosh\left(\sqrt{\frac{k_1}{D}}\,a\right) + 2A \sinh\left(\sqrt{\frac{k_1}{D}}\,a\right) \quad (14.20b)$$

$$M_0 - 2A\frac{\sqrt{k_1 D}}{v}\left[1 - \cosh\left(\sqrt{\frac{k_1}{D}}\,a\right)\right] = G \qquad \text{Solid} \quad (14.21)$$

5. At the lower boundary $z = b$, we can also write the continuous steady-state conditions as in the case at the boundary $z = a$:

$$\frac{v}{D}\left[J_1 \exp\left(-\frac{k_{21}}{v}b\right) + J_2 \exp\left(-\frac{k_{22}}{v}b\right)\right] = E \qquad \text{Flux} \quad (14.22)$$

$$L_\infty - \frac{v^2}{Dk_{21}} J_1 \exp\left(-\frac{k_{21}}{v}b\right) + \frac{v^2}{Dk_{22}} J_2 \exp\left(-\frac{k_{22}}{v}b\right)$$

$$= Eb + F \qquad \text{Liquid} \qquad (14.23)$$

$$J_1 \exp\left(-\frac{k_{21}}{v}b\right) + J_2 \exp\left(-\frac{k_{22}}{v}b\right) + M_\infty = G \qquad \text{Solid} \quad (14.24)$$

Finally, the following solutions are obtained by eliminating the integral constants:

First layer ($0 \leq z \leq a$):

$$L = 2A \sinh\left(\sqrt{\frac{k_1}{D}}\,z\right) = \frac{v(M_0 - M_\infty)}{\sqrt{k_1 D}} \sinh\left(\sqrt{\frac{k_1}{D}}\,z\right) \quad (14.25)$$

$$M = M_\infty + (M_0 - M_\infty) \cosh\left(\sqrt{\frac{k_1}{D}}\,z\right) \qquad (14.26)$$

Second layer ($a \leq z \leq b$):

$$L = \frac{v(M_0 - M_\infty)}{2\sqrt{k_1 D}}\left[\sqrt{\frac{k_1}{D}} \cosh\left(\sqrt{\frac{k_1}{D}}\,a\right)(z - a) + \sinh\left(\sqrt{\frac{k_1}{D}}\,a\right)\right]$$

$$(14.27)$$

$$M = M_\infty + (M_0 - M_\infty)\cosh\left(\sqrt{\frac{k_1}{D}}\,a\right) = M_{max} \qquad (14.28)$$

Third layer ($z > b$):

$$L = L_\infty - \frac{v^2}{Dk_{21}}J_1\exp\left(-\frac{k_{21}}{v}z\right) - \frac{v^2}{Dk_{22}}J_2\exp\left(-\frac{k_{22}}{v}z\right) \qquad (14.29)$$

$$M = M_\infty + J_1\exp\left(-\frac{k_{21}}{v}z\right) + J_2\exp\left(-\frac{k_{22}}{v}z\right) \qquad (14.30)$$

For convenience we have left the integral constants J_1 and J_2, but they can be eliminated by solving equations 14.18 to 14.24 simultaneously.

Numerical Calculations and Discussion

The following values are assumed from the observed data: water content, 60 percent; density of dry sediments, 2.5 g/cm³; L_∞ = 16 ppm in water = 12.6 µg/cm³; M_0 = 4,100 ppm in dry sediments = 2,170 µg/cm³; M_∞ = 800 ppm in dry sediments = 420 µg/cm³; the sedimentation rate, 1.5×10^{-3} cm/yr. The observed mean values with standard deviations are 60 ± 3 percent and 2.49 ± 0.05 g/cm³ for the water content and density of dry sediments, respectively. The sedimentation rate is estimated from the geomagnetic data observed by Tonouchi and Kobayashi (personal communication) and from the ^{230}Th/^{232}Th method by Tsunogai and Yamada (1977).

1. First Layer

As shown in figure 14–4, the best-fit curves with the forms of equations 14.25 and 14.26 drawn through the observed values of the concentrations of manganese in the liquid and solid phases are

$$L = 5.0 \times 10^{-2} \sinh(0.107z) \quad \text{in µg/cm}^3 \qquad (14.25')$$

$$M = 420 + 1750 \cosh(0.107z) \quad \text{in µg/cm}^3 \qquad (14.26')$$

In the preceding calculation we first obtained the parameter $\sqrt{k_1/D}$ in equation 14.26 from manganese in the solid phase and then calculated the parameter $2A$ in equation 14.25 from manganese in the liquid phase, applying the parameter $\sqrt{k_1/D}$. From the two parameters estimated, we have calculated the apparent diffusivity of interstitial manganese (D) and the specific removal rate of manganese from the liquid phase k_1 to be 1.5×10^{-5} cm²/s and 5.6 yr^{-1}, respectively.

The apparent diffusivity obtained is 10 times larger than that estimated by Bender (1971): 1.5×10^{-6} cm²/s as an upper limit based on the molecular diffusivity and tortuosity. A tenfold larger diffusivity agrees with the fact that the vertical profile of manganese in the liquid phase increases more gradually with depth compared to the profile of manganese in the solid phase. Therefore, we must investigate another mechanism besides molecular diffusion, such as bioturbation or physical mixing in the surface sediments, for the explanation of the large apparent diffusivity.

The specific removal rate of manganese corresponds to the half-life of 45 days for the oxidation of manganous ion in water in the oxidizing layer, if the removal is the result of oxidation of dissolved manganese. The rate is nearly the same as that discussed by Elderfield (1976), but we have no directly measured rate of Mn(II) oxidation in the sediments. The flux of manganese from the sediments to the bottom water is calculated to be

$$D(\partial C/\partial z)_{z=0} = v(M_0 - M_\infty) = 2.6 \times 10^{-2} \text{ g/(m}^2 \cdot \text{yr)} \qquad (14.18')$$

Although the rate is less than 10 percent of the rate of manganese dissolution in the sediments, the flux is about one order of magnitude larger than the rate of supply of manganese from the continents to the ocean. That supply rate [3.6×10^{-3} g/(m²·yr)] corresponds to a 2,200-year residence time for manganese in the oceans. The flux also indicated that the mean time of the removal of manganese in particulate form from seawater is a few hundred years, which is similar to the residence times of 150 years for iron, aluminum, or titanium. We anticipate, therefore, that manganese behaves similarly to iron or aluminum during particulate removal from seawater.

2. Third Layer

It is expected that the diffusivity of manganese in the third layer may be different from that in the first layer because of the extremely large apparent diffusivity in the first layer and the fact that the depth where $\partial L^2/\partial^2 z) = 0$ is much deeper than the layer where $\partial M/\partial z = 0$ (figure 14–4). Therefore, we have determined the parameters in the third layer independently from the preceding calculation in the first layer.

A more accurate distribution for manganese is necessary for the estimate of the second term on the right side of equation 14.30. Fortunately, we have supplemental data on total manganese from 1-cm slices in part of the core. We have estimated the third term in equation 14.30 from data for sixteen

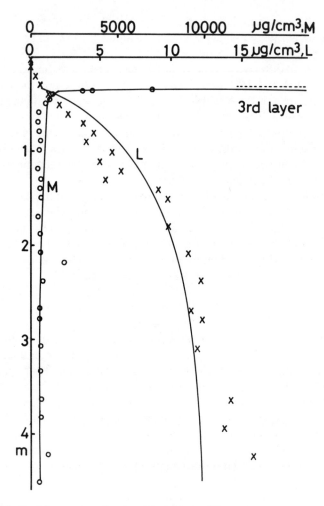

Figure 14–5. Manganese in the third layer. The calculated curves, L and M, for manganese in the liquid and solid phases given, respectively, in the equations 14.29' and 14.30' are compared to the observed concentrations of manganese in the liquid ($\times$) and solid ($\bigcirc$) phases. Vertical scale is in meters of sediment.

sections of the core from $z = 41$ to $z = 57$ cm and the assumption that $M_2 = 40$ ppm at $z = 400$ cm, and we have estimated the second term from data for six sections of the core from $z = 35$ to $z = 41$ cm after subtracting the first and third terms from the observed concentration. The result of the calculation is shown in figure 14–5. After introducing the numerical values into equation 14.30, we obtain

Migration of Manganese

$$M = 420 + 9{,}100 \exp(-0.87z') + 540 \exp(-8.8 \times 10^{-3}z') \quad (14.30')$$

where $z' = z - 35$. For the liquid phase, the remaining degree of freedom is only 1, namely, the apparent diffusivity, and the equation 14.29 with numerical values is given by

$$L = 12.6 - 1.8 \exp(-0.87z') - 10.1 \exp(-8.8 \times 10^{-3}z') \quad (14.29')$$

The calculated parameters are $k_{21} = 4.1 \times 10^{-11}$ s^{-1} for a half-life of 530 years for the rapid dissolution of manganese from the solid phase; $k_{22} = 4.2 \times 10^{-13}$ s^{-1} for a half-life of 5.2×10^4 years for the slow dissolution; and $D = 5.4 \times 10^{-7}$ cm^2/s. The apparent diffusivity in the third layer is only 0.04 of that in the first layer, but it is close to the value (4×10^{-7} cm^2/s) used by Li et al. (1969), which seems to be probable for water-clay mixtures. The deviation of the calculated curve from observed data for manganese in the liquid phase is rather large in the upper part of the third layer. That may be due to a larger diffusivity in the upper part of the third layer. We have no data on the dissolution of manganese from the solid phase for comparison, and that dissolution may vary with many factors, such as the organic carbon content, the biological activity, the diffusivity in interstitial water, and so forth.

The color of the reducing layer of this core is not so different from that of the oxidizing layer. The cause seems to be that the reduction of manganese is less dramatic than that of iron, which changes the color of sediments from brown to blue.

3. Second Layer

The thickness of the second layer cannot be determined precisely because the thickness is less than 5 cm, i.e., within the 30-to-35-cm depth zone. For practical purposes, we have considered the second layer as a moderate layer connecting the first and third layer continuously. The following values, therefore, are chosen arbitrarily: $a = 30.5$ cm for the upper boundary; $b = 34.0$ cm for the lower boundary; and 43,000 ppm Mn for dry sediments (23,000 μg/cm^3) for the maximum concentration. Those values satisfy the continuous conditions at the both boundaries and the condition that the observed concentration of manganese in the section from $z = 30$ to 35 cm is equal to tha calculated from the preceding solutions equations.

Concluding Remarks

The model used here is rather simple. We have found a few peaks of total manganese for some cores. That may be chiefly due to the organic carbon content in sediments, which is influenced by the depth, by the variations of biological activity in seawater, and by sedimentation rate. Thus some other indexes such as pE, organic carbon content, isotopic data, etc. are necessary to make a model for complicated distributions of manganese in other cores.

We have assumed a steady-state condition in our calculations. Although we do not have positive evidence for steady state, a catastrophic change in sedimentation does not seem to have occurred in the pelagic ocean during the last 100,000 years. Thus the calculated flux and rate constants may be realistic for the western North Pacific.

If the calculated flux of manganese from the sediments to the bottom water in the region is applicable to all oceans, the many questions on manganese in the ocean stated earlier can be quantitatively resolved. However, although fluxes of manganese are much larger in coastal areas or shallow seas (Bender, 1971; Tsunogai and Uematsu, 1978), those in sterile subtropical regions of red clay may be much smaller because red clay contains little of the organic matter which reduces manganese oxide. Our future work, therefore, is to examine more extensively and quantitatively the fact of migration of manganese in various parts of the ocean.

References

Anikouchine, W.A. 1967. Dissolved chemical substances in compacting marine sediments. *J. Geophys. Res.* 72:505–509.

Bender, R.A. 1971. Does upward diffusion supply the excess manganese in pelagic sediments? *J. Geophys. Res.* 76:4212–4215.

Bonatti, E., Fisher, D.E., Joensuu, O, and Rydell, H.S. 1971. Post-depositional mobility of some transition elements, phosphorus, uranium and thorium in deep sea sediments. *Geochim. Cosmochim. Acta* 35: 189–201.

Brewer, P.G. 1975. Minor elements in sea water. In J.P. Riley and G. Skirrow (Eds.), *Chemical Oceanography* Vol. 1, 2nd Ed., London: Academic Press, pp. 415–496.

Chester, R., and Hughes, M.J. 1967. A chemical technique for the separation of ferromanganese minerals, carbonate minerals and adsorbed trace elements from pelagic sediments. *Chem. Geol.* 2:249–262.

Elderfield, H. 1976. Manganese fluxes to the oceans. *Mar. Chem.* 4:103–132.

Goldberg, E.D., and Arrhenius, G.O.S. 1958. Chemistry of Pacific pelagic sediments. *Geochim. Cosmochim. Acta* 13:153–212.

Horn, M.K., and Adams, J.A.S. 1966. Computer derived geochemical balance and element abundance. *Geochim. Cosmochim. Acta* 35:279–297.

Landergren, S. 1964. On the geochemistry of deep sea sediments. *Reports of the Swedish Deep Sea Expedition X.* Goteborg, Sweden: Elanders Boktrycheri Aktiefolag, pp. 57–154.

Li, Y.-H., Bischoff, J., and Mathieu, G. 1969. The migration of manganese in arctic basin sediments. *Earth Planet. Sci. Lett.* 7:265–270.

Lynn, D.C., and Bonatti, E. 1965. Mobility of manganese in diagenesis of deep-sea sediments. *Mar. Geol.* 3:457–474.

Michard, G. 1971. Theoretical model for manganese distribution in calcareous sediments cores. *J. Geophys. Res.* 76:2179–2186.

Nasu, N. (Ed.). 1975. *Preliminary Cruise Report of KH-74-4.* Geological and geophysical researches in the Northwest Pacific Ocean Research Institute, University of Tokyo.

Tsunogai, S., and Kusakabe, M. 1973. The state of chemical elements in seawater and a new definition of residence time. Paper presented at annual meeting of the Geochemical Society of Japan, Akita.

Tsunogai, S., and Yamada, M. 1977. Th and Ra in the deep sea sediments and the migration of Ra. Paper presented at annual meeting of the Geochemical Society of Japan, Tokyo.

Tsunogai, S., and Uematsu, M. 1978. Particulate manganese, iron and aluminum in coastal water, Funka Bay, Japan. *Geochem. J.* 12:47–56.

Van Der Weijden, C.H., Schuiling, R.D., and Das, H.A. 1970. Some geochemical characteristics of sediments from the North Atlantic Ocean. *Mar. Geol.* 9:81–99.

15 On the Mechanisms of the Formation of Ferromanganese Concretions in Recent Sediments

I.I. Volkov

Abstract

Previous studies of marine ferromanganese concretions have often lacked important information about the abundance of the ore elements Fe and Mn and the abundance of trace elements such as Cu in the surrounding sediment. Thus the diagenetic mechanisms of formation of the concretions from the surrounding sediments have been difficult to depict. Numerous analyses of concretions and associated sediments have been conducted in the Black Sea, the Pacific Ocean, and the Indian Ocean. A combined regression indicates that the total Fe/Mn ratio in concretions is frequently correlated with the ratio of reactive Fe to total Mn in the surrounding sediment. Thus the reactive Fe in the surrounding sediment could serve as the source for the Fe in concretions. Similar regressions of trace-element ratios between concretions and sediments have suggested that surrounding sediments may be the source for those elements in concretions (especially in the Black Sea), although the correlations are not always as strong as that for Fe.

Introduction

Fe-Mn concretions are the most vivid and widespread illustration of diagenetic processes under the oxidizing conditions of marine and oceanic sediments. In addition to presenting scientific problems that remain to be solved, Fe-Mn concretions may constitute ore from which important metals could be recovered in the near future.

Reserves of oceanic polymetallic ore in the form of Fe-Mn concretions are enormous and estimated at hundreds of billions of tons. According to Strakhov et al. (1968), manganese accumulations in the ocean exceed all other sources of manganese ore put together.

Hundreds of chemical analyses of concretions from lake, sea, and ocean sediments are now available. Most of these cover only the major ore elements

---

This abstract was prepared by the editors.

like Fe and Mn; complete analyses are rare. The diagenetic origin of these nodules seems well established, but details of the mechanisms of formation are often incomplete. One reason is that data on the composition of the enclosing sediments are scarce, so conclusions often are derived from comparisons between *average* data for concretions and sediments. That problem exists for comparisons based on the abundance of ore elements in concretions, on the fractionation of elements during the formation of concretions, and on other types of information.

The chemical composition of concretions from recent marine sediments varies both for the main ore-forming elements (Fe and Mn) and for trace elements, particularly Ni, Co, Cu, and Mo. Usually the concretions in shallow-water sediments of inland and border seas are poorer in manganese and trace metals than those in deep-sea pelagic sediments. Also, within the group of concretions from border areas there seems to be more variation in composition than within the group of pelagic concretions. Therefore, an improved understanding of the causes of the variabilities in nodule composition is one of the principal objectives of this chapter.

I believe that the formation of concretions, whether in shallow marine or oceanic areas, is a two-stage process. In shallow sea sediments, such as the Black Sea, the Baltic Sea, the Russian Arctic border-sea areas, and other continental margin seas where reducing conditions exist in the sediments, the first step is a diagenetic fractionation of elements between the reduced deeper sedimentary layers and the oxidized surface layers. The upper oxidized layer is enriched in Fe, Mn, Ni, Co, Mo, P, and other elements, with iron and manganese occurring as freshly precipitated, reactive hydroxides that incorporate trace elements.

Then, in shallow-water sediments, the second stage is a diagenetic aggregation of colloidal amorphous hydroxides in the oxidized sediment layer into nodules and concretions. Thus the second stage is the formation of concretions proper, and it encompasses the processes of contraction, dehydration, and subsequent crystallization of authigenic colloidal iron and manganese hydroxides. Trace elements, which make no mineral phases of their own in the sediments, are carried into concretions by accumulating there together with iron and manganese hydroxides. During the first stage, the redistribution occurs vertically in the sediments, while, in the second stage, horizontal migration of materials in the surface layer of oxidized sediments becomes predominant.

In the pelagic sediments, where there are usually no reducing conditions except at considerable depth in the sediments, the condition prior to concretion formation is sedimentation. Material washed into the ocean is fractionated by sedimentary differentiation so that the geochemically most mobile elements (Mn, Ni, Co, Cu, Mo, and so on) are accumulated in higher concentration in deep-sea clays. As in the shallow seas, they mostly occur in association with Fe and Mn hydroxides. The second stage in oceanic sediments, i.e., formation of concretions proper, is identical with that described earlier for shallow-sea concretions (Volkov and Sevastyanov, 1968; Strakhov et al., 1968). Sedimentation rates for the deep ocean are much slower, so concretions there may take much longer to form than those on continental margins.

Given that Fe-Mn concretions are diagenetic units formed by the colloidal chemical processes just sketched, the causes of their varying compositions must lie with the varying compositions of the enclosing sediments. In an earlier paper (Volkov and Sevastyanov, 1968), we argued that the ratio of Fe and Mn in concretions, as well as their trace-element contents, must be connected with the relative contents of reactive forms of those elements in the enclosing sediment. At the time we could not, however, support that hypothesis with factual material, which was then confined to limited data on Black Sea concretions and surrounding sediments.

Studies conducted during subsequent years in the Pacific and Indian Oceans by the chemical section of the Institute of Oceanology of the USSR Academy of Sciences make it possible to discuss in more detail the likely causes of variation in the chemical composition of concretions and to take a fresh look at the formation mechanisms behind them.

This study is based on data on the chemical compositions of concretions in the Black Sea (Volkov and Sevastyanov, 1968), the Pacific Ocean from the coast of Japan to the American shore (Glagoleva, 1972, Volkov et al. 1976), and four samples from the Indian Ocean (Isaeva, 1967). Only those samples are considered where a parallel investigation of the enclosing sediment has been conducted. The investigation of concretions and sediments included determinations of total concentrations of Fe, Mn, Ni, Co, Cu, and Mo. In addition, the concentration of reactive Fe (by extraction in 3N H_2SO_4), and quadrivalent Mn (by determination of active oxygen with the aid of $FeSO_4$ in a sulfuric acid medium) were studied in sediments. Also studied were reactive (hydroxide) forms of Mn and Fe, along with Ni, Co, and Cu contained in those hydroxides, for which the reagent of Chester and Hughes (1967) was used (extraction by 30% $NH_2OH \cdot HCl$ in 30% CH_3COOH). Determination of elements in the extract was by atomic absorption. Total concentrations of elements in the samples were evaluated by conventional chemical techniques.

Proceeding from what was just said about the formation mechanisms of concretions, I assumed that Fe, Mn, Co, Cu, and Mo now contained in the concretions were originally reactive forms of those elements in the sediments. Therefore, all comparisons were conducted in terms of total concentrations of the elements in the concretions and their ratios. Some of the elements in concretions are in fact contained in detrital clay or other non-ore matter, but that portion is negligible. Trace elements chosen for consideration are, first, those elements which become part of manganese minerals and have a strong correlation with manganese in concretions (table 15-1) and, second, those elements which are strongly enriched in the surrounding oxidizing sediments, either because of redistribution between oxidized and reduced sediments or as a result of oceanic sedimentation. In this study it is assumed that the enrichment occurs because the reactive (as opposed to the detrital) forms of the elements are increased in the sediments. Those reactive forms may often prevail in the sediments or may often be equated with the total amounts of the elements present in the sediments.

Results and Discussion

Manganese and iron oxides make up at least 90 percent of the ore matter in the concretions. In concretions, Fe and Mn occur in abundances such that a decreased content of one in the ore matter of a concretion is matched by an increased content of the other (table 15-1). This seemingly "universal" regularity was established for pelagic concretions (Strakhov et al., 1968) and for shallow-water concretions (Sevastyanov and Volkov, 1966). The Fe/Mn or Mn/Fe ratio may vary in a broad range, even among oceanic concretions, but the Fe + Mn sum in the ore remains virtually constant. We have pointed out elsewhere (Volkov and Sevastyanov, 1968) that Mn and Fe are equivalent in the concretion-forming process, competing as they do in the chemical composition of the concretions. We also proposed that the ratio of Fe and Mn in a concretion must depend on the varying ratio of the reactive forms of those elements in the enclosing sediments. That hypothesis can now be supported with data on concretions and their surrounding sediments from the Pacific Ocean and the Black Sea. The data on reactive iron are either the content of sulfuric-acid-soluble iron in ooze from the Pacific Ocean and Black Sea or the content of amorphous iron hydroxides in Pacific samples extracted by the Chester and Hughes (1967) reagent. For manganese in the sediments around nodules, data on Mn^{4+} and on total manganese (Mn_{tot}) were equally useful, since they are linked by direct dependence with a high correlation ($r = +0.96$ for thirty-one samples of concretion-enclosing sediments from the Pacific and Black Sea). Although most of the manganese in concretions is found as Mn^{4+} compounds in the form of $MnO(OH)_2$, I used the data on the total Mn content in concretions for calculation. Data given in table 15-1 show that there is a direct dependence between Mn^{4+} and Mn_{tot} in concretions ($r = +0.98$).

Table 15-2 presents the relationship between the Fe/Mn ratio in concretions and in the enclosing sediments. In the majority of cases, the two ratios show a positive correlation. By way of illustration, one case from table 15-2 is elaborated upon in figure 15-1. The distribution of data points seems

Table 15-1
Coefficients of Correlation (r) between the Contents of Fe, Mn, and Some Trace Elements in Concretions

Ocean Concretions		Black Sea Concretions	
Element Pair	r	Element Pair	r
Mn-Fe	−0.52	Mn-Fe	−0.94
Mn_{tot}-Mn^{4+}	+0.98	Mn_{tot}-Mn^{4+}	+0.98
Mn-Ni	+0.37	Mn-Ni	+0.92
Mn-Co	+0.35	Mn-Co	+0.81
Mn-Cu	+0.34	Mn-Cu	+0.83
Mn-Mo	+0.51	Mn-Mo	+0.92

to show a regular pattern of the two ratios increasing concurrently. In the lowermost part of the field (with Fe/Mn in concretions less than 1) are points from samples of pelagic deep-sea red clays of the northwestern and eastern troughs of the Pacific. These are samples of typical oceanic concretions with a prevalence of manganese over iron in the chemical composition. Above this area are plotted samples taken near the Hawaiian Islands. The composition of the concretions here in part reflects the introduction of iron into the sediments by the weathering of basaltic material. Next, in the part of the diagram where the Fe/Mn ratio of concretions ranges from 2 to 6, are data points for hemipelagic and transitional clay samples from the Northwestern depression, where reducing conditions are observed within the sedimentary series and where Fe greatly prevails over Mn in the sediment. Therefore, those concretions are similar to the ones from shallow seas such as the Black Sea. The upper right portion of the diagram is occupied by Black Sea concretions, with enclosing sediments having high concentrations of Fe_{react} and high Fe_{react}/Mn_{tot} (up to 22) in the concretions.

The frequent dependence of the Fe/Mn ratio in concretions on the Fe/Mn ratio in the enclosing sediment seems clear for the samples studied. It is therefore reasonable to assume that a similar dependence applies to most Fe-Mn concentrations in marine sediments, while keeping in mind that exceptions may exist.

It is also possible that a similar relationship holds for trace elements in ferromanganese concretions. These trace elements, e.g., Ni, Co, and Cu, are sometimes accumulated in relatively high concentrations but form no mineral phases of their own. Instead, they occur in concretions as a sorption complex or are chemically co-precipitated or otherwise held in Fe and Mn hydroxides. General regularities of trace-element accumulations in concretions have already been reported for Black Sea nodules (Volkov and Sevastyanov, 1968). In shallow-water sediments that are usually rich in biogenous and terrigenous compounds, developing Fe/Mn concretions are poor in trace

Table 15-2
Correlation Coefficients (r) for Fe/Mn Ratio in Concretions and Reactive Fe and Mn in Sediments under Various Conditions (The Forms of Fe and Mn are described in the text.)

Ratio in Sediments	Ratio in Concretions	Basin	Number of Samples	Correlation Coefficient (r)
Fe_{react}/Mn^{4+}	Fe_{tot}/Mn_{tot}	Pacific	27	0.67
Fe_{react}/Mn^{4+}	Fe_{tot}/Mn_{tot}	Pacific and Black Sea	31	0.85
Fe_{react}/Mn_{tot}	Fe_{tot}/Mn_{tot}	Black Sea	8	0.42
Fe_{react}/Mn_{tot}	Fe_{tot}/Mn_{tot}	Pacific	27	0.47
Fe_{react}/Mn_{tot}	Fe_{tot}/Mn_{tot}	Pacific and Black Sea	35	0.69
Fe/Mn in the Chester and Hughes (1967) extract	Fe_{tot}/Mn_{tot}	Pacific	25	0.70

elements because trace-element concentrations in the sediments are low and exceed the Clarke value only slightly through diagenetic redistribution. In contrast, oceanic concretions are strongly enriched in trace elements because they develop in an environment of very slow accumulation of sediment richer in trace elements. For example, the red clays of the central part of the northwestern Pacific depression are 5.3 times richer in Ni, 7.8 times richer in Co, 3.6 times richer in Cu, and 4.8 times richer in Mo than the coastal terrigenous sediments. Similar enrichments are found in sediments from the eastern depression between the Hawaiian and Mexican shores (Volkov et al., 1974; Glagoleva et al., 1975). The proportion of reactive forms of trace elements in terrigenous coastal sediments is small because the greater part of them are associated with detrital minerals. In the deep ocean, reactive forms prevail overwhelmingly over detrital ones; they accumulate there at the expense of hydrogenous forms. The main forms in which trace elements occur in the deep ocean are as sorbed phases or chemically bound components of hydrated manganese and iron oxides.

The content of trace elements in the concretions in this study show a positive correlation to the manganese contents (table 15-1). The correlations are significantly large for Black Sea concretions, suggesting that the trace elements in these concentrations are present inside manganous minerals. Apparently when concretions develop from the enclosing ooze, these trace elements follow the manganese with which they were initially associated.

That kind of association of trace elements with Mn phases in concretions may be more widespread, as seen in table 15-3, where I have evaluated the ratios of Ni, Co, Cu, and Mo to Mn in concretions and compared them with the corresponding Me/Mn ratios in the enclosing sediments. For a combined suite of samples from the Pacific Ocean, the Indian Ocean, and the Black

Ferromanganese Concretions

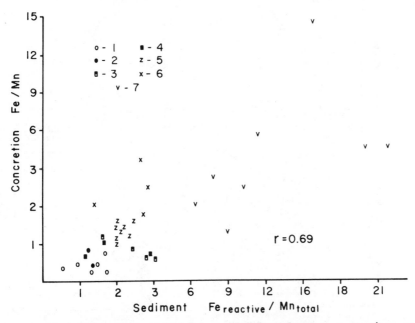

Figure 15-1. The relationship between Fe/Mn ratio in concentrations and Fe_{react}/Mn_{tot} ratio in Pacific and Black Sea sediments: 1 = red clays with zeolites, Eastern depression; 2 = red clays with ash, Eastern depression; 3 = red clays with zeolites, Northwestern depression; 4 = red clays with ash, Northwestern depression; 5 = sediments of the sampling area near Hawaii; 6 = hemipelagic and transitional clays, Northwestern depression; 7 = Black Sea sediments; 8 = calcareous ooze, Indian Ocean; and 9 = radiolarian ooze, Indian Ocean.

Sea, there is clear correlation between the ratio of all four trace elements to manganese in sediments and in concretions. The relationship is illustrated graphically in figure 15-2.

As with the Fe/Mn ratio (figure 15-1), the arrangement of the points on the diagram follows a distinct pattern. For all the elements, the bottom left part of the diagram, with minimal Fe/Mn in concretions and in sediments, is occupied by Black Sea samples. The top right part, with high Me/Mn in sediments and concretions, is occupied by deep-sea red clays of the Pacific and occasionally by pelagic globigerina and radiolarian oozes of the Indian Ocean. Sediments and concretions of the area near Hawaii and the transitional Pacific clays hold an intermediate position on the diagrams. In all cases (except for Cu), the points on the diagrams are scattered on both sides of a 45° line. For Cu, they all lie below that line. This indicates that Cu/Mn

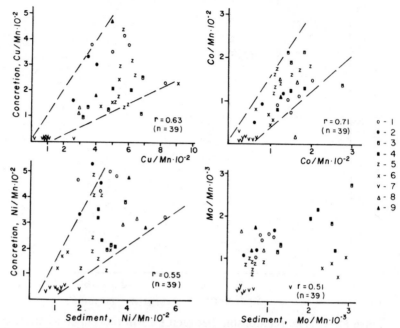

Figure 15-2. Relationship between Me/Mn ratio in concretions and in surrounding sediments for sediments from the Pacific Ocean, Indian Ocean, and the Black Sea. Symbols for data points are the same as in figure 15-1: (a) = Cu/Mn; (b) = Ni/Mn; (c) = Co/Mn; and (d) = Mo/Mn.

in enclosing sediments is consistently higher than in concretions. So it is possible that Cu in oceanic sediments is being held by forms other than hydrated manganese dioxide.

From the reasonably large correlation coefficients in these comparisons it may be concluded that the trace elements under study in oxidized oceanic sediments quite frequently are associated with hydrated manganese dioxide. It also appears that Ni, Co, Cu, and Mo are accumulated together with Mn during the formation of the concretions. The content of trace elements in Fe-Mn concretions is often influenced by their content in the enclosing sediment.

The observed data thus show that the chemical composition of the ore-forming part of Fe-Mn concretions can be correlated with reactive forms of elements in the enclosing sediments. This idea confirms the concretion-forming mechanisms described earlier; further purposeful research in that direction should be continued to encompass a broader range of elements. Mero (1965) proposed that any comprehensive theory of manganese concretion formation must account for the occurrence and distribution of all

Table 15-3
Correlation Coefficients (r) of Me/Mn Ratios in Concretions and in Enclosing Sediments

Ratio in Sediments	Ratio in Concretions	Basin	Number of Samples	Correlation Coefficients (r)			
				Ni/Mn	Co/Mn	Cu/Mn	Mo/Mn
Me_{tot}/Mn_{tot}	Me_{tot}/Mn_{tot}	Pacific and Indian Oceans	31 (27 + 4)	0.20	0.43	0.18	0.33
Me_{tot}/Mn_{tot}	Me_{tot}/Mn_{tot}	Pacific and Indian Oceans and Black Sea	39 (31 + 8)	0.55	0.71	0.63	0.60
Me_{tot}/Mn_{tot}	Me/Mn in Chester and Hughes (1967) extract	Pacific Ocean	24	0.42	0.65	0.20	—

elements in the nodules. In agreement with that suggestion, further research extending the present results to a broader range of elements should be conducted.

References

Chester, R., and Hughes, M.J. 1967. A chemical technique for the separation of ferromanganese minerals, carbonate minerals and absorbed trace elements from pelagic sediments. *Chem. Geol.* 2:249–262.

Glagoleva, M.A. 1972. Regularities of variation of the chemical composition of Fe-Mn concretions in Northwestern Pacific sediments. (Russ.) *Litol. i polezn. iskop.* 4:40–49.

Glagoleva, M.A., Volkov, I.I., Sokolov, V.S., and Yagodinskaya, T.A. 1975. Chemical elements in Pacific sediments in profile from the Hawaiian to the Mexican shore. (Russ.) *Litol. i polezn. iskop.* 5:16–28.

Isaeva, A.B. 1967. Chemical composition of concretions of the Indian Ocean. (Russ.) *Litol. i polezn. iskop.* 3:43–56.

Mero, J. 1965. *The Mineral Resources of the Sea.* New York: Elsevier.

Sevastyanov, V.F., and Volkov, I.I. 1966. Chemical composition of Fe/Mn concretions of the Black Sea. (Russ.) *Dokl. AN SSSR* 166(3):701–704.

Strakhov, N.M., Shterenberg, L.E., Kalinenko, V.V., and Tikhomirova, E., 1968. *Geokhimiya osadochnogo margantsovorudnogo protsessa.* Moscow: Nauka.

Volkov, I.I., and Sevastyanov, V.F. 1968. Redistribution of chemical elements in the diagenesis of the Black Sea sediments. (Russ.) In *Geokhimiya Osadochnykh Porod i Rud.* Moscow: Nauka, pp. 134–182.

Volkov, I.I., Fomina, L.S., and Yagodinskaya, T.A. 1976. Chemical compositions of Fe-Mn concretions of the Pacific along the profile of Wake Atoll–Mexican shore. (Russ.) In *Biogeokhimiya diageneza osadkov okeana.* Moscow: Nauka, pp. 186–204.

Volkov, I.I., Sokolov, V.A., Sokolova, E.G., and Pilipchuk, M.F. 1974. Rare and trace elements in northwestern Pacific sediments. (Russ.) *Litol. i polezn. iskop.* 2.

16 Trace Metals in Interstitial Waters from Central Pacific Ocean Sediments

M. Hartmann and *P.J. Müller*

Abstract

Transition metal concentrations in interstitial waters of siliceous ooze, red clay, and calcareous ooze from the Central Pacific Ocean were found to be considerably higher than in deep oceanic waters. Enrichment factors relative to seawater of 12 to 19 for Mn, 5 to 28 for Cu, 3 to 4 for Zn, and 2 to 5 for Ni were found. Most sediment cores showed a tendency to have the highest interstitial Cu concentrations close to the upper sediment surface. Positive correlations existed between dissolved Mn, Cu, and Ni, and between Ni and Zn.

The results showed that trace-metal mobilization and upward migration in highly oxic sediments (Eh $>$ +400 mV) are restricted to the sediment surface (0 to 2 cm). Flux estimates suggested that the amounts of trace metals released from the sediments are sufficient to act as the source for manganese nodule growth.

Introduction

The question of trace-metal supply for deep-sea manganese nodule formation, either directly from the seawater or from the interstitial water of underlying sediments, remains unresolved. In anoxic sediments, reductive dissolution of manganese and iron oxides, upward diffusion of their reduced species (Mn^{2+}, Fe^{2+}) and reprecipitation in near-surface oxic sediments are well-studied processes (Hartmann, 1964; Lynn and Bonatti, 1965; Presley et al, 1967; Bischoff and Ku, 1971; Elderfield, 1976; Hartmann et al. 1976; and others). It appears unlikely that a similar mechanism of mobilization and upward migration exists for trace metals in pelagic red clay and siliceous ooze because of the high redox potentials (Eh $>$ +400 mV) throughout those sediments.

The purpose of this chapter is to describe the heavy-metal distribution in the interstitial waters of highly oxidized sediments and to discuss possible mobilization mechanisms. Sediments studied were subsampled from eight "Kastengreifer" cores and three 6- to 10-m-long box cores taken in four small investigation areas (labeled 1 to 4, each about 1,400 km^2) between the Clarion and Clipperton fracture zones southeast of Hawaii during cruise VA 08-1/1974 of the German research vessel Valdivia (see table 16–1). Preliminary results of geophysical, sedimentological, biostratigraphical, soil mechanical, and geochemical properties of sediments, interstitial water, and manganese nodules from the four areas were published in the cruise report (Beiersdorf et al., 1974). Therefore, only a brief description of a few important characteristics of the sediments is presented here. All cores from areas 1, 2, and 4 contained red clay and siliceous ooze. They were free of carbonates and had manganese nodules on the sediment surfaces. Organic carbon contents in the surface sediments were 0.3 to 0.4 percent dry weight and decreased exponentially with depth to values areound 0.1 percent or less (Müller, 1975, 1977, and unpublished data). In contrast, the cores from area 3 consisted mainly of carbonates (40 to 90 percent $CaCO_3$) and contained less than 0.2 percent organic carbon. As expected from the low organic matter contents, the sediments showed high redox potentials (between +400 and +550 mV).

Methods

Sediment cores were subsampled on board ship immediately after core retrieval. Interstitial waters were squeezed through 30-cm membrane filters (0.45 μm nominal pore size) with N_2 under 10 atm pressure exerted across a rubber membrane separating the sediment from the gas. The PVC-lined filter press (modified after Hartmann, 1965) had a capacity of about 1,000 ml wet sediment from which up to 500 ml of interstitial water were recovered. Pore waters were extracted in a temperature-controlled room at about 7°C to minimize the "temperature of squeezing effect" (Mangelsdorf et al., 1969). Sediments from "Kastengreifer" cores did not warm before squeezing since they were exposed to temperatures higher than 15°C for 30 min or less. The long box-core sediments, however, were exposed to open-air temperatures for several hours before they could be stored in the cold lab for interstitial water extraction. Immediately after interstitial water recovery, 200 to 300 ml of each sample were passed through a Chelex-100 ion exchange column (modified after Riley and Taylor, 1968) to extract and preserve the heavy metals for analyses at home. Heavy metals were determined by atomic absorption after elution with 2 N HNO_3.

Results

Table 16–2 lists transition-metal concentrations of seventy-five interstitial water samples from the eight "Kastengreifer" cores. Results for fifty-two

Trace Metals

Table 16–1
Sampling Stations. All cores except those from area 3 contained manganese nodules on top of the sediment. Geographical positions not released for publication

Station No.	Ship Board No.	Area	Water Depth (m)	Core Length (cm)
10140–1	VA08–1–13	4	5144	33
10141–1	VA08–1–19	4	5189	38
10144–1	VA08–1–35	4	5113	32
10145–1	VA08–1–36	3	4599	32
10147–1	VA08–1–41	3	4619	31
10149–1	VA08–1–44	2	5205	25
10175–1	VA08–1–58	2	5164	29
10178–1	VA08–1–80	1	5101	38
10142–1	VA08–1–20	4	5195	983
10148–1	VA08–1–43	3	4609	592
10176–1	VA08–1–70	2	5214	989
Water samples:				
10138–1	VA08–1–04	1	5293	
10143–1	VA08–1–34	4	5154	
10146–1	VA08–1–38	3	4619	
10150–1	VA08–1–55	2	5246	

Note: See cruise report by Beiersdorf et al., 1974.

additional samples from the three 6- to 10-m-long box cores are given in table 16–3. The results from seven deep oceanic water samples, taken with 30-L PVC Niskin samplers 1 and 100 m above bottom, are listed in table 16–4 along with mean values for interstitial water and interstitial enrichment factors relative to seawater.

Most of the interstitial Fe, Mn, and Cu concentrations fell within the range of 1 to 10 μg/L. Zn concentrations were considerably higher and showed relatively large fluctuations around a mean value of about 20 μg/L. Ni concentrations were generally lower than 2 μg/L. Co concentrations were found to be below the detection limit (0.2 μg/L) in most of the samples except those with high Ni contents. One should expect smooth gradients of the trace-metal concentrations within the cores. The scattering of the trace-metal concentrations encountered despite careful treatment may, in some cases, result from handling artifacts during pore water extraction or from analytical scatter. In other cases, the cause could be transitional changes of microenvironments (due probably to benthic organisms) that produce metal mobilization (see subsequent discussion).

Comparison with the mean concentrations found in seawater reveals a considerable enrichment of trace metals in the interstitial waters. The mean enrichment factors (table 16–4) were higher for Mn (12 to 19) and Cu (5 to 28) than for Zn (3 to 4) and Ni (2 to 5). Because of high Fe blank readings, no reliable iron values were obtained for seawater.

Table 16-2
Heavy Metal Concentrations (μg/L) in Interstitial Waters from Short Box Cores ("Kastengreifer")

10178-1 (Area 1)

Depth (cm)	Eh (mV)	pH	Fe	Mn	Cu (μg/L)	Zn	Ni
0-2	—	—	9.9	1.6	4.5	10.9	1.3
2-4	—	—	6.7	3.9	2.5	14.7	1.1
4-6	—	—	13.4	1.1	2.0	9.1	0.7
6-9	—	—	8.2	3.7	3.7	18.7	2.0
9-13	—	—	14.2	2.0	1.2	12.7	0.7
13-18	—	—	15.8	2.5	3.4	16.1	1.0
18-23	—	—	9.7	2.6	2.3	13.1	0.9
23-28	—	—	2.3	1.3	1.1	7.0	0.7

10175-1 (Area 2)

Depth (cm)	Eh (mV)	pH	Fe	Mn	Cu (μg/L)	Zn	Ni
0-2	+490	7.61	4.2	2.0	2.4	7.4	0.9
2-4	+490	7.63	3.6	3.6	3.0	9.2	1.4
4-6	+490	7.66	2.3	2.2	2.1	8.1	0.9
6-9	+505	7.61	0.3	2.0	1.7	7.3	0.8
9-12	+495	7.44	2.0	2.2	2.2	7.4	1.2
12-16	—	7.48	3.8	2.2	1.3	6.5	1.0
16-20	+505	7.44	3.6	3.4	0.9	7.8	1.3
20-24	+520	7.45	—	1.3	0.5	4.6	0.8
24-28	—	7.48	1.0	2.2	0.7	6.2	1.1

10149-1 (Area 2)

Depth (cm)	Eh (mV)	pH	Fe	Mn	Cu (μg/L)	Zn	Ni
0-2	+510	7.64	0.4	5.2	3.6	6.8	1.2
2-4	+490	7.59	0.4	10.9	2.5	12.7	1.8
4-6	+495	7.57	0.2	7.1	3.4	5.6	1.9
6-9	+465	7.56	0.7	0.6	0.6	4.9	0.3
9-12	+495	7.48	0.5	1.8	1.2	7.8	0.9
12-16	+535	7.50	13.1	7.7	1.5	15.7	2.5
16-20	+535	7.50	0.5	3.9	2.5	23.6	1.5
20-24	+545	7.43	0.5	2.3	1.2	9.3	1.1

10140 (Area 4)

Depth (cm)	Eh (mV)	pH	Fe	Mn	Cu (μg/L)	Zn	Ni
0-2	—	—	10.4	2.5	1.2	71.4	1.7
2-3	+395	7.51	4.9	0.8	0.6	24.4	0.5
3-5	+415	7.42	5.4	1.8	0.9	41.8	1.4
5-7	+435	7.45	5.8	1.1	0.8	93.6	1.2
7-10	+445	7.44	6.6	2.7	1.6	126.4	2.7
10-13	+450	7.44	3.0	62.4	2.3	17.8	4.9
13-16	+490	7.40	2.9	10.2	2.0	16.6	2.9
16-20	+495	7.40	3.8	13.5	4.3	44.6	4.2
20-24	+485	7.42	2.8	2.8	4.1	15.8	2.3
24-28	+475	7.32	5.8	5.8	4.2	27.0	4.3
28-32	—	7.36	1.2	1.3	8.7	18.3	1.3

Trace Metals

10141-1 (Area 4)

Depth	Eh	pH					
0–2	+470	7.48	8.1	6.4	6.0	34.5	1.6
2–4	+455	7.49	4.0	3.2	3.1	27.5	1.3
4–6	—	—	0.9	0.6	0.7	4.5	0.5
6–10	+455	7.55	8.2	2.3	0.8	15.6	0.5
10–14	+470	7.54	4.0	1.7	0.8	11.4	0.2
14–18	—	—	6.6	2.9	0.9	23.9	0.7
18–23	+480	7.45	6.2	2.6	1.0	19.3	0.7
23–28	+495	7.47	3.9	1.4	0.6	12.7	0.2
28–33	+490	7.47	2.6	2.4	0.8	18.5	0.6
33–37	+500	7.45	—	1.3	0.8	19.6	0.5

10144-1 (Area 4)

Depth	Eh	pH					
0–2	+545	7.66	2.5	6.3	5.2	20.5	1.4
2–4	+530	7.64	2.3	6.4	2.8	15.1	1.5
4–6	+525	7.62	1.7	3.5	1.8	16.2	1.5
6–9	+495	7.62	1.7	3.6	1.8	15.5	1.4
9–12	+495	7.61	1.3	2.2	1.0	15.4	1.0
12–18	+495	7.60	—	10.9	5.2	26.1	3.4
18–22	+505	7.43	2.4	2.1	2.3	11.9	1.0
22–26	+515	7.45	4.2	17.6	5.4	32.4	4.3
26–30	+515	7.49	2.4	4.3	0.7	14.5	0.9

10145-1 (Area 3)

Depth	Eh	pH					
0–2	+525	7.55	—	1.8	7.5	15.8	1.6
2–4	+475	7.62	—	1.0	2.0	9.8	0.5
4–6	+480	7.67	—	—	—	—	—
6–8	+465	7.65	—	3.4	2.1	37.6	1.1
8–11	+460	7.69	—	—	—	—	—
11–14	+445	7.71	—	1.2	1.4	22.7	0.6
14–18	+445	7.70	—	—	—	—	—
18–22	+440	7.61	—	4.8	1.8	20.0	0.6
22–26	+425	7.65	—	—	—	—	—
26–30	+425	7.70	—	—	—	—	—

10147-1 (Area 3)

Depth	Eh	pH					
0–2	+445	7.71	—	—	8.3	19.6	7.6
2–4	+445	7.75	—	—	—	—	—
4–6	+445	7.79	10.3	2.8	3.9	10.3	1.4
6–8	+445	7.80	9.2	3.2	3.3	15.5	0.9
8–11	+450	7.75	13.5	3.1	2.1	9.2	1.8
11–14	+445	7.76	—	—	—	—	—
14–18	+450	7.76	14.0	2.4	1.8	18.0	1.3
18–22	+480	7.75	—	—	—	—	—
22–26	+470	7.81	17.6	—	2.6	58.3	3.1
26–30	+470	7.83	—	—	—	—	—

— means not determined.

Table 16-3
Heavy Metal Concentrations (μg/L) in Interstitial Waters from Long Box Cores ("Kastenlot")

Depth (cm)	Eh (mV)	pH	Fe	Mn	Cu	Zn	Ni
					(μg/L)		
10176-1 (Area 2)							
8–22	+525	7.53	16.0	2.5	1.1	25.5	1.0
35–49	+520	7.50	—	0.9	0.4	15.8	0.5
61–75	—	7.53	2.0	1.1	0.3	20.9	0.1
76–90	—	7.48	2.2	1.0	0.3	56.0	22.1
125–139	—	—	6.6	1.0	0.3	8.9	0.1
184–198	—	—	2.5	1.3	0.5	60.8	15.4
262–276	—	—	9.8	1.7	0.7	24.5	0.6
310–324	—	—	11.3	8.7	1.1	58.3	1.7
368–382	—	—	—	1.7	—	12.7	—
516–530	—	—	19.5	11.3	3.0	24.1	2.3
650–664	—	—	3.3	2.9	1.9	14.5	1.8
830–844	—	—	9.5	2.0	2.3	28.6	1.2
920–934	—	—	12.3	11.4	—	—	4.9
10148-1 (Area 3)							
11–25	+495	7.61	—	1.9	1.2	20.6	0.8
20–40	+495	7.66	—	2.2	0.4	24.9	0.3
45–49	+490	7.64	—	—	—	—	—
60–74	+475	7.71	—	—	—	—	—
81–95	+480	7.69	0.7	3.0	0.9	19.4	0.6
106–120	+505	7.71	—	—	—	—	—
136–150	+460	7.69	0.2	5.5	2.0	10.3	3.2
212–226	+485	7.77	0.7	1.6	1.2	7.8	0.8
318–332	+505	7.65	—	2.6	1.1	5.4	0.9
386–400	+520	7.73	—	1.9	1.6	13.4	1.6
416–430	+525	7.75	16.3	4.0	2.0	47.0	3.1
431–445	+520	7.76	—	0.4	—	2.3	0.5
467–481	+480	7.72	1.5	1.0	0.7	10.5	0.8
483–497	+540	7.68	4.2	1.7	0.9	6.6	0.8
498–512	+495	7.71	0.5	1.7	1.6	8.7	1.1
543–557	+505	7.64	4.3	1.1	1.2	6.2	1.0
559–573	+510	7.64	4.7	1.5	0.9	6.7	1.5
575–589	+480	7.79	4.3	2.0	1.1	6.4	1.1

— means not determined.

Trace Metals

Depth (cm)	Eh (mV)	pH	10142-1 (Area 4)				
			Fe	Mn	Cu	Zn	Ni
			(µg/L)				
38–52	+495	7.44	—	3.7	—	31.7	—
62–76	+505	7.37	7.3	3.0	0.7	24.0	0.8
80–94	+525	7.36	5.2	1.5	0.4	14.9	0.4
121–135	+525	7.27	4.9	2.7	0.5	23.2	1.0
232–246	+475	7.37	5.5	2.0	0.3	15.7	0.4
262–276	+505	7.35	0.8	—	—	—	—
277–291	+505	7.35	7.0	3.4	0.4	19.5	0.7
337–351	+535	7.30	—	1.5	0.3	23.3	0.5
392–406	+505	7.28	—	1.6	0.8	91.3	0.5
440–454	+490	7.25	—	1.8	1.6	26.0	0.7
519–533	+515	7.13	8.4	3.2	0.6	23.1	1.0
574–588	+490	7.25	5.9	1.5	0.3	13.7	0.5
597–612	+505	7.27	1.6	2.7	0.5	14.6	0.9
614–628	+505	7.26	1.2	3.5	0.9	23.0	1.1
629–643	+490	7.27	—	1.1	0.1	6.9	0.4
644–658	+495	7.22	4.5	3.2	0.5	20.5	0.8
659–673	+515	7.34	2.6	2.4	0.5	—	0.8
719–733	+505	7.34	3.4	2.9	0.8	17.0	0.8
779–793	+525	7.32	5.6	3.2	0.7	28.1	0.9
899–913	+515	7.30	2.1	4.4	0.9	25.1	2.1
959–973	+550	7.27	2.7	3.9	0.6	17.6	1.4

Table 16-4
Heavy Metal Concentrations (μg/L) in Deep Oceanic Water, Mean Values for Interstitial Waters from Tables 16–2 and 16–3, and Mean Interstitial Water/Seawater Enrichment Factors.

Station No.	Water Depth (m)	Sediment Section (cm)	Fe	Mn	Cu	Zn	Ni
			(concentration in μg/L)				
Seawater:							
10138–1	5292		—	0.4	0.3	7.7	0.5
	5192		—	0.2	0.2	5.6	0.6
10143–1	5153		—	—	0.1	6.7	0.4
	5053		—	—	0.1	8.5	0.5
10146–1	4618		—	—	(1.6)	1.3	0.5
	4518		—	(16.0)	0.2	3.5	(1.9)
10150–1	5245		—	0.1	0.1	4.8	0.3
Average[a]			—	0.23	0.17	5.4	0.47
Interstitial water (mean values ± s[b]):							
		0–2	5.9 ± 4.1	3.7 ± 2.2	4.8 ± 2.4	17.3 ± 20.7	2.2 ± 2.2
		2–38	5.1 ± 4.6	3.6 ± 3.2	2.1 ± 1.5	19.9 ± 20.3	1.4 ± 1.1
		38–973	5.1 ± 4.4	2.8 ± 2.4	0.9 ± 0.6	19.6 ± 17.2	1.1 ± 0.9
Mean enrichment factors (interstitial water/seawater):							
		0–2	—	1.86	28.4	3.2	4.6
		2–38	—	15.2	12.4	3.7	3.0
		38–973	—	12.2	5.4	3.6	2.3

Note: — means not determined.
[a]Values in parentheses excluded from calculation.
[b]s = standard deviation.

Table 16–5
Correlation Matrix for Heavy Metals in Interstitial Waters. Number of samples = 87; correlation coefficients indicated by crosses (+) were significant at $\alpha = 0.01$. Extraordinary high single values (> 50 µg/L Fe, > 20 µg/L Mn, and > 10 µg/L Ni, four samples in total) were omitted from the calculation.

	Fe	Mn	Cu	Zn	Ni
Fe		0.072	0.064	0.254	0.076
Mn	0.072		0.429+	0.167	0.740+
Cu	0.064	0.429+		0.023	0.510+
Zn	0.254	0.167	0.023		0.346+
Ni	0.106	0.740+	0.510+	0.346+	

Mn, Cu, and Ni showed positive correlations with each other ($\alpha = 0.01$, table 16–5), as illustrated in figure 16–1 for Mn and Ni. The only significant correlation for Fe was with Zn ($\alpha = 0.05$).

Table 16–4 shows that all metals except Zn tended to have their highest

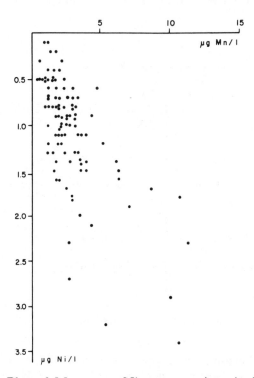

Figure 16–1. Plot of Mn versus Ni concentrations in interstitial water samples.

concentrations in interstitial waters near the sediment surface. However, because of the low number of samples from the uppermost core sections (six to eight samples), that feature proved significant only for Cu at a level of significance $\alpha = 0.01$ (figure 16–2). We have, at present, no explanation for the opposite trend of the Cu concentration in core 10140-1.

Our results compare favorably with results reported by Callender (1976), who found Mn concentrations between 1 and 9 µg/L and Cu concentrations between 1 and 8 µg/L in interstitial waters of surface (0 to 30 cm) siliceous oozes and red clays from the northeastern equatorial Pacific Ocean. Most of his Cu concentrations clearly decreased with increasing sediment depth, but in a few cases, he reported an increase in interstitial water Cu concentration below 20 cm, similar to the trend found in core 10140-1 (figure 16–2). In addition, he found trace metals in interstitial waters appreciably enriched in comparison to Pacific Ocean water (Mn by a factor of 35, and Cu by a factor of 13). It should be mentioned, however, that Orren and Sayles (1976) reported considerably higher interstitial trace-metal concentrations in Pacific siliceous oozes using sampling by *in situ* probe (10 to 200 µg/L for Fe, 5 to 50 µg/L for Mn, 10 to 60 µg/L for Cu, and 1 to 6 µg/L for Ni).

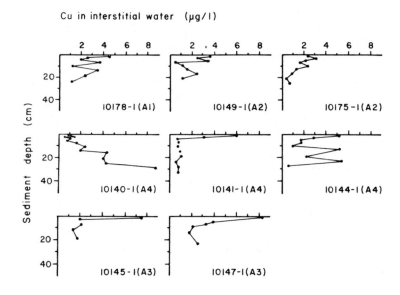

Figure 16–2. Interstitial water concentrations of Cu in sediment profiles from "Kastengreifer" cores (A1, A2, A3, and A4 mark the four sampling areas).

Discussion

Decreasing concentrations of interstitial trace-metals with depth are the opposite of what one usually finds in anoxic sediments, where Mn and Fe concentrations, and perhaps those of other trace-metals, increase with depth. Deep-sea siliceous ooze and red clay sediments, however, are deposited with extremely low sedimentation rates (a few millimeters or less per 1,000 years; Heath and Moore, 1965; Chester and Aston, 1976). The resulting sediments have high redox potentials and little organic matter (generally 0.3 to 0.5 percent organic carbon in surface sediments and about 0.1 percent below a few decimeters of sediment depth; Volkov et al., 1975; Müller, 1977). Furthermore, the resistant organic matter fraction of the total organic matter increases with depth since it seems to be associated with clay minerals (Müller, 1977) or show a high degree of metal complexation (Degens and Mopper, 1976). Both conditions protect the resistant organic fraction from microbial attack. Consequently, the redox potentials remain high in those sediments (tables 16-2 and 16-3). As a consequence of increased organic matter decay at the sediment surface, most cores even show a slight but significant Eh decrease toward the uppermost core sections (disregarding the 0- to 2-cm sections, which are influenced by prolonged exposure of the sediment surface to air after sampling). Redox potentials of the upper 15 cm of the cores are significantly lower than those from the deeper sediment sections. Thirty samples gave a mean value of +468 mV compared to sixty-six deeper samples which had a mean value of +498 mV. The two mean values differed at a level of significance of $\alpha = 0.001$. The high redox potentials are consistent with nitrate concentrations increasing with core depth, suggesting the presence of free oxygen within the interstitial water (Hartmann and Müller, 1974; Müller, 1975).

Hence the following mechanism might explain the interstitial trace-metal distribution in oxic sediments. Trace metals originally bound to organic matter (see, eg., Boström et al., 1974; Degens and Mopper, 1976), to biogenic silica (e.g. Arrhenius, 1963; Greenslate et al., 1973), or to sedimentary metal oxides are released to the interstitial water by diagenetic reactions (decomposition or dissolution of solids) and are subsequently reprecipitated as oxides or sorbed on to clay minerals. The metal release rate is highest in surface sediments due to their higher content of degradable organic matter and to the fact that the dissolution rate of silica rapidly decreases with depth (Hurd, 1973). In addition, the presence of burrowing organisms or their organic remains in the upper sediment layers may locally cause slightly lowered redox conditions, enhancing the dissolutin of manganese oxides (and associated metals) in certain micro-environments.

Recent studies of light brown to yellow spots of concentric or tubelike structures found in the brown or dark red brown surrounding sediment of most Pacific cores support that conclusion (Hartmann et al., 1975; Hartmann 1979). The spots have significantly lower Mn, Cu, and Ni contents than the surrounding (darker) material, while Fe and Zn remain somewhat constant. The finding suggests a mobilization of Mn and associated heavy metals (particularly Cu and Ni) in certain restricted microenvironments within the sediment by reductive dissolution of oxides, possibly caused by the decay of locally buried polychaetes or the organic substances secreted by them. The heavy metals preferentially mobilized by this mechanism are the same as those experimentally mobilized (Hartmann et al., 1975) by leaching sediments under mild acid reducing conditions (after Chester and Hughes, 1967).

From the foregoing, it follows that the heavy-metal release rate must be highest in the surface sediments and decrease with depth because it is closely related to the availability of degradable organic matter or soluble silica frustules. Surface sediments contain more of those solids than deeper sediments.

Conclusions

The decreasing Cu concentrations with increasing sediment depth show that there exists no diffusive upward flux of Cu, at least from sediment sections more than 2 cm below the sediment surface. However, the differences in trace-metal concentrations betweeen pore water and seawater (see table 16–4) indicate that upward diffusive fluxes of trace metals must exist within the upper 2 cm of the sediments investigated. The fluxes are difficult to estimate since the maximum trace-metal concentrations and the depths where they occur are difficult to sample.

The lower limit of each flux, however, can be estimated from the mean values for seawater and near-surface interstitial water listed in table 16–4, assuming a linear concentration gradient between 0 and 1 cm of sediment depth. In this special case, the following equation may be used (from Berner, 1971):

$$J_x = -D_s \phi \frac{(C_{x'} - C_0)}{x'} \tag{16.1}$$

where J_x = diffusional flux (mass/area/time)

D_s = diffusion coefficient in sediment (area/time)

ϕ = porosity of sediment

Trace Metals

$C_{x'}$ = concentration at $x = x'$ (mass/volume)

C_0 = concentration at $x = 0$ (mass/volume)

x' = a shallow depth within the linear gradient (length)

With the mean trace-metal concentrations listed in table 16-4 for seawater (C_0) and near-surface interstitial water ($C_{x'}$ assuming $x' = 1$ cm), a porosity of 0.84 (calculated after data from Kögler, 1974), and diffusion coeffcients taken from Li and Gregory (1974, corrected for tortuosity), we obtained the diffusional flux rates listed in table 16-6. For comparison, the accumulation rates of the respective trace metals are included in table 16-6. They were calculated from unpublished sediment data and an assumed sedimentation rate of 1 mm/1,000 yrs. The comparison shows that the loss of metals from the sediments by diffusion is appreciably higher for Cu, Zn, and Ni than their respective accumulation rates. In view of the assumptions made, the fluxes could conceivably be larger than those calculated due to a steeper concentration gradient at the uppermost skin of the sediment.

From the loss of metals from the uppermost sediment section one would expect that the uppermost core sections should exhibit higher metal contents within the sediment compared to the deeper core sections. The initial excess of trace metals is lost by diffusion to the water column during diagenesis. Increased trace-metal contents in the upper sediment section, however, have not as yet been found. Thus one has to conclude that the residence time of mobilizible trace metals is very short, resulting in, at most, a very thin layer enriched in trace metals just beneath the sediment/water interface (probably < 2 mm). Moreover, only a very low portion (probably $\ll 1$ percent) of the trace metals deposited on the sediment surface will finally be incorporated in the sediments, and the overwhelming portion will return to the free ocean water through dissolution and diffusion. Another possibility is that the content of mobilizible trace metals in freshly deposited sediment is initially low and thereafter remains low because the mobilizable metals dissolve at the same rate as other sedimentary components. The amorphous silica in frustules of diatoms and radiolarians follows the second possiblity, and Hurd (1973) calculated that only a low fraction of the biological opal production is ever actually incorporated into the sediments.

Diffusion rates of heavy metals can be compared with their accretion rates to the surface of manganese nodules. Assuming a manganese nodule growth rate of 0.5 cm/10^6 yr. (mean value estimated from Heye, 1975) and a dry nodule density of 1.4 g/cm^3 (Halbach and Özkara, 1975), a nodule accretion rate of 0.7 g/cm^2 per million years [$= 0.7$ μg/(cm$^2 \cdot$yr)] was calculated. A mean Mn content of 27 percent (Friedrich and Plüger, 1974)

Table 16-6
Rates of Diffusive Flux from the Sediments and Rates of Metal Accumulation to the Sediments [Values in $\mu g/(cm^2 \cdot yr)$]

	Fe	Mn	Cu	Zn	Ni
Rate of diffusion		0.18	0.23	0.58	0.08
Rate of accumulation	3-4	0.3-0.7	0.01-0.06	0.01-0.02	0.01-0.02

results in a Mn accretion rate of 0.19 $\mu g/(cm^2 \cdot yr)$. That is the same value found for the Mn diffusion rate from the sediment (table 16-6). Assuming a 1:1 sediment-surface/nodule-surface ratio, corresponding to about 25 percent sediment-surface coverage, the value suggests that the diffusive flux of Mn from the sediment is sufficient to be the source for the growth of manganese nodules. Corresponding values of accretion rates of Cu, Ni, and Zn to nodule surfaces are appreciably lower than the respective rates of diffusive flux; i.e., those metals are supplied in excess.

The preceding results indicate that the most intensive heavy-metal reactions occur in the sediment within a few millimeters of the sediment/seawater interface. That is the zone where manganese nodules are found and are growing. More precise results from the uppermost sediment section require more refined sampling methods with less disturbance of this boundary layer.

Acknowledgments

The authors wish to thank M. Whiticar for his critical comments and help with the English text and I. Dold, H. Lass, and H. Hensch for their technical assistance. We are also indebted to captain and crew of R/V Valdivia for help during sampling operations. The financial support by the Deutsche Forschungsgemeinschaft and the Bundesministerium für Forschung und Technologie is gratefully acknowledged.

References

Arrhenius, G. 1963. Pelagic sediments. In M.N. Hill (Ed.), *The Sea*, Vol. 3. New York: Wiley-Interscience, pp. 655-727.

Beiersdorf, H., Dürbaum, H.J., Friedrich, G., Hartmann, M., Kögler, F.O., Müller, P.J., Plüger, W., Schlüter, H.U., and Wolfart, R. 1974. Geophysikalische, geologische and geochemisch-lagerstättenkundliche Untersuchungen im Bereich von Manganknollenfeldern im zentralen Pazifik. *Meerestechnik* 5:187-206.

Berner, R.A. 1971. *Principles of Chemical Sedimentology.* New York: McGraw-Hill.

Bischoff, J.L., and Ku, T.L. 1971. Pore fluids of recent marine sediments. II. Anoxic sediments of 35° to 45°N Gibraltar to Mid-Atlantic ridge. *J. Sed. Petrol.* 41:1008–1017.

Boström, K., Joensuu, O., and Brohm, I. 1974. Plankton: Its chemical composition and its significance as a source of pelagic sediments. *Chem. Geol.* 14:255–271.

Callender, E. 1976. Transition metal geochemistry of interstitial fluids extracted from manganese-nodule-rich pelagic sediments of the Northeastern Equatorial Pacific Ocean. In *Abstracts of the Joint Oceanographic Assembly 1976.* Edinburgh.

Chester, R., and Aston, S.R. 1976. The geochemistry of deep-sea sediments. In J.P. Riley and R. Chester (Eds.), *Chemical Oceanography,* Vol. 6, 2nd Ed. London: Academic Press, pp. 281–390.

Chester, R., and Hughes, M.J. 1967. A chemical technique for the separation of ferro-manganese minerals, carbonate minerals and adsorbed trace elements from pelagic sediments. *Chem. Geol.* 2:249–262.

Degens, E.T., and Mopper, K. 1976. Factors controlling the distribution and early diagenesis of organic material in marine sediments. In J.P. Riley and R. Chester (Eds.), *Chemical Oceanography,* Vol. 6, 2d Ed., London: Academic Press, pp. 59–113.

Elderfield, H. 1976. Manganese fluxes to the oceans. *Mar. Chem.* 4:103–132.

Friedrich, G., and Plüger, W. 1974. Die Verteilung von Mangan, Eisen, Kobalt, Nickel, Kupfer und Zink in Manganknollen verschiedener Felder. In Beiersdorf et al., Geophysikalische, geologische und geochemisch-lagerstättenkundliche Untersuchungen im Bereich von Manganknollenfeldern im zentralen Pazifik. *Meerestechnik* 5:203–206.

Greenslate, J.L., Frazer, J.Z., and Arrhenius, G. 1973. Origin and deposition of selected transition elements in the seabed. In M. Morgenstein (Ed.), *The Origin and Distribution of Manganese Nodules in the Pacific and Prospects for Exploration.* Honolulu: pp. 45–70.

Halbach, P., and Özkara, M. 1975. Stoffliche Eigenschaften mariner Manganknollen. In G. Friedrich (Ed.), Geowissenschaftliche Untersuchungen auf dem Gebiet der Manganknollenforschung, BMFT-Forschungsbericht No. M 75–02. Deutsche Forschungs- und Versuchsanstalt für Luft- und Raumfahrt, München, pp. 41–59.

Hartmann, M. 1964. Zur Geochemie von Mangan und Eisen in der Ostsee. *Meyniana* 14:3–20.

Hartmann, M. 1965. An aparatus for the recovery of interstitial water from recent sediments. *Deep-Sea Res.* 12:225–226.

Hartmann, M. 1979. Evidence for early diagenetic mobilization of trace metals from discolorations of pelagic sediments. *Chemical Geol.* 26: in press.

Hartmann, M., and Müller, P.J. 1974. Geochemische Untersuchungen an Sedimenten und Porenwassern. In Beiersdorf et al., Geophysikalische, geologische und geochemisch-lagerstättenkundliche Untersuchungen im Bereich von Manganknollenfeldern im zentralen Pazifik. *Meerestechnik* 5:201–202.

Hartmann, M., Kögler, F.C., Müller, P.J., and Suess, E. 1975. Ergebnisse der Untersuchungen zur Genese von Manganknollen. In Friedrich (Ed.), Geowissenschaftliche Untersuchungen auf dem Gebeit der Manganknollenforschung, BMFT-Forschungsbericht No. M 75–02. Deutsche Forschungs- und Versuchsanstalt für Luft- und Raumfahrt, München, pp. 60–113.

Hartmann, M., Müller, P.J., Suess, E., and Van Der Weijden, C.H. 1976. Chemistry of Late Quaternary sediments and their interstitial waters from the NW African continental margin. *"Meteor" Forsch.-Ergebn.* C(24):1–67.

Heath, G.R., and Moore, T.C., Jr. 1965. Subbottom profile of abyssal sediments in the Central Equatorial Pacific. *Science* 149:744–746.

Heye, D. 1975. Wachstumsverhältnisse von Manganknollen. *Geologisches Jahrbuch* Reihe E(5).

Hurd, D.C. 1973. Interactions of biogenic opal, sediment and seawater in the Central Equatorial Pacific. *Geochim. Cosmochim. Acta.* 37:2257–2282.

Kögler, F.C. 1974. Sediment-physikalische Eigenschaften von drei Tiefseekernen des zentralen Pazifischen Ozeans. In Beiersdorf et al., Geophysikalische, geologische und geochemisch-lagerstättenkundliche Untersuchungen im Bereich von Manganknollenfeldern im zentralen Pazifik. *Meerestechnik* 5:199–201.

Li, Y.H., and Gregory, S. 1974. Diffusion of ions in sea water and in deep-sea sediments. *Geochim. Cosmochim. Acta* 38:703–714.

Lynn, D.C., and Bonatti, E. 1965. Mobility of manganese in diagenesis of deep-sea sediments. *Mar. Geol.* 3:457–474.

Mangelsdorf, P.C., Jr., Wilson, T.R.S., and Daniell, E. 1969. Potassium enrichment in interstitial waters of recent marine sediments. *Science* 165:171–173.

Müller, P.J. 1975. Diagenese stickstoffhaltiger organischer Substanzen in oxischen und anoxischen marinen Sedimenten. *"Meteor" Forsch.-Ergebn.* C(22):1–60.

Müller, P.J. 1977. C/N ratios in Pacific deep-sea sediments: Effect of inorganic ammonium and organic nitrogen compounds sorbed by clays. *Geochim. Cosmochim. Acta* 41:765–776.

Orren, M., and Sayles, F.L. 1976. Trace metal determinations in interstitial water collected in situ in deep-sea sediments. Paper presented at the Joint Oceanographic Assembly 1976, Edinburgh.

Presley, B.J., Brooks, R.R., and Kaplan, J.R. 1967. Manganese and related elements in the interstitial water of marine sediments. *Science* 158:906–909.

Riley, J.P., and Taylor, D. 1968. Chelating resins for the concentration of trace elements from sea water and their use in conjunction with atomic absorption spectrophotometry. *Anal. Chim. Acta* 40:479–485.

Volkov, I.I., Rozanov, A.G., and Sokolov, V.S. 1975. Rodox processes in diagenesis of sediments in the northwest Pacific Ocean. *Soil Sci.* 119:28–35.

Part VI
Transition Metals in Nearshore Sediments

Part VI

17 Diagenetic and Environmental Effects on Heavy-Metal Distribution in Sediments: A Hypothesis with an Illustration from the Baltic Sea

Rolf O. Hallberg

Abstract

Experiments have shown that chelating agents in a dynamic sediment system become concentrated in layers above the H_2S zone (redoxcline) and may react with heavy metals there and sweep them out of the system before they have time to be trapped and fixed as sulfides. Even though chelators are not very competitive for heavy metals in the presence of significant sulfide, they can play a significant role, particularly under more oxidizing conditions.

The $(Cu + Mo)/Zn$ ratio is proposed as a paleoecological indicator for ancient redox conditions. It increases sharply during anoxic conditions and decreases during oxidizing conditions at ancient sediment/water interfaces. A ^{210}Pb-dated 40-cm core from the eastern Gotland Basin of the Baltic Sea was sliced into millimeter-thick layers corresponding to 1 to 2 years. Variation in the $(Cu + Mo)/Zn$ ratio showed detailed correspondence with hydrologic conditions in the Baltic during the twentieth century and indicated fluctuations of much greater amplitude during the seventeenth and eighteenth centuries. Values of $(Cu + Mo)/Zn$ were as high as 6 and as low as 0.1.

Introduction

During an early stage of sediment diagenesis, the initial compaction results in collapse of the more unstable of the original sedimentary structures, producing closer packing of the sediment particles. Increased packing results in porosity decreases, and interstitial water is thereby expelled from the sediment. For the upper part of the sediment, where the sediment/seawater interaction is most predominant, steady-state diagenesis is an acceptable

---

This abstract was prepared by the editors.

assumption since porosity and compaction change predictably during burial (see figure 17–1). Thus compaction could produce a transport of ions from the sediment to the overlying water; another transport mechanism is diffusion.

As sediments accumulate, the oxidized overlying sediment layer is buried and gradually becomes more deficient in oxygen because of oxygen consumption by infauna and the limited contribution of oxygen to the sediment by diffusion. Consequently, reducing reactions become predominant in the subsurface part of the sediment where H_2S is produced by the sulfate-reducing bacteria. The H_2S moves upward and produces a gradual increase in sulfur-bearing compounds in the oxidized layer. During that process, some compounds (e.g., hydroxides, oxyhydroxides, and various carbonates) have lower stability than the sulfides and are therefore redissolved.

In sediments, stages of the microbiological decomposition of organic matter are not synchronized and result in an increase of intermediate

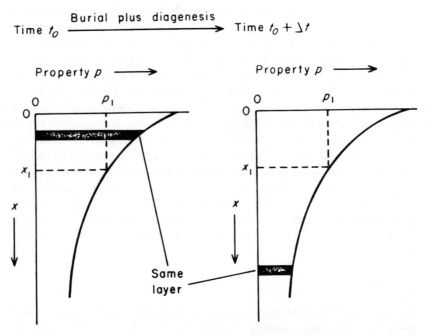

Source: R.A. Berner, *Principles of Chemical Sedimentology*. Copyright © 1971 by McGraw-Hill, Inc. Used with the permission of McGraw-Hill Book Company.

Figure 17–1. Diagrammatic representation of steady-state diagenesis. Note that at a given depth x_1, the sediment property $p = p_1$ does not change with time but that p for a given *layer* changes as it is buried.

Diagenetic and Environment Effects

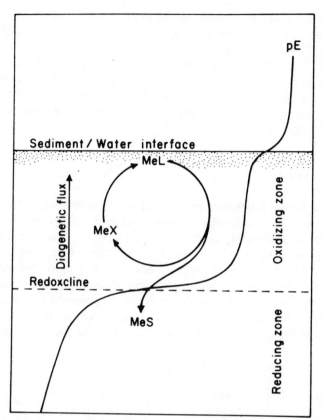

Source: Schippel et al., *OIKOS Supplementum* 15:64–67, 1973. Reprinted with permission.
Figure 17–2. The cycle of chelated heavy metals above the redoxcline.

compounds which may act as metal chelates. Therefore, there will be a competition for the heavy metals between chelating agents and hydrogen sulfide.

Initially, we can expect most of the H_2S to be precipitated as metal sulfides, and chelating agents will therefore be transported upward into the oxidizing zone to react with metals buried there. If a metal-chelating compound has an upward transport rate which is greater than that of the H_2S, chelated metal will be deficient in the H_2S zone. Thus a mineral cycle exists above the redoxcline where the chelated metals are transported upward and the process is further enchanced by compaction of the sediment (figure 17–2).

As the sediments accumulate, the metals can climb with them. They become concentrated in the uppermost layers and escape from being trapped and fixed as stationary sulfides (see Elderfield and Hepworth, 1975). Metals differ in their ability to chelate, so the more easily chelated ones tend to become enriched in the uppermost layer. In contrast to Gardner (1974) and Emerson (1976), I assume a dynamic system which is more relevant for natural processes. In this dynamic system, the metals are chelated and/or complexed *before* they meet the sulfide environment. Thus, even if the chelators are not very competitive for heavy metals in the presence of significant amounts of sulfide, they can play a major role under relatively more oxidizing conditions.

Chemical Model

A chemical model (computer program Haltafall according to Ingri et al., 1967) was used to describe the behavior of copper in a system with various concentrations of chelating agents and a fixed amount of sulfur. The redox potential, given as pE, was between 6 and −9, while pH was kept constant at 7. Since we do not know what chelating or complexing agents we are dealing with, we have to assume values for stability constants. Amino acids have values around 10^{-10}. Peptides and porphyrins are generally believed to have greater affinity for metals. In this chemical model, a value of $K = 10^{-10}$ was used, and the following reactions were assumed to be the predominant:

$$H_2S \rightleftharpoons HS^- + H^+ \qquad K = 10^{-7} \qquad (17.1)$$

$$9H^+ + 8e^- + SO_4^{2-} \rightleftharpoons HS^- + 4H_2O \qquad K = 10^{34} \qquad (17.2)$$

$$8H^+ + 8e^- + HSO_4^- \rightleftharpoons HS^- + 4H_2O \qquad K = 10^{36} \qquad (17.3)$$

$$Cu^{2+} + 3HS^- \rightleftharpoons [Cu(HS)_3]^- \qquad K = 2.82 \times 10^{-26} \qquad (17.4)$$

$$Cu^{2+} + HS^- \rightleftharpoons CuS + H^+ \qquad K = 1.58 \times 10^{-21} \qquad (17.5)$$

$$Cu^{2+} + \text{chelate} \rightleftharpoons \text{Cu Chelate} \qquad K = 10^{-10} \qquad (17.6)$$

The logarithms of the concentrations of $[Cu(HS)_3]^-$ and Cu chelate against variations of pE and log concentration of chelate are shown in figure 17–3. The redoxcline is shown and the point at which CuS starts to precipitate is pE = −1.5. The intersections between the graphs for equations 17.4 and 17.6 fall along a line identical with a pE ≈ −3. Thus the reaction between Cu and the chelating agent is effective even at redox conditions below the redoxcline. The redox environment where Cu chemistry seems to be controlled by chelating processes is shown as a shaded area at the bottom of the

Diagenetic and Environment Effects

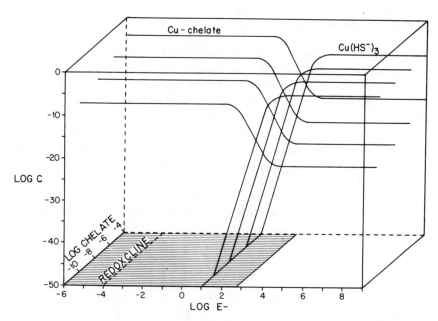

Figure 17-3. Three-dimensional diagram showing the equilibria between Cu $(HS^-)_3^-$ and a Cu-chelate at various chelate concentrations and electron activities. Remaining chemical conditions are: total sulfur concentration in formulae (1-5), 10^{-3}M; total Cu-concentration 10^{-5}M, and pH = 7. CuS starts to precipitate at pE = -1.5. As noticed, the chelate can easily compete with the sulfides down to pE ≈ -3 (the shaded area) which is far below the redoxcline at a pE ≈ 4. For further information see text.

figure. In a dynamic system where the diagenetic flux can transport the chelated products upward, we can thus expect the relative abundances of metals in the uppermost part of the sediment to be determined by their respective ability to be chelated. Under such circumstances, the stability constants for the metal organic compounds need not be especially low. Therefore, the chelating process should be active in the pE range common to surface sediments like those of the Baltic Sea (Bågander and Niemistö, 1978). That kind of environment would also have the highest bioactivity in the sediments. Meiofauna and microorganisms are known to take an active part in metal uptake from their environment and to enhance the transportation processes in the sediment.

Cu/Zn Ratios in Sediments

In a simulated sedimentary system (Hallberg et al., *in press*), experimental evidence was obtained for the involvement of organic matter in the distribution of trace metals in sediments by examining the mechanisms of incorporation of Cu and Zn at the redoxcline of a sulfide-rich sediment. The presence of sulfide and the absence of organic compounds created a common precipitation mechanism for Cu and Zn—presumably as CuS and ZnS. Thus, when a sulfide-bearing brine came in contact with a solution of Cu and Zn, the metals were precipitated and fixed in a common sedimentary layer (figure 17–4). A linear correlation between the concentrations of Cu and Zn give $R = 0.57$, which is significant at the 1 percent level. When organic compounds were present, and especially when they were metabolized with sulfate by sulfate-reducing bacteria to produce H_2S, the precipitation and

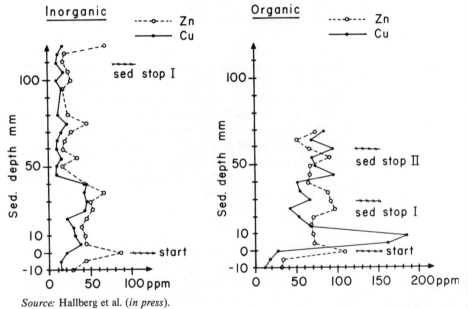

Source: Hallberg et al. (*in press*).

Figure 17–4. Precipitation of Cu and Zn in two experimental systems with continuous sedimentation. In the inorganic system, sulfide was contributed as a sulfide solution in the absence of organic matter. In the organic system, sulfide was produced by sulfate-reducing bacteria during their degradation of added organic matter. Note the correlation of Zn and Cu precipitation for the inorganic system ($R = 0.57$) in contrast to that for the organic system ($R = 0.13$).

fixation mechanism for metals was more complex. Figure 17–4 shows that under such conditions Cu and Zn are separated and fixed in different sedimentary layers.

Thus the Cu/Zn ratio in a sediment should reflect redox conditions of the ancient environment. The ratio will increase when reducing conditions prevail in the entire sedimentary column, and it will decrease when oxidizing conditions exist in the topmost part of the sediment and bottom water. In the oxidizing case, Cu is not fixed as a sulfide but is transported upward. Theoretically, the Cu/Zn ratio is shown to be a representative environmental indicator (Hallberg, 1972). In one of his geochemical profiles from the Baltic, Manheim (1961) has demonstrated how Cu and Zn behave

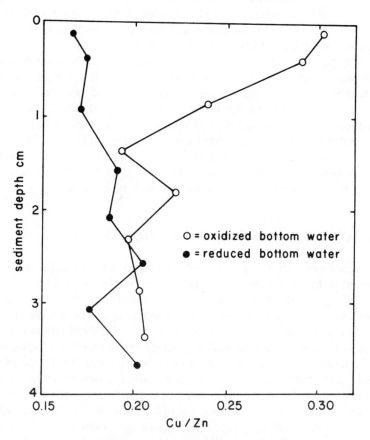

Source: Schippel et al., *OIKOS Supplentum* 15:64–67, 1973. Reprinted with permission.

Figure 17–5. Cu/Zn ratio of a sediment before (○) and after (●) a redox turnover (analysis taken from a nearshore Baltic sediment).

in a reduced basin. The Cu/Zn ratio markedly increases above unity in the intensely reduced sediments at great water depths.

On the continental shelf in and around Walvis Bay, South West Africa, there is an increase in the Cu/Zn ratio toward the deeper and more H_2S-rich regions (Calvert and Price, 1970). Unfortunately, no redox values are given for the sediment samples. Modern Baltic sediments have been investigated in situ where the transformations of an oxidizing to a reducing environment occurred under controlled conditions in a closed system (Hallberg et al., 1972). The Cu/Zn ratio increased in the topmost centimeters above the redoxcline of the sediment during oxidizing conditions in the bottom water (figure 17-5). After reducing conditions had been established in the system, copper was leached from the sediment and gave rise to a vertical Cu/Zn ratio profile with no significant variation in the uppermost part of the sediment (figure 17-5). However, in this case, the leached Cu was not precipitated as copper sulfide, probably because the closed-system technique did not permit sufficient quantities of hydrogen sulfide to be produced.

Molybdenum, like copper, is significantly more abundant in intensely reduced environments compared to other metals. It has been found to be concentrated only in the deeper reduced basins of fjords (Gross, 1967). Gross did not find any correlation between Mo and total carbon in the sediment, but he did find a good correlation with the redox potential. In contrast, Curtis (1966) suggested a good correlation between Mo and total carbon. Since the deeper parts of a basin generally contain more organic matter than the rest of the basin, the deeper parts will be reduced first; therefore, the correlation just mentioned is to be expected.

The (Cu + Mo)/Zn Ratio as a Paleoecological Indicator

In accordance with the preceding discussion, both the Cu/Zn ratio and the content of Mo in sediments are likely indicators for ancient conditions. The ratio (Cu + Mo)/Zn is consequently proposed as an even better indicator than the Cu/Zn ratio for the redox conditions in the ancient uppermost sedimentary layer. In times past, the (Cu + Mo)/Zn ratio should have been high when conditions were reducing and low when conditions were oxidizing.

The ratio $R = $ (Cu + Mo)/Zn for a sediment core of the Eastern Gotland Basin of the Baltic is presented in figure 17-6. The uppermost 130 mm of the core coresponds approximately to the twentieth century. The graph has some peaks in this part of the core worthy of further examination. The individual peaks are indicated by years according to age determinations using ^{210}Pb measurements (Niemistö and Viopio, 1974). The thickness of the sediment slices corresponded to a period of about 2 years, except for the uppermost part of the core where each slice coresponded to a shorter period (< 1 year).

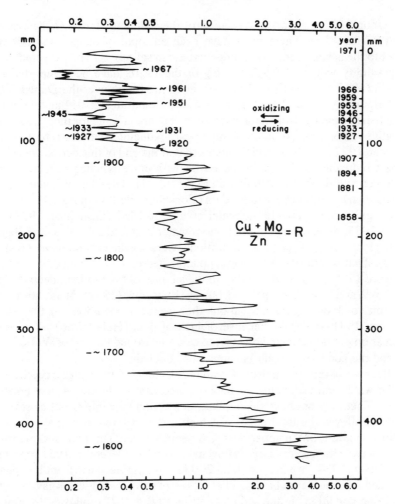

Source: L.E. Bågander, A.G. Engvall, M. Lindström, S. Odén, and F.A. Schippel. *Contributions from the Askö Lab.* 2:106, 1973. Reprinted with permission.

Figure 17-6. Variation of $R = (Cu + Mo)/Zn$ in a sediment core from the Baltic. Oxygen concentration observed in the bottom water (Fonselius, 1969) coincides with peaks in the R values as indicated by years written beside the profile in the upper-most part of the graph. The mean values for the ^{210}Pb dating are given in the right margin.

It must be noted that the analyses gave integrated average values for redox conditions of time periods during which the redox conditions may have changed considerably.

Around 1920, a marked oxidation of the sediment started (R decreases from 0.6 to 0.25). There must have been an input of oxygenated bottom water to the basin. A corresponding increase in salinity is to be expected, and high salinity values at the beginning of the 1920s have been reported by Soskin (1963). The inflow of new water to the Baltic over the Danish sills started some years earlier (Fonselius and Rattanasen, 1970) but cannot be established from lightship data since such data are missing for those years. A negative redox trend (from oxidizing toward reducing conditions) took place between 1927 and 1931, which means that at the end of this period we could expect reducing conditions in the basin, with corresponding occurrence of hydrogen sulfide in the bottom water ($R = 0.55$). The idea of a stagnation period then is verified by salinity data (Fonselius and Rattanasen, 1970) and by the existence of H_2S in the water column in 1931 (Granquist, 1932).

In 1933, an exchange of the bottom water in the Eastern Gotland Basin took place, and oxygenated conditions were again predominant (Kalle, 1943). That is indicated by a corresponding decrease in R from 0.55 to 0.25 in figure 17–6. A general stability in the salinity of the Gotland deep during World War II has been pointed out by Fonselius (1969). It is, however, difficult to draw conclusions about the chemical conditions in the basin between 1941 and 1947 because of lack of data (Hela, 1966). A second stagnation period seemed, however, to cease at around the end of World War II, and the conditions again became more oxidizing.

Because of the mechanism of such redox changes, the better oxygenation of the water and sediment can be explained best by inflow of new bottom water. That may not have taken place as a continuous inflow but as several smaller inflows, the total effect of which was registered as a change in the redox conditions of the uppermost sedimentary layer. The small inflows may have ended with the very large inflow in November-December 1951 reported by Wrytki (1954), when more than 200 km^3 of marine water from Kattegatt intruded into the Baltic basins. That inflow of oxygenated water gave rise to oxidizing conditions in the sediment surface ($R = 0.22$) followed by a new redox turnover during the succeeding stagnation period which lasted till about 1961 ($R = 0.55$). The salinity in the bottom water continuously decreased during this period. Oxygen measurements were carried out until the value zero was recorded at the beginning of 1958 (Fonselius and Rattanasen, 1970). Unfortunately, no data for H_2S are available for the period 1958–1961, but the smell of H_2S from bottom-water samples was noticeable during observations in 1958 (Engstorm, personal communication, in Fonselius, 1962).

The inflows of oxygenated water into the area after 1961 were of such quantities that they did not affect the redox conditions for too long. Now the conditions in the basin seem to be such that the reducing conditions are very rapidly re-established after each new inflow of more saline water. One reason may be the small size of the inflows, which means that only small amounts of oxygen would be injected. Another reason may be the increasing pollution of the Baltic, since decomposable organic matter is the driving force for the oxygen consumption and production of hydrogen sulfide.

In the twentieth century there was a fluctuation between oxidizing and reducing conditions in the Gotland Basin according to the graph in figure 17-6. The fluctuations seem to have been of a greater amplitude during the seventeenth and eighteenth centuries than during the twentieth century.

In none of the preceding cases, however, have specific metal-organic complexes been identified, and the possibility that inorganic ions or phases were involved in the equilibria cannot be discounted.

References

Bågander, L.E., Engvall, A.G., Lindström, S., Odén, S., and Schippel, F.A. 1973. Chemical microbiological dynamics of the sediment-water interface. *Contr. Askö Lab.* 2: 106.
Bågander, L.E., and Niemistö, L. 1978. An evaluation of the use of redox measurements for characterizing recent sediments. *Est. Coast. Mar. Sci.* 6: 127–134.
Berner, R.A. 1971. *Principles of Chemical Sedimentology.* New York: McGraw-Hill.
Calvert, S.E., and Price, N.B. 1970. Minor metal contents of recent organic-rich sediments of South West Africa. *Nature* 227: 593–595.
Curtis, C.D. 1966. The incorporation of soluble organic matter into sediments and its effect on trace-element assemblages. Advances in organic geochemistry. *Proc. Int. Meeting in Rueil-Malmaison 1964.* London: Pergamon, pp. 1–13.
Elderfield, H., and Hepworth, A. 1975. Diagenesis, metals and pollution in estuaries. *Marine Poll. Bull.* 6(6): 85–87.
Emerson, S. 1976. Early diagenesis in anaerobic lake sediments: Chemical equilibria in interstitial waters. *Geochim. Cosmochim. Acta* 40: 925–934.
Fonselius, S.H. 1962. Hydrography of the Baltic Deep Basins. *Fishery Board of Sweden, Series Hydrography* 13:41.
Fonselius, S.H. 1969. Hydrography of the Baltic Deep Basins. III. *Fishery Board of Sweden, Series Hydrography* 23:97.
Fonselius, S.H., and Rattanasen, Ch. 1970. On the water renewals in the Eastern Gotland Basin after World War II. *Meddelande Havsfiskelab. Lysekil* 90:11.
Gardner, L.R. 1974. Organic versus inorganic trace metal complexes in

sulfidic marine waters—Some speculative calculations based on available stability constants. *Geochim. Cosmochim. Acta* 38:1297–1302.
Granquist, G. 1932. Croisière thallassologique et observations en bateaux routiers en 1931. *Merentutkimuslait. Julk./Havsforskningsinst. Skr.* 81:24.
Gross, M.G. 1967. Concentration of minor elements in diatomaceous sediment of a stagnant fjord. In G.H. Lauff (Ed.), *Estuaries.* Washington: A.A.A.S., pp. 273–282.
Hallberg, R.O. 1972. Sedimentary sulfide mineral formation—An energy circuit system approach. *Mineral. Deposita* 7:189–201.
Hallberg, R.O., Bågander, L.E., Engvall, A.G., and Schippel, F.A. 1972. Method to study geochemistry of sediment-water interface. *Ambio* 2:71–72.
Hallberg, R.O., Bubela, B., and Ferguson, J. Metal chelation in sedimentary systems. Geomicrobiol. Jour. (in press).
Hela, I. 1966. Secular changes in the salinity of the upper waters of the northern Baltic Sea. *Comm. Phys. Mathem. Soc. Sci. Fennica* 31 (14):21.
Ingri, N., Kakolowicz, W., Sillén, L.G., and Warnqvist, B. 1967. High speed computers as a supplement to graphical methods. V. *Talanta.* 14:1261–1286.
Kalle, K. 1943. Die grosse Wasserumschichtung in Gotlandtief vom Jahre 1933–34. *Ann. hydrogr.* 71(4–6): 142.
Manheim, F.T., 1961. A geochemical profile from the Baltic Sea. *Geochim. Cosmochim. Acta* 25:52–70.
Niemistö, L., and Viopio, A. 1974. Studies on the recent sediment of the Gotland Deep. *Merentutkimuslait. Julk./Havsforskningsinst. Skr.* 238: 17–32.
Schippel, F.A., Hallberg, R.O., and Oden, S. 1973. Phosphate exchange at the sediment-water interface. *OIKOS Supplementum* 15:64–67.
Soskin, I.M. 1963. *Continuous Changes in the Hydrochemical Characteristics of the Baltic Sea.* Leningrad: Hydrometeorological Press (Ref. Fonselius and Rattanasen, 1970) (in Russian).
Wyrtki, K. 1954. Der grosse Salzeinbruch in die Ostsee im November und Dezember 1951. *Kieler Meeresforsch.* 10(1):19.

18 Depth Distributions of Copper in the Water Column and Interstitial Waters of an Alaskan Fjord

David T. Heggie and *David C. Burrell*

Abstract

The concentrations and fluxes of copper in the deep basin of an Alaskan fjord were found to be controlled by reactions in the water column and surface sediments and by exchanges across the sediment/seawater interface. Copper is removed from the deep water onto particulates and transported to the sediments where it is remobilized within approximately the upper 7 cm. Approximately 20 percent of copper removed from the water column and carried to the sediments is returned to the overlying water. Copper is removed from the interstitial waters adjacent to but below the remobilization zone, probably by precipitation as sulfides. The remobilization and removal reactions in those sediments occur in narrow zones close to the sediment/seawater interface.

Introduction

Trace metallic elements dissolved in the oceans exist at concentrations of the order of 10^{-9} mole L^{-1} Those concentrations in seawater are not generally believed to be controlled by known precipitation equilibria. Of more importance is adsorption onto and incorporation into suspended particulate matter, which eventually settles to the sea floor. The metals' ultimate fate is burial with the sediments. However, demonstrations of non-conservative behavior in the water column, the elucidation of sources and sinks (other than by examinations of solid phases), and the computation of fluxes for those elements is a difficult task because (1) the extremely low concentrations require sensitive and precise analytical techniques, few of which are available, and (2) the effects of physical transport processes must be distinguished from in situ processes such as reactions within the water column and exchanges across boundaries.

Contribution No. 346, Institute of Marine Science, University of Alaska, Fairbanks, Alaska 99701.

Some of the latter difficulties have been overcome by making measurements in deep-water environments where mixing processes are relatively well defined and appear to be confined to the vertical direction only. Craig (1974) applied a vertical advection-diffusion mixing model to the distribution of copper in the northeast Pacific Ocean (GEOSECS I intercalibration station) using the data of Spencer et al. (1970), and Spencer and Brewer (1971) applied a similar model to the vertical distribution of manganese in the Black Sea. Steady-state conditions in those models required only a single observation of the distribution in time. At continental boundaries, and particularly within estuaries, the time and length scales over which reactions and processes proceed are shorter than those in the deep sea, and steady-state conditions cannot be assumed. Therefore, for naturally occurring non-radioactive elements, demonstrations of non-conservative behavior and flux computations require spatial measurements through real time.

Measurements of copper in interstitial waters are not common, and little is known of the processes which control depth distributions or concentrations in sediment pore waters. Of the few data reported in the literature (Brooks et al., 1968; Presley et al., 1972; Duchart et al., 1973), one observation is common: concentrations in the interstitial waters are higher than in the overlying waters. Marine sediments may well be an important sink for copper and other trace metals entering the ocean, but the data suggest that interstitial waters must also be considered a potential source of copper for oceanic waters.

Precise measurements of copper have been made with a sensitive electrochemical technique (differential pulsed anodic stripping voltammetry) in both the water column of a fjord basin, Resurrection Bay (south central Alaska), and in the interstitial water to investigate the processes controlling copper concentrations and distributions in that estuary. This chapter examines the seasonal variations in copper concentrations in the water column with respect to sources and sinks and develops a mass balance between the sediments and overlying waters.

Methods and Area of Sampling

Sample Collection

Samples of seawater were collected in 1.7-L Niskin bottles and filtered immediately after collection through 0.4-μm Nuclepore filters (washed with $\sim$ 10% v/v hydrochloric acid and rinsed with deionized double-distilled water) into 250-ml pre-cleaned polyethylene bottles (soaked 24 hours in $\approx$ 30% v/v hydrochloric acid, washed with $\sim$ 10% v/v acid, and then rinsed with deionized double-distilled water). The filtered samples were acidified to pH 2.0 to 3.0 with 0.5 ml of ~ 4 N HCl (addition of acid contributed < 0.02 μg L^{-1} to the copper concentration) and stored frozen prior to analysis.

The sediment cores were collected with a Benthos gravity corer, and 5- to 10-cm segments were extruded directly into Reeburgh sediment squeezers (Reeburgh, 1967). No significant differences in copper concentrations were detected between a core loaded into the squeezers in a glove bag under a nitrogen atmosphere and those loaded directly on deck. Several milliliters of interstitial water were discarded before samples were collected. The interstitial water obtained was passed successively through two sets of 0.4-μm Nuclepore filters (prewashed in ~ 10% v/v HCl and deionized double-distilled water) directly into a few drops of ~ 10% v/v hydrochloric acid. The final pH of those samples was ~ 2.0.

Measurements

Water column and interstitial water samples were analyzed by differential pulsed anodic stripping voltammetry, using a PAR model 174 polarographic analyzer (Heggie, 1977). Copper was concentrated onto a mercury-coated, rotating, glassy carbon electrode for a fixed period of time. A suitable voltage range was scanned in an anodic (oxidizing) direction and the peak current, which resulted from the oxidation of the metal at the electrode surface, was recorded. Standard operating conditions (except for plate time, which is variable depending on sample concentration) are shown in figure 18–1. The peak current is proportional to the concentration of copper in the bulk of the solution, which is determined by the method of standard additions (figure 18–1a). Because the differential pulsed technique discriminates between background (capacitative) and desired signal (faradaic) currents, for equal deposition times it provides two-fold to five-fold greater sensitivities than linear dc anodic stripping methods (figure 18–1b). Including shipboard sampling, the overall precision of the differential pulsed technique is about 10 percent (table 18–1).

Resurrection Bay Hydrography

In order to examine the copper data it is necessary to understand something of the seasonal water fluxes through this fjord. Resurrection Bay (figure 18–2) opens onto the Gulf of Alaska and is divided into two basins (250 m outer, 290 m inner) by a sill at about 185 m. During oceanographic summer (June through October), the deep waters (> 150 m) of the fjord are replaced each year by more dense water that is advected up onto the adjacent continental shelf and subsequently penetrates the inner reaches of the fjord. That water

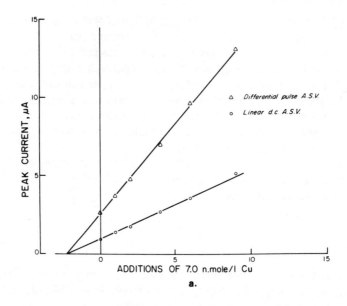

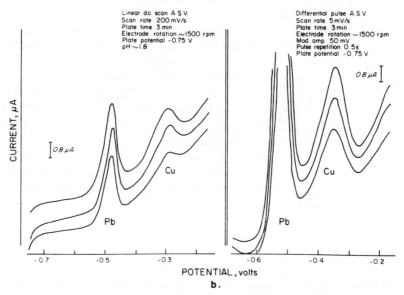

Figure 18–1. (a) Linear response of anodic stripping techniques to standard additions of copper. (b) Comparison of differential pulsed anodic stripping with linear dc stripping.

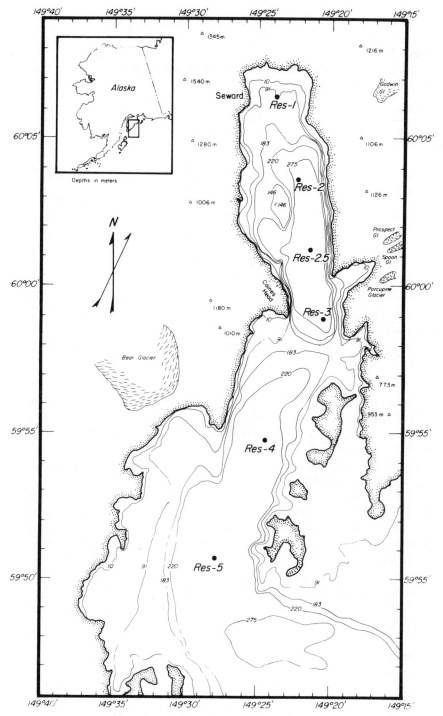

Figure 18-2. Resurrection Bay showing station locations and outer and inner basin bathymetry.

Table 18–1
Analyses of Precision of Copper Determinations by Differential Pulsed Anodic Stripping Voltammetry

Individual Measurement[a]		Within Niskin Bottle[b]		Between Niskin Bottle[c]	
Sample	Cu (nmoles/L)	Sample	Cu (nmoles/L)	Sample	Cu (nmoles/L)
1	9.60	S_1	4.25	A_4	9.61
2	8.98	S_2	3.78	B_4	11.81
3	8.82	S_3	4.41	C_4	11.02
4	9.45	S_4	4.25	D_4	10.08
5	9.76	S_5	3.78	F_4	12.13
6	9.13	—	—	N_{16}	11.65
	$\bar{x} = 9.29$		$\bar{x} = 4.09$		$\bar{x} = 11.02$
	$\sigma = 0.38$		$\sigma = 0.30$		$\sigma = 1.00$
Percent coefficient of variation = 4.0%		Percent coefficient of variation = 7.3%		Percent coefficient of variation = 9.2%	

[a] Samples from 250 ml aged seawater pH ≈ 2.2.
[b] Samples drawn from the same Niskin bottle.
[c] Single sample drawn from the same Niskin bottle lowered to the same depth in the "isolated" basin over a 14-hr period.

remains behind the sill of the inner basin during the ensuing oceanographic winter months (November through April) to exchange, predominately *via* turbulent mixing, with the less dense water found above sill depth. During the study period, vertical eddy mixing coefficients of 3 to 5 cm^2/s were calculated for water near the sill depth of the central region of the inner basin, and volumes of water between 10 and 90 percent of the inner basin volume were exchanged by advection of more dense water across the sill (Heggie et al., 1977). Because of the sluggish exchange below the sill during the winter, there is a seven-month period during which the effects of reactions within the deep water and exchanges across the boundaries of the deep basin are relatively well preserved in the water column. Station Res-2.5 (figure 18–2) was chosen as a representative of the inner basin, and other inner-basin stations, although sampled in less detail, confirmed our choice. Station Res-4 represented the outer basin and documented variations in the source waters of the fjord.

Results

Water Column

Concentrations of copper varied between 0.15 and 3.13 $\mu g\ L^{-1}$. From about June (1974), the deep water of the inner basin had been undergoing

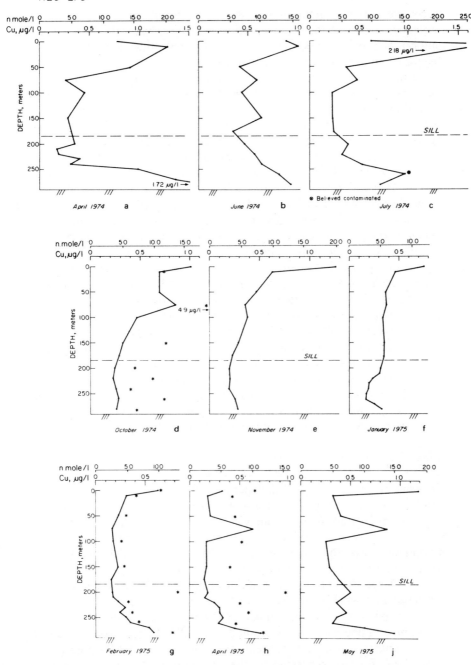

Figure 18-3. Vertical profiles of copper at Station Res-2.5 of the inner basin (April 1974–May 1975). Asterisks (*) refer to copper concentrations of unfiltered acidified samples.

successive advective replacements of water across the sill (figure 18–3b through d), and by October concentrations of copper below the sill were very uniform at ~ 0.26 µg L^{-1} (figure 18–3d), reflecting concentrations in the fjord source water around the sill depth. During the succeeding months (November through April), concentrations generally decreased in the surface and intermediate waters (figure 18–3e through h), but increased toward the basin floor until the concentration at 280 m was 0.74 µg L^{-1} by April of 1975. The increases at Res-2.5 in the intermediate waters and just below sill depth (~ 100 to ~ 220 m) during January can be explained by the advection of a core of water from beyond the sill into the inner reaches of the fjord. The advected water had copper concentrations between about 0.24 and 0.68 µg L^{-1}. That water was not dense enough to penetrate below the sill depth, but dissipation of the core by lateral and vertical mixing resulted in the slight increases in copper observed. An advective influx into the inner basin, sufficient to replace about 90 percent of the resident deep basin water below the sill, occurred sometime between April and May of 1975. The concentrations of the incoming source water varied between 0.40 and 0.56 µg L^{-1}, which explain the elevated concentrations observed in the inner basin during May (figure 18–3j). A similar profile was observed during the previous spring (April 1974, figure 18–3a); the gradient directed into the sediments is strongly developed. During oceanographic summer (June through October), the deep water of the inner basin was subject to successive advective influxes across the sill. The deep and bottom waters were displaced upward in the water column, while freshwater input at the surface and particulate fluxes (USGS, unpublished data) through the water column were near their seasonal maxima. During this period, decreases in the copper content were generally observed throughout the intermediate and deep waters.

Interstitial Waters

Concentrations of copper varied between 9.98 and 1.02 µg L^{-1}, and a persistent type of profile (figure 18–4) was measured in six cores taken over a twelve-month period. A maximum was always found at the sediment surface (0 to ~ 7 cm), and the concentrations here were about one order of magnitude greater than those in the overlying waters at 280 m. However, soluble contents decreased sharply to the 2 µg L^{-1} level between about 7 and 20 cm. Deeper within the sediment, down to about 120 cm, copper levels increased somewhat to between 3 and 4 µg L^{-1}. The sediments below the surface-maximum zones were anoxic. Addition of dilute hydrochloric acid to raw sediments liberated hydrogen sulfide from all but the top few centimeters of the sediments. The high concentrations of iron, ~ 20,000 ppm (in the "reducible" fraction generated by the procedure of Chester and Hughes, 1967), on the sediments and the apparent absence of free hydrogen sulfide

Depth Distributions of Copper

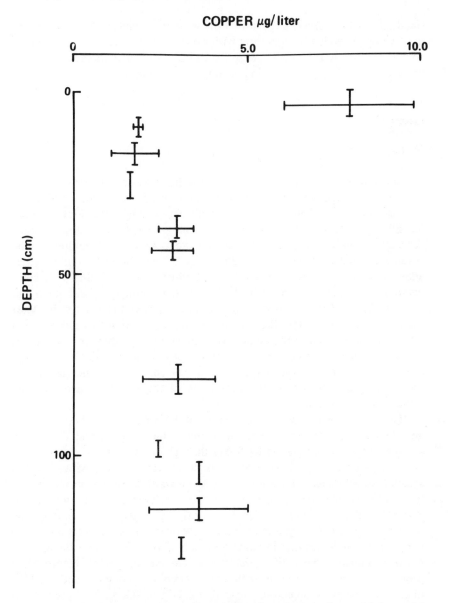

Figure 18–4. Composite vertical profile of copper in interstitial waters. Data from six cores collected at Res-2.5 between July 1974 and May 1975. Horizontal bars represent one standard deviation of measured concentrations.

(method of Cline, 1969) suggest that the interstitial water sulfide concentrations were controlled by equilibria with iron minerals. The concentrations of copper ($\sim 10^{-8}$M), however, were far in excess of their theoretical values had they been controlled by simple sulfide precipitation.

Discussion

Water Column

A copper budget for the deep water of the inner basin was determined by comparing changes in copper in the basin with advective and/or diffusive contributions. The changes in copper were obtained by integration between the sill depth and the basin floor over the time period between observations. Advective contributions to the budget were obtained by multiplying the volume of water exchanged by the difference between the copper in the water added to and displaced from the basin. Diffusive contributions were computed as the product of the vertical eddy mixing coefficient at sill depth and the vertical gradient in copper concentration across the sill depth. After accounting for the latter, the remainders represent losses or gains which result from reactions in the column (nonconservative behavior), exchanges across the sediment boundary, or combinations of those. The results of that exercise are listed in table 18-2. The ranges cited reflect uncertainties in analytical precision and in the evaluation of the advective and diffusive terms.

The data show net losses of copper from the water column between June and January, i.e., summer and midwinter, which are highest during the summer when particulate matter fluxes through the fjord are at a maximum. Net increases in copper were found between January and May, and the seasonal development of a gradient directed toward the sediments indicates a flux from the surface sediments into the overlying fjord bottom waters. Assuming that the increases in the overlying water originate from the sediment surface, a simple mass balance between the sediments and overlying waters indicates that it would take only about 10 days for the upper 7 cm of sediment to be entirely depleted of interstitial copper. Hence some process must be operating to maintain the surface maximum and the high interstitial copper concentrations (~ 8 μg L^{-1} in the surface zone. High concentrations of copper (~ 20 ppm)—about three orders of magnitude greater than the adjacent interstitial waters—were found on the surface sediments (method of Chester and Hughes, 1967). It is proposed that copper is remobilized within the surface sediments from solid phases.

The supply of copper to the sediments is maintained by particles raining through the water column. Several samples of seawater were acidified but not

Table 18-2
Computed Rates of Change of Copper Concentrations in the Water Column below Sill Depth at Station Res-2.5 of Resurrection Bay (April 1974–May 1975)

Observation	Rate of Change of Copper Concentration $\mu g\ Cu\ L^{-1}\ yr^{-1}$
21 April 1974	
28 June 1974	+ 0.19 (0.12–0.26)[a]
25 July 1974	− 0.99 (2.24–0.73)[b]
18 October 1974	Unaccountable advective replacements of deep basin water
26 November 1974	− 0.41 (0.31–0.55)[b]
21 January 1975	− 0.26 (0.15–0.36)[b]
27 February 1975	+ 0.91 (0.99–0.84)[a]
9 April 1975	+ 0.33 (0.24−0.42)[a]
13 May 1975	+ 0.39 (0.03–0.75)[a]

[a]Mean of net increases 0.46 $\mu g\ Cu\ L^{-1}\ yr^{-1}$
[b]Mean of net losses 0.55 $\mu g\ Cu\ L^{-1}\ yr^{-1}$

filtered, and all had higher concentrations of copper than acidified *and* filtered samples collected from the same depths (figure 18–3). The nature of that filterable material has not been determined; neither is it known how copper is associated with it (e.g., adsorbed onto surfaces or incorporated into the structure). However, the data do indicate that copper is associated with particulate material which must eventually reach the sediment surface. That material may be responsible for the observed losses of copper from the water column during the summer months (table 18–2).

Sediments and Interstitial Waters

The decreased concentrations of copper in interstitial waters below approximately 10 cm can be explained only by removal following burial. The removal reaction is probably precipitation as copper sulfide ($pK = 35.2$), but concentrations have probably been modified by reactions with "humic" material (Rashid and Leonard, 1973). It is not necessary to invoke any further removal reaction below about 20 cm, and deeper concentrations can be explained by simple burial alone (or burial with some secondary remobilization from a solid phase) if the apparent, albeit slight, gradient from 20 to 120 cm in the sediments is real. The interstitial water data indicate at least two processes operate within narrow zones in the surface sediments to control the interstitial water copper concentrations: remobilization at the sediment surface (0 to $\sim$ 7 cm), and removal over the 7- through $\sim$ 20-cm depth zone.

Berner (1971, 1974) has presented general kinetic models of dissolved chemical species in interstitial waters. However, the discontinuous nature of the copper profiles presented here cannot be adequately described by such models. The rate of removal of interstitital copper has been computed from a steady-state equation which considers only diffusion from the sediment surface down into the removal zone and from deeper within the sediment up into the removal zone. The advective effects of burial, compaction, and extrusion of interstitial water were small and therefore neglected. The flux into the removal zone was dominated by that from the sediment surface, and for a diffusion coefficient of 3×10^{-6} cm^2/s the flux is 0.08 μg/(cm$^2 \cdot$ yr) of Cu. If that supply is removed by precipitation occurring mainly within a narrow 20-cm zone, then the estimated downward flux of copper is equivalent to a removal rate of 4 μg/(L$\cdot$yr) of Cu.

For a steady-state interstitial copper profile, the rate of remobilization in the surface segment of sediments must balance the sum of all the apparent removal processes—i.e., that which is lost by precipitation as sulfide in the anoxic zone below, plus that which is returned to the overlying water. If the remobilization is assumed to occur in a narrow 7-cm zone at the sediment surface and all the flux into the overlying water is assumed to originate from that zone, the remobilization rate is 668 Cu L^{-1} yr^{-1}, 98 percent of which is accounted for by loss to the overlying water. The calculation implies that copper is being returned to the overlying water almost as fast as it is being remobilized. The definition of the extent of the remobilization horizon is limited by the sampling interval used, and the computations of the remobilization and removal rates are sensitive to the depth intervals over which the reactions are assumed to occur.

The Copper Sink

For 7 months of the year, copper was removed from the water column of the inner basin below sill depth at a mean net rate of 0.55 μg Cu L^{-1} yr^{-1} (table 18-2). Application of that rate to the total 300-m water column yielded an integrated net annual removal of 9.6 μg of Cu per cm^2 of water. Over the remaining five months, the deep-basin copper budget was dominated by the return of copper to the overlying water at a mean net rate of 0.46 μg Cu L^{-1} yr^{-1} (table 18-2). Assuming that all the increase in the water column emanated from the sediments, the integrated net annual return of copper across 1 cm^2 of the sediment/seawater interface was calculated to be 1.9 μg. If that figure is compared to the amount of copper removed annually from the water column (9.6 μg), then approximately 20 percent of the copper removed from the water column is returned from the sediments to the overlying water.

The calculations depend on the concentrations and gradients observed in interstitial and overlying waters. Subsequent to the remobilization of copper but prior to its burial and removal reactions within the surface sediments could modify the interstitial water copper concentrations, so that the effects of remobilization would not be fully reflected. Any such reactions that tend to remove interstitial copper may therefore act as a control on the flux to the overlying water.

Manganic oxide phases are found in high concentrations in surface oxidized sediments and are known to be effective scavengers of transition metals in natural waters (Goldberg, 1954; Krauskopf, 1956; Jenne, 1968). The diagenetic remobilization of manganese by reduction of the oxides at depth in sediments results in the enrichment of pore waters with Mn(II), which can migrate to the sediment/seawater interface and undergo reoxidation (Lynn and Bonatti, 1965; Calvert and Price, 1972). As long as potentially reducible manganese in the sediment is not exhausted, a mechanism exists for continuously providing fresh surfaces of oxidized manganese at the sediment/seawater interface. It is therefore possible that the scavenging of remobilized copper by precipitating manganese oxides could act as a control on the flux of copper to bottom waters of the fjord. Maximum concentrations of extractable manganese (4,200 ppm) were found in the upper 0.5 cm of sediments of Resurrection Bay. Extractable Mn decreased to about 350 ppm at 10 cm of depth—a distribution best explained by diagenetic remobilization of manganese and its reprecipitation near the sediment/seawater interface. Maximum concentrations of copper also would be expected on the surfaces of grains in the sediment if the capture of remobilized copper by those oxides were effective. However that was not found to be the case here. Concentrations of extractable copper on sediments were uniform with depth ($\sim$ 20 ppm), implying that any such scavenging was unimportant. Copper has been shown to be strongly associated with organic matter in surface sediments (Nissenbaum and Swaine, 1976), and that association may inhibit the affinity of copper for the manganese oxides. However, the apparent absence of a copper maximum on the sediments may also be misleading, since the effects of the acid reducing agent used (Chester and Hughes, 1967) on sulfide phases and organic matter in these sediments may promote reactions which, together with the possibility of some oxidation of the sediments prior to analysis, may obsure any copper maximum. The importance of a potential scavenging effect by manganese oxides therefore remains unresolved at present.

Sediment Reaction Zones

Copper carried to the sediment surface by particulate material was found to be remobilized in the surface sediments and removed deeper within the sediments to a different solid phase. The different locations of the remobilization horizon, in environments of different sedimentation rates and fluxes of particulate biogenic matter, ought to give some indication of the nature of the particles that are responsible for transport to the sediment surface (i.e., primarily whether they are biogenic or inorganic). Moreover, those locations ought to provide some test of whether or not the remobilization and removal reactions are confined to narrow zones in the sediments or occur over extended sediment depths.

The chemical environment of a sediment changes largely as a result of the oxidation of organic matter such that the oxidation intensity (pE) decreases with depth (figure 18–5a through c). Within that redox gradient, there is a reaction sequence for many elements (Stumm and Morgan, 1970; Breck, 1974), as schematically illustrated in figure 18–5b. In deep oceanic environments, where sedimentation rates and organic loads are both relatively low, the gradient in redox intensity generally would be expected to be slight, and oxidation-reduction reactions of the types shown in figure 18–5b are probably separated by relatively large vertical distances. That environment may be contrasted with continental borderland and estuarine sediments which have relatively high sedimentation rates and organic loads. Here the redox-intensity gradient would be greater, with the same reaction sequence compressed into a relatively thin surface sediment layer. Those controls on the redox gradient by sedimentation rate and organic load are illustrated in figures 18–5a and c.

Because of its widespread association with copper in the sea, the geochemistry of manganese in different sediments with respect to the preceding is of interest. Deep-sea oceanic sediments are generally oxidizing, and manganese oxides are stable to extended depths (Lynn and Bonatti, 1965); but on continental borderlands where hydrogen sulfide may be found at shallow depths, they are unstable and reduced to Mn(II). Sediment and interstitial water profiles of manganese in those types of environments have been discussed by Calvert and Price (1972), who found maximum interstitial manganese concentrations in the very surface of sediments in a Scottish loch. In contrast, Li et al. (1969) found maximum interstitial manganese concentrations at about 130 cm of depth in the Arctic Ocean. Copper in the sea may be associated with mixed oxide phases of manganese and iron, e.g., see Goldberg (1954), and/or with biogenic particulate matter, e.g., see Martin (1970), Lowman et al. (1971), and Martin and Knauer (1973). Within the context of sediment redox reactions at different pE levels, the remobilization and removal reactions in sediments reported in this chapter should occur at

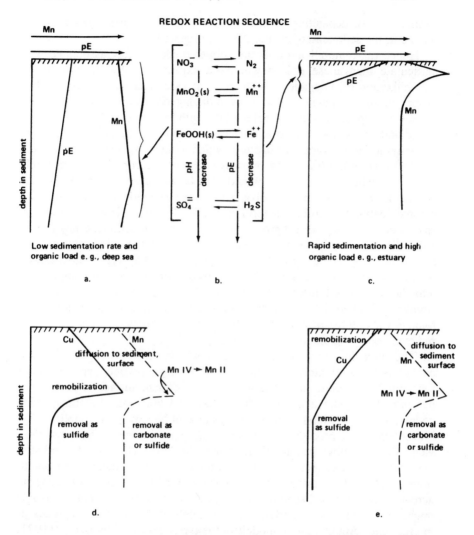

Figure 18–5. Schematic representations of (a) redox gradient and interstitial manganese profiles for an environment of low sedimentation rate and organic load; (b) the sequence of redox reactions according to redox intensity; (c) redox gradient and interstitial water manganese profiles for an environment of high sedimentation rate and organic load; (d) interstitial water copper profile associated with the reduction of manganese oxides; and (e) interstitial water copper profile associated with the degradation of biogenic matter.

various depths, depending on the particle type transporting copper to the sediment, the sedimentation rate, and the organic load.

Hence the remobilization of the copper predominately associated with particulate manganese oxides would be expected to follow closely the remobilization of manganese, with removal as copper sulfides proceeding deeper in the sediment under more reducing conditions (figure 18–5d). However, copper associated principally with oxidizable biogenic matter would undergo remobilization in the surface sediments (figure 18–5e). There the activities and numbers of bacteria which oxidize organic matter are at a maximum (Kriss, 1963). The remobilization of the copper predominately associated with siliceous tests would also be expected within the surface sediments, where the tests dissolve in response to removal of the protective organic coating by aerobic bacterial and animal activity. However, removal as copper sulfide would still proceed deeper within the sediment, beyond the redox discontinuity in the anoxic region. Where sedimentation rates and the particulate biogenic flux are relatively high, profiles of the types illustrated by figures 18–5d and e would be indistinguishable because of their compression into the surface sediments. Resurrection Bay appears to be that type of situation. Conversely, where sedimentation rates and biogenic loads are relatively low, those profiles should be extended, with the remobilization and removal zones separated in the sediments.

The preceding are simplistic developments based on the supposition that copper is predominately associated with a single particle type. Because suspended particulate matter falling to the sediment surface is probably a mixture of biogenic and inorganic matter and both may be transporting copper, actual interstitial copper profiles cannot be viewed so simplistically. Nevertheless, the preceding models provide some framework from which to view copper profiles in interstitial waters. More closely spaced interstitial measurements of copper, together with indicators of sediment redox conditions (nitrate, manganese, iron, and sulfate) and indicators of bacterial activity across a broad spectrum of sedimentation rates and organic loading might clarify the location and nature of remobilization and removal processes in sediments. Application of models of the types described by Berner (1971, 1974) should then provide a useful test for zonation of reactions. If remobilization and removal reactions do occur in separate narrow zones, then the interstitial profiles in the layers of sediment between the zones should closely fit an advection-diffusion profile with no reaction terms involved.

Summary

The measurements of copper in the overlying water column and interstitial waters have illustrated several processes controlling the seasonal concentration variations and fluxes of copper through this fjord:

1. Copper is removed from the water column and transported to the sediments by particulate matter.
2. Copper is remobilized in the upper 7 cm of sediment.
3. Approximately 20 percent of the copper removed from the water column and transported to the sediments is returned, subsequent to remobilization, across the sediment/seawater interface to the overlying water.
4. Copper is removed from interstitial waters, below 7 cm depth in the anoxic region of sediments, probably by precipitation as copper sulfide.

The remobilization and removal processes appear to take place in narrow zones within the surface sediments. It is proposed that capture of copper, subsequent to remobilization, by precipitating manganese oxides at the sediment surface could act as a potential control on the concentration of copper in the surface interstitial waters and hence on the flux into the overlying waters. That mechanism did not, however, appear important in this particular fjord, probably because of an association of copper with organic matter. Such a control on the flux of remobilized metal across the sediment/seawater interface may be quantitatively more important in the deep sea because of the relative absence of organic matter in pelagic sediments. The remobilization of copper associated with biogenic material would always be expected at the sediment surface, whereas that associated with manganese and iron oxides should be at various sediment depths depending on the gradient in redox intensity.

Acknowledgments

This work was supported in part by the U.S. Department of Energy under grant number E(45-1)-2229. The assistance of the crew of the R/V *Acona* and E.R. Dieter and D.W. Boisseau during field work is acknowledged.

References

Berner, R.A. 1971. *Principles of Chemical Sedimentology*. New York: McGraw-Hill.

Berner, R.A. 1974. Kinetic models for the early diagenesis of nitrogen, sulfur, phosphorus and silicon in anoxic marine sediments. In E.D. Goldberg (Ed.), *The Sea*, Vol. 5: *Marine Chemistry*. New York: Wiley-Interscience, pp. 427–449.

Brooks, R.R., Presley, B.J., and Kaplan, I.R. 1968. Trace elements in the interstitial waters of marine sediments. *Geochim. Cosmochim. Acta* 32:397–414.

Breck, W.G. 1974. Redox levels in the sea. In E.D. Goldberg (Ed.), *The Sea*, Vol. 5: *Marine Chemistry*. New York: Wiley-Interscience, pp. 153–179.

Calvert, S.E., and Price, N.B. 1972. Diffusion and reaction profiles of dissolved manganese in the pore waters of marine sediments. *Earth Planet. Sci. Lett.* 16:245–249.

Chester, R., and Hughes, M.J. 1967. A chemical technique for the separation of ferro-manganese minerals, carbonate minerals and adsorbed trace elements from pelagic sediments. *Chem. Geol.* 2:249–262.

Cline, J.D. 1969. Spectrophotometric determination of hydrogen sulfide in natural waters. *Limnol. Oceanogr.* 14:454–458.

Craig, H. 1974. A scavenging model for trace elements in the deep sea. *Earth Planet. Sci. Lett.* 23:149–159.

Duchart, P.S., Calvert, S.E., and Price, N.B. 1973. Distribution of trace metals in the pore waters of shallow water marine sediments. *Limnol. Oceanogr.* 18:605–610.

Goldberg, E.D. 1954. Marine Geochemistry. I. Chemical scavengers of the sea. *J. Geology* 62:249–265.

Heggie, D.T. 1977. Copper in the sea: A physical-chemical study of reservoirs fluxes and pathways in an Alaskan fjord. Ph.D. dissertation, Institute of Marine Science, University of Alaska, Fairbanks.

Heggie, D.T., Boisseau, D.W., and Burrell, D.C. 1977. Hydrography, nutrient chemistry and primary productivity of Resurrection Bay, Alaska. Report R77-2, Institute of Marine Science, University of Alaska, Fairbanks.

Jenne, E.A. 1968. Controls on Mn, Fe, Co, Ni, Cu and Zn concentrations in soils and water. The significant role of hydrous Mn and Fe oxides. In R.F. Gould (Ed.), *Trace Inorganics in Water*. Advances in Chemistry Series 73. Amer. Chem. Soc., Washington, D.C.

Krauskopf, K.B. 1956. Factors controlling the concentration of thirteen rare metals in seawater. *Geochim. Cosmochim. Acta* 9:1–32.

Kriss, A.E. 1963. *Marine Microbiology* (Deep Sea). London: Oliver and Boyd.

Li, Y.H., Bischoff, J., and Mathieu, G. 1969. The migration of manganese in the Arctic Basin sediment. *Earth Planet. Sci. Lett.* 7:265–270.

Lowman, F. G., Rice, T.R., and Richards, A. 1971. Accumulation and redistribution of radionuclides by marine organisms. In *Radioactivity in the Marine Environment*. Washington: National Academy of Sciences.

Lynn, D.C., and Bonatti, E. 1965. Mobility of manganese in the diagenesis of deep-sea sediments. *Mar. Geol.* 3:457–474.

Martin, J.H. 1970. The possible transport of trace metals via moulted copepod exoskeletons. *Limnol. Oceanogr.* 15:756–761.

Martin, J.H., and Knauer, G.A. 1973. The elemental composition of plankton. *Geochim. Cosmochim. Acta* 37:1639–1653.

Nissenbaum, A., and Swaine, D.G., 1976. Organic matter metal interactions in recent sediments: the role of humic substances. *Geochim. Cosmochim. Acta* 40:809–816.

Presley, B.J., Kolodny, T., Nissenbaum, A., and Kaplan, I.R. 1972. Early diagensis in a reducing fjord, Saanich Inlet, British Columbia. II. Trace element distribution in interstitial water and sediment. *Geochim. Cosmochim. Acta* 36:1073–1090.

Rashid, M.A., and Leonard, G.D. 1973. Modifications in the solubility and precipitation behaviour of various metals as a result of their interactions with sedimentary humic acid. *Chem. Geol.* 11:89–97.

Reeburgh, W.S. 1967. An improved interstitial water sampler. *Limnol. Oceanogr.* 12:163–165.

Spencer, D.C., and Brewer, P.G. 1971. Vertical advection diffusion and redox potentials as controls on the distribution of manganese and other trace metals dissolved in waters of the Black Sea. *J. Geophys. Res.* 76:5877–5892.

Spencer, D.W., Robertson, D.E., Turekian, K.K., and Folsom, T.R. 1970. Trace element calibrations and profiles at the GEOSECS test station in the northeast Pacific ocean. *J. Geophys. Res.* 75:7688–7696.

Stumm, W.W., and Morgan, J.J. 1970. *Aquatic Chemistry*. New York: Wiley-Interscience.

Part VII
The Influence of Seawater on Freshwater Deposits

19 Authigenic Barite in Varved Clays: Result of Marine Transgression over Freshwater Deposits and Associated Changes in Interstitial Water Chemistry

Erwin Suess

Abstract

Barite micronodules in varved clay sediments of the southern Baltic Sea are thought to have formed diagenetically when freshwater deposits were permeated by marine waters following postglacial transgression of the Atlantic Ocean. Interstitial exchange reactions between the ion complement of river clays, dominated by Ca, and the major seawater ions Na, K, and Mg result in the displacement of sorbed Ba from exchange sites and its subsequent precipitation as $BaSO_4$. That hypothesis is supported by the Ba-ion exchange behavior of suspensions of montmorillonite in waters of various salinities, as experimentally determined by Puchelt (1967). It is further supported by the nonconservative behavior of Ba during mixing, as reported by Hanor and Chan (1977) for waters from the Mississippi River and Gulf of Mexico.

Introduction

After the deposition of glacial varved clays in the wake of the receding continental ice sheet of northern Europe, inundations of saline waters covered and permeated the freshwater deposits at least twice (Sauramo, 1925, 1958; Kolp, 1966). Delay in glacial rebound of the Scandinavian Shield behind eustatic sea-level rise facilitated marine transgressions in the western and southern part of the present-day Baltic Sea. By comparison, in the central and northern parts, the postglacial deposits were either not reached at all by the marine ingressions of Atlantic Ocean waters, as in the case of varved clay deposits on land, or did not come into contact with saline waters until very recently (Sandegren, 1934; Pratje, 1948; Kullenberg, 1954; Masicka, 1963; Nilsson, 1970; Ericsson, 1972, 1973).

Thus the southern Baltic Sea is ideal for studying a number of chemical processes associated with marine transgression over freshwater clay deposits, e.g., cation-exchange reactions, diffusive mass transport, and uptake of cations by clays into structural positions. A detailed account of those processes is found in Suess (1976) along with descriptions of methods, distributions of major and minor elements in interstitial waters and sediments, and lithostratigraphic correlations of the postglacial deposits in the Bornholm Basin area of the southern Baltic.

One consequence of seawater/terrigenous clay interaction is the proposed interstitial formation of barite micronodules, which is the subject of this chapter. Pertinent methodological information is summarized below and precedes the descriptions of micronodules and the interstitial water chemistry and ion-exchange characteristics of the varved clays. Finally, Ba-exchange behavior is used in speculations on the mode of formation of the barite micronodules.

Methods

A series of Kasten cores representing a composite cross section of the Bornholm Basin area were taken for chemical analysis of sediment and interstitial water (figure 19-1). Pressure filtration of interstitial water for the study reported here (core 13438-1) was performed on board ship within 6 h of core retrieval. Determinations of concentrations of dissolved nutrients and major ions were performed by modified standard seawater methods, as outlined by Hartmann et al. (1976).

Sediment radiographs were used in locating core sections containing barite micronodules (figure 19-2). Isolation of those nodules by sieving and heavy liquid separation was then performed immediately following the cruise.

Determination of exchangeable ion complements was done on splits of squeezed sediment samples remaining after the extraction of interstitial water. A 2-g sample of moist squeezer cake was twice suspended at room temperature in 100 ml of 1 M $BaCl_2$ solution to exchange all sorbed ions for Ba^{2+} (Kretzschmar, 1972; Schlichting and Blume, 1966). Sodium and potassium concentrations were measured by flame-emission spectroscopy, and magnesium and calcium contents were determined by atomic absorption spectrophotometry.

In the process of ion exchange with $BaCl_2$ solution, the major seawater ions of the entrapped interstitial solution were removed as well. Correction for the interstitial ions was possible using the porosity of the sediments and the interstitial concentrations of dissolved ions. In this analytical approach, the clay-mineral exchange complement was not affected by shifts in the exchange equilibria associated with the usual rinsing of the interstitial electrolyte solution prior to the exchange reaction (Sayles and Mangelsdorf, 1977). The total cation-exchange capacity of the sediments from the varved clay section was then separately determined by back exchange of Ba with Mg—a standard procedure in analyses of soils (Jackson, 1956). The presence of calcite and dolomite in the samples suggested that the exchange solution of $BaCl_2$ dissolved some carbonates. Therefore, the sum of the individual exchangeable cations was probably somewhat larger than the total cation-exchange capacity derived from the Ba back exchange. An attempt was made to partition the excess between $MgCO_3$ and $CaCO_3$ sources based on the relative amounts of calcite and dolomite in the samples (Suess, 1976).

Results and Discussion

Barite Micronodules

The nodules were first observed in the coarse-grained size fraction of varved deposits underlying the modern marine sediment of the Bornholm Basin (Suess, 1976). However, nodular barite concentrations with shapes and sizes similar to those more common sediment constituents have also been discovered in other areas of the Baltic Sea (Blashchishin, 1976). The particular arrangement of barite micronodules within the varved clays of the Bornholm Basin is illustrated by the x-ray radiograph in figure 19-2a. A close-up view of one of the micronodules is given in figure 19-2b. They commonly range in size from 0.1 to 1.0 mm, have a smooth surface with circular bumps, and are composed of massive, elongated, radially oriented barite crystals, as shown in figures 19-2b and 2c. X-ray diffraction and chemical analyses established that the major constituent of the micronodules is the mineral barite (table 19-1). a significant percentage of SiO_2 was also present, yet almost no other crystalline phase besides barite was found. However, in one barite sample, a quartz grain appeared as an inclusion.

The barite micronodules were most likely formed in situ sometime after the sediment was deposited. There are three physical reasons for that conclusion: (1) the random distribution of the micronodules throughout the varved clays; (2) their large size compared to surrounding sediment grains; and (3) their smooth surfaces structures, which indicated a lack of transport.

Interstitial Water Chemistry

The conditions and mechanism for in situ formation of the nodules appear to be related to permeation of the varved clays by marine waters, causing displacement of sorbed Ba^{2+} and Ca^{2+} by Na^+, K^+, and Mg^{2+} with

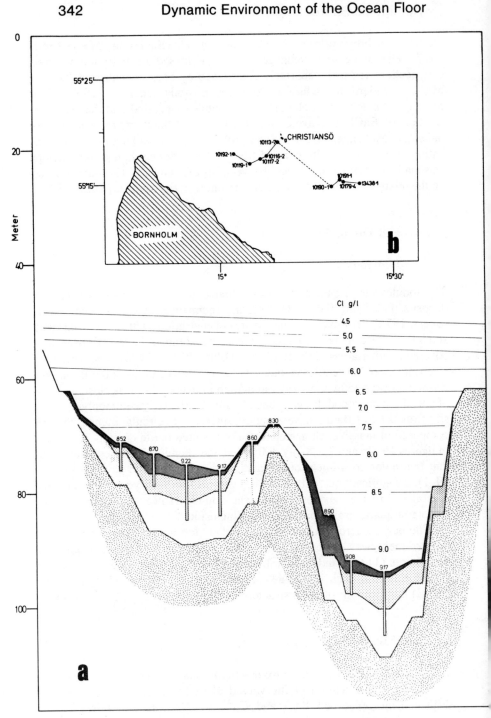

Authigenic Barite

Figure 19–1. Generalized distribution of glacial, postglacial, and recent marine sediments of the Bornholm Basin. (a) The numbers indicate typical chlorosities of interstitial water samples from core locations along the composite profile 10192-1 to 13438-1. (b) Locations 10116-2 and 13438-1 mark the sites where interstitial waters were extracted (Fig. 19–3) and barite micronodules found in the varved clay section. The cross section is based on data by Kögler (1976).

---

subsequent precipitation of $BaSO_4$. That is inferred from the composition of the interstitial waters and the exchange complement of the varved clay section of the core. In figure 19–3, the composition of present-day Baltic Sea bottom water is shown for all major constituents (Na, Mg, Ca, K, and SO_4) in relation to chloride. The scaling of the concentration axis is such that for interstitial waters of "true" Baltic Sea composition all ion/chloride ratios coincide with the chloride concentration as shown for the bottom-water sample near the sediment/water interface. That approach has the advantage of more readily showing deviations from the standard seawater composition. It should be pointed out that the SO_4/Cl, Na/Cl, K/Cl, and Mg/Cl ratios used here are the same as those for oceanic waters according to Millero (1974) and that, for the Ca/Cl ratio, the well-known positive Ca anomaly of Baltic Sea water was taken into account, i.e., $Ca/Cl = 0.0236$ instead of 0.0212 (Gripenberg, 1937; Wittig, 1941; Kremling, 1969).

The chloride concentration gradient is thought to be the result of the diffusive downward flux of seawater ions; the seven-fold increase in dissolved Ca^{2+} over that of Baltic Sea bottom water is due to desorption of Ca^{2+} from exchangeable sites on clays with simultaneous uptake of Na^+, K^+, and Mg^{2+}. A charge balance for the added and lost cations of the dissolved load appears

Table 19–1
Composition of Barite Micronodules from Bornholm Basin Sediments; Core 13428-1 between 1,040 and 1,050 cm of depth

	Sample 1 (wt. %)	Sample 2 (wt. %)
Loss on ignition, 800°C	11.4	Not ignited
SiO_2	16.1	16.3
$BaSO_4$	72.9 ± 0.9[a]	67.7 ± 0.9[b]
	100.4	84.0

[a] Total $BaSO_4$ recovered gravimetrically after HF treatment.
[b] $BaSO_4$ calculated from Ba analysis; data from Suess (1976).

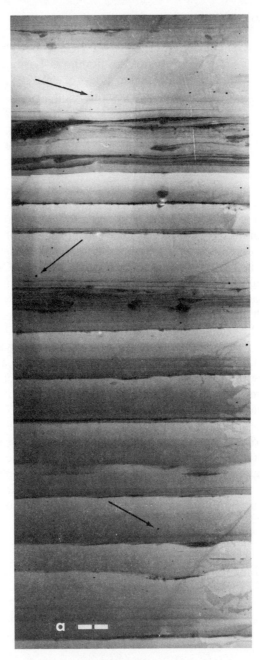

Figure 19–2. (a) Radiography of section from varved clay facies of Bornholm Basin; arrows mark three of numerous barite micronodules throughout the sediment; scale = 1 cm. (b) Barite micronodule, 1.2-mm long; scale = 0.5 mm. (c) Surface and fracture plane of micronodule with massive, elongated, and radially oriented barite crystallites; scale = 5 μm.

Authigenic Barite 345

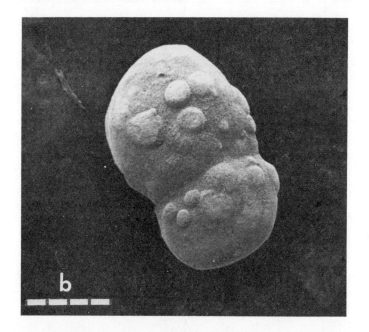

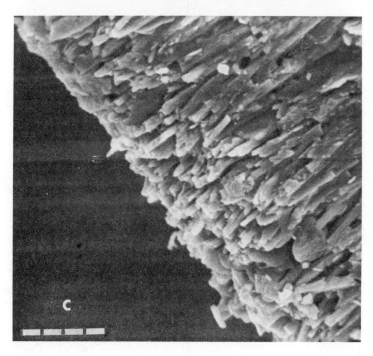

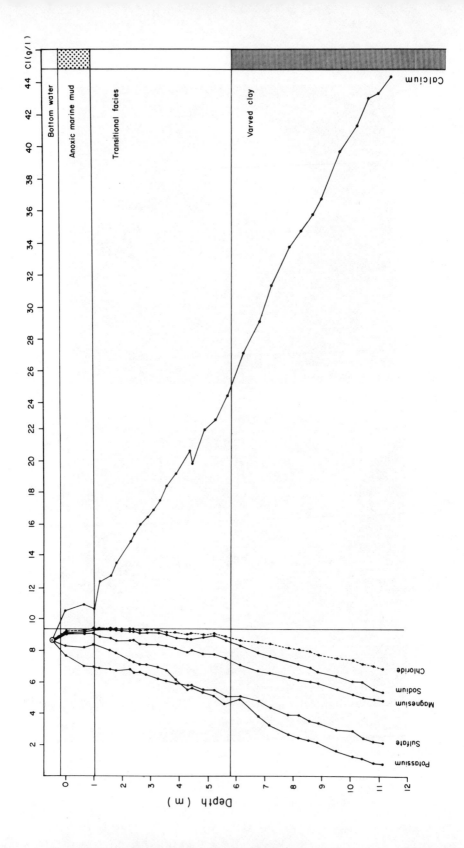

Figure 19-3. Depth distribution of major ions of interstitial waters from core 13438-1; facies boundaries and sediment/water interface are indicated by horizontal lines; the scaling of the concentration axis is: chloride = as labeled; sodium = 0.5556*·Cl (g/L); potassium = 0.0206* ·Cl (g/L); magnesium = 0.0668*·Cl (g/L); sulfate = 0.140*·Cl (g/L); calcium = 0.0236**·Cl (g/L). Any deviation in ionic composition from that of Baltic Sea bottom water (⊙) is shown by the degree of deviation from the Cl-concentration profile, most notably for Ca and K; * = ion-to-chloride ratio of the principal ions in seawater (Millero, 1974); ** = anomalous ion-to-chloride ratio of Ca in Baltic Sea water (Gripenberg, 1937).

---

to confirm ion-exchange as the reaction responsible for the major ion changes observed (figure 19-4). The variations in charge balance around the 1:1 slope probably reflect changes in anion concentrations, or they may be due to the exchange of other cations, such as Ba^{2+}, Sr^{2+}, or H^+, along with Ca^{2+}. However, they were not determined in the interstitial water solutions.

The titration alkalinity of the interstital waters decreases with depth, thus excluding dissolution of $CaCO_3$ as one possible source for the Ca increase. A second source for the increase of Ca, could have been early-diagenetic dolomitization described by Manheim and Sayles (1974) in terms of the reaction

$$2\ CaCO_3 + Mg^{2+} \rightarrow CaMg\,(CO_3)_2 + Ca^{2+}.$$

That possibility can also be excluded since the Mg decrease compensates for only a very small fraction of the Ca increase in the interstitial waters of the varved clays.

The distributions of interstitial nutrients and alkalinity in the varved clays do not support bacterial sulfate reduction as the mechanism for the loss of sulfate shown in figure 19-3 (Hartmann et al., 1973, 1976; Suess, 1976). Instead, barite precipitation may be a mechanism for the control of interstitial SO_4 in this particular core section. Further, the major ion-concentration gradients indicate that little or no Na^+ was removed from the interstitial water solution within the transitional facies, but that a significant depletion of Mg^{2+} and K^+ occurred there. In the varved clay facies, however, more Na^+ was lost from the interstitial water than Mg^{2+}. Diffusive fluxes along the concentration gradients tend to obscure changes at other facies boundaries.

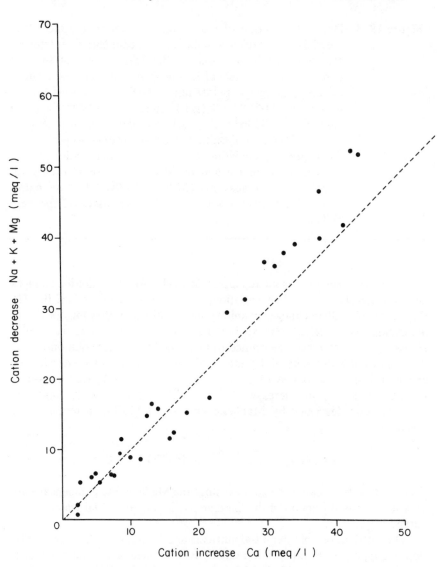

Figure 19–4. Combined changes in cation composition of interstitial waters of core 13438-1 calculated from the following relationships, where (*) marks the equivalent fractions of the respective ions in seawater (Millero, 1974): cation increase (mEq/L) = 2 Ca (mmol/L) − 1.18*·Cl (g/L); cation decrease (mEq/L) = 24.17*·Cl (g/L) − Na (mmol/L) − 0.527*·Cl (g/L) − K (mmol/L) + 5.497*·Cl − 2 Mg (mmol/L).

Authigenic Barite

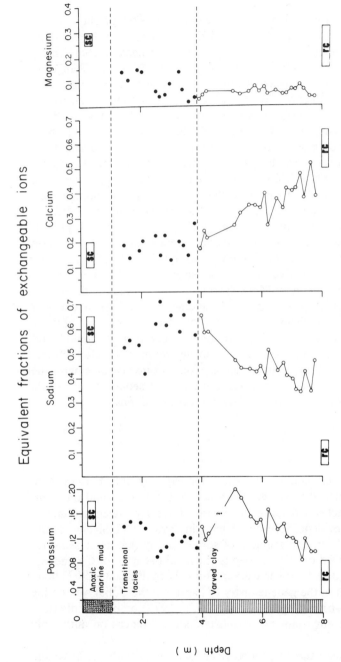

Figure 19-5. Changes in equivalent fractions of exchangeable ions of Bornholm Basin sediments from core 10116-2; note the large changes within the varved clay facies from values like those of seawater-saturated clays at the top of the section to values like those of river clays. The ranges for the equivalent fractions of river clay (*rc*) and seawater clay (*sc*) are from Sayles and Manglesdorf (1977). It is suggested that cation-exchange reactions also displace Ba from exchangeable sites of varved clays—in analogy with the displacement of Ca—which eventually results in the precipitation of $BaSO_4$; total cation-exchange capacities ranged from 16 to 23 mEq/100 g.

Exchangeable Ion Complement of Varved Clays

The displacement of sorbed Ca^{2+} from exchangeable sites of clays by the major seawater cations, notably by Na and K as inferred from the changing interstitial water composition, is also seen in the exchangeable complement of the varved clays of core 10116-2. Figure 19-5 illustrates the changes in sorbed ion equivalent fractions for two of the three sedimentary facies of the Bornholm Basin. The thicknesses and the depth distribution of the transitional and varved clay facies of core 10116-2 are comparable to those encountered in the eastern part of the Bornholm Basin (core 13438-1) where the interstitial water data were obtained (figure 19-3). Of particular interest here is the shift of exchangeable Ca^{2+} from the $>$ 50 percent of the total cation-exchange capacity (CEC) in the varved clay at 800 cm of sediment depth to $<$ 20 percent at the boundary with the overlaying transitional facies. Correspondingly, the exchangeable Na^+ and K^+ fractions increase from $\sim$ 30 percent to $\sim$ 60 percent and from $\sim$ 10 percent to $\sim$ 14 percent respectively. The exchangeable Mg^{2+} fraction remains essentially constant at $<$ 10 percent of the total CEC. The ranges of equivalent fractions of Na^+, K^+, Mg^{2+}, and Ca^{2+} for river clays and seawater-saturated sediments (Zaytseva, 1966; Sayles and Mangelsdorf, 1977) strongly indicate the freshwater origin of the varved clays and the diagenetic effects of seawater penetration on the ion-exchange reactions. More detailed accounts of those processes are found in Suess (1976) and are in preparation. It should be reemphasized that the analytical procedure in determining the clay-mineral exchange complement in the varved deposits did not involve rinsing of the interstitial electrolyte solutions but was done on wet sediments of known water contents and dissolved interstitial cation compositions.

Ba Ion Exchange

As pointed out earlier, neither dissolved nor exchangeable barium were determined at the time the interstitial waters were extracted and the equivalent exchange fractions determined on the Bornholm Basin sediments. Consequently, the evidence presented here in favor of authigenic barite formation as a result of displacement of sorbed Ba by the major seawater cations is largely based on the analogous chemical behavior of Ca and Ba. However, there are data by other workers, notably by Puchelt (1967) in an exhaustive treatment of the geochemistry of barium in the global weathering cycle (see figure 19-6) and by Hanor and Chan (1977) on barium distribution in an estuarine environment, to substantiate inferences on authigenic barite formation.

In a situation where the suspended river load mixes with ocean water,

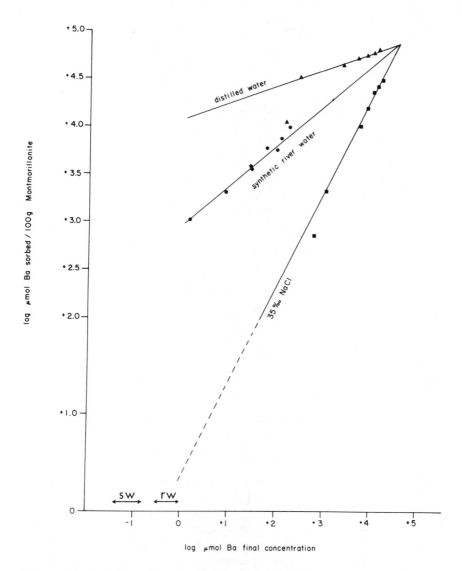

Figure 19–6. Adsorption isotherms from the aqueous system Ba and montmorillonite as a function of increasing ionic strength. The isotherms converge at ~ 125 mEq Ba/100 g of montmorillonite, which probably corresponds to the total cation-exchange capacity of the clays use in these experiments. Mixing of a Ba-saturated montmorillonite suspension in river water (mean Ba concentration $10^{0.2}$ μmol/L) and a 35‰ NaCl solution would result in the displacement of sorbed Ba from ~ 1,000 μmol/100 g to ~ 2 μmol/100 g; sw = range of Ba concentration in seawater; rw = range of Ba concentration in river water. The isotherms are calculated from experiments by Puchelt (1967).

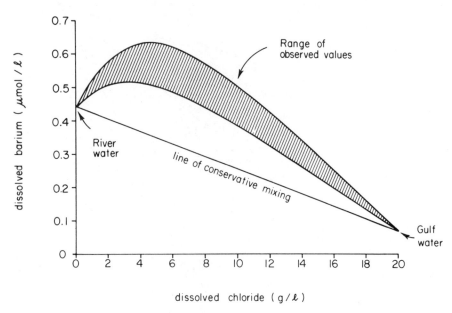

Source: Modified from Hanor and Chan (1977).

Figure 19-7. Distribution of dissolved Ba as a function of salinity in the Mississippi River and Gulf of Mexico mixing zone. The deviation from the line of conservative mixing indicates introduction of dissolved Ba from suspended river clays by cation exchange.

barite probably forms as finely divided crystals or even becomes directly precipitated on clays, as Puchelt suggested, provided that nucleation is induced. Hanor and Chan (1977) have shown that in the zone of mixing of Mississippi River and Gulf of Mexico waters, dissolved Ba behaves nonconservatively and suggested desorption from suspended river clays in exchange for the major seawater cations as a mechanism for the introduction of an excess of dissolved Ba (figure 19-7). In the waters of the mixing zone, the ion-activity product of Ba and SO_4 greatly exceeds the solubility-product constant of barite, and it is not obvious when and how barite nucleation and precipitation might proceed. In the marine transgressive situation encountered here, however, where the Baltic Sea waters permeate the varved clays after deposition, much more favorable conditions are created for the nucleation and subsequent formation of barite. In particular, deposited sediment has an exceedingly large ratio of clay to seawater compared to the Mississippi River plume. Such a favorable ratio provides a large reservoir of sorbed Ba for reaction with SO_4. In sediments there is an abundance of preferred nucleation sites as well as stable physical conditions. Thus the nodular form of the barite encountered in the southern Baltic Sea deposits might be favored. Evaluation of the usefulness of these barite micronodules as *indicators* of marine transgression will be an interesting subject for future research.

Acknowledgments

Field work for this study was supported by the German Research Society. The information on sediment facies distribution and the radiographs of the varved clays were kindly made available by F.C. Kögler (Kiel). The scanning electron micrographs were taken by V. Reimann (Kiel). F.T. Manheim (Woods Hole) drew attention to the work by A.I. Blashchishin on nodular barite in Baltic Sea sediments. Final evaluation of the data and publication was supported by the National Science Foundation under Grant OCE77-20376.

References

Blashchishin, A.I. 1976. Mineral'nyi sostav donnykh osadkov (mineral composition of bottom sediments). In V.K. Gudelis and E.M. Emelyanova (Eds.), *Geologiya Baltiiskogo morya*. Vilnius: Izdat. Mokslas, pp. 221–254.

Ericsson, B. 1972. The chlorinity of clays as a criterion of the paleosalinity. *Geol. Förens. Förhandl.* 94:5–21.

Ericsson, B. 1973. The cation content of Swedish post-glacial sediments as a criterion of paleosalinity. *Geol. Förens. Förhandl.* 95:181–220.

Gripenberg, S. 1937. The calcium content of Baltic water. *J. du Cons.* 12:293–304.

Hanor, J.S., and Chan, L.-H. 1977. Non-conservative behavior of barium during mixing of Mississippi River and Gulf of Mexico waters. *Earth Planet. Sci. Lets.* 37:242–250.

Hartmann, M., Müller, P., Suess, E., and Van Der Weijden, C.H. 1973. Oxidation of organic matter in recent marine sediments. *"Meteor"-Forsch. Ergebn.* C(12):74–86.

Hartmann, M., Müller, P. Suess, E., and Van Der Weijden, C.H. 1976. Chemistry of Late Quaternary sediments and their interstitial waters from the NW African continental margin. *"Meteor"-Forsch. Ergebn.* C(24):11–67.

Jackson, M.L. 1956. Soil Chemical Analysis—Advanced Course. Univ. of Wisconsin, Dept. of Soils Publ.

Kögler, F.-C. 1976. *Wissenschaftliche und technische Anwendung von*

Echolot und Sonaranlagen. Herausgeber: ELAC, D-2300 Kiel, Bericht 76/2.
Kolp, O. 1966. Rezente Fazies der westlichen und südlichen Ostsee. *Peterm. geograph. Mittl.* 110:1–18.
Kremling, K. 1969. Untersuchungen über die chemische Zusammensetzung des Meerwassers aus der Ostsee vom Fühjahr 1966. *Kieler Meeresforsch.* 29:81–104.
Kretzschmar, R. 1972. *Kulturtechnisch-bodenkundliches Praktikum.* Inst. für Wasserwirtschaft und Meliorationswesen der Universität Kiel.
Kullenberg, B. 1954. On the presence of seawater in the Baltic ice lake. *Tellus* 6:221–118.
Larsen, B., and Kögler, F.-Ch. 1975. A submarine channel between the deepest parts of the Arkona and the Bornholm Basins in the Baltic Sea. *Dt. Hudrogr. Z.* 28(6):274–276.
Manheim, F.T., and Sayles, F.L. 1974. Composition and origin of interstitial waters of marine sediments, based on deep sea drill cores. In E.D. Goldberg (Ed.), *The Sea*, Vol. 5: *Marine Chemistry*. New York: Wiley-Interscience, pp. 527–566.
Masicka, H. 1963. Essai de définition stratigraphique ainsi que l'âge de la carotte prélevée de la Baié Gdansk. *Baltica* 2:61–70.
Millero, F.J. 1974. Seawater as a multicomponent electrolyte solution. In E.D. Goldberg (Ed.), *The Sea*, Vol. 5: *Marine Chemistry*. New York: Wiley-Interscience, pp. 3–80.
Nilsson, E. 1970. On the late-Quaternary history of southern Sweden and the Baltic Basin. *Baltica* 4:11–32.
Pratje, O., 1948. Die Bodenbedeckung der südlichen und westlichen Ostsee und ihre Bedeutung für die Ausdeutung fossiler Sedimente. *Dt. Hydrogr. Z.* 1:45–61.
Puchelt, H. 1967. *Zur Geochemie des Bariums im exogenen Zyklus.* Sitzungsber. Heidelberger Akad., Wiss. math.-nat. Kl., 4 Abhandl., pp. 85–205.
Sandegren, R., 1934. Über das geologische Alter der polnischen Bändertone. *Geolog. Förens. Förhandl.* 56:624–628.
Sauramo, M. 1958. Die Geschichte der Ostsee. *Ann. Acad. Sci. Fennicae, Ser. A, III, Geolog.-Geograph.* 51:6–522.
Sauramo, M. 1925. Über die Bändertone in den ostbaltischen Ländern vom geochornologischen Standpunkt. *Fennia* 45(6):1–10.
Sayles, F.L., and Mangelsdorf, P.C., Jr. 1977. The equilibration of clay minerals with seawater: Exchange reactions. *Geochim. Cosmochim. Acta* 41:951–960.
Schlichting, E.H., and Blume, P. 1966. *Bodenkundliches Praktikum.* Hamburg/Berlin: Parey Verlag.
Suess, E. 1976. Porenlösungen mariner Sedimente—Ihre chemische Zusam-

mensetzung als Ausdruck frühdiagenetischer Vorgänge. Habilitationsschrift,, Universität, Kiel.
Wittig, H. 1940. Über die Verteilung des Kalziums und der Alkalinität in der Ostsee. *Kieler Meeresforsch.* 3:460–496.
Zaytseva, E.D. 1966. Capacity of exchange and exchange cations of sediments of the Pacific Ocean. In S.V. Brujewicz (Ed.), *Chemistry of the Pacific Ocean* Moscow: Izd. Nauka (in Russian).

Part VIII
Hydrothermal Interactions

20 "Heated" Bottom Water and Associated Mn-Fe-Oxide Crusts from the Clarion Fracture Zone Southeast of Hawaii

H. Beiersdorf, H. Gundlach, D. Heye, V. Marchig, H. Meyer, and C. Schnier

Abstract

Metal-enriched crusts with "baked"-appearing sediment and anomalously warm bottom waters (9 and 28°C) were found in a depression where the Hawaiian Ridge meets the Clarion Fracture Zone in the Pacific Ocean. Average composition of the crusts was approximately: SiO_2, 32 percent; Mn, 8 percent; Fe, 11 percent; Ni, 0.36 percent; Cu, 0.32 percent; Co, 0.1 percent; and Zn, 0.045 percent. It was similar to nodules from nearby siliceous ooze areas. Sedimentary clay layers were interrupted by volcanic sequences having paleomagnetic ages of approximately 1 million years. The authors conclude that volcanic activities can provide material for the formation of manganese deposits, including nodules.

Preliminary Report

During two cruises of the German research vessel Valdivia in 1976, unusual observations of deep-sea sediments and near-bottom seawater were made in the region where the Hawaiian Ridge meets the Clarion Fracture Zone (14°N and 153°W). The two cruises were VA 13/1 and VA 13/2.

In order to compare whole sediment, manganese nodules, and seawater from the siliceous ooze belt between the Clarion and Clipperton Fracture Zones with those of the red clay zone north of the Clarion Fracture Zone, samples were taken at seven stations in the area (figure 20–1). Before sampling, the area was mapped bathymetrically using 1-mi by 1-mi grids. The samples were taken in a northwest- to southeast-trending trough between a comparatively low abyssal hill in the west and a plateau in the east, both

---

This abstract was prepared by the editors.

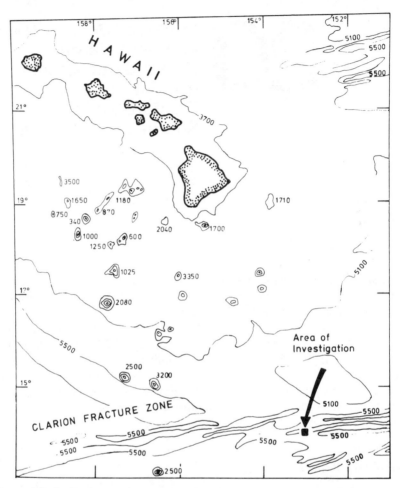

Figure 20-1. Position of the area of investigation (near the intersection of the Hawaiian Ridge with the Clarion Fracture Zone).

rising not higher than 150 to 200 m above the bottom of the trough (figure 20-2). The trough has a length of approximately 4 miles and a width of about 1 mile. Water depths in the area range from 5,600 to 5,840 m.

After an unsuccessful water sampling station (150, figure 20-2) during the first cruise (VA 13/1), a bottom-water sample at station 152 showed an unusually high temperature of 28°C measured immediately after the sampler came on deck. At first we assumed that the sealing mechanism had failed to keep warm surface water out of the sampling bottle. However, that assumption did not explain why the bottom-water sample was at 28°C while the surface water was only 23 to 24°C.

"Heated" Bottom Water   361

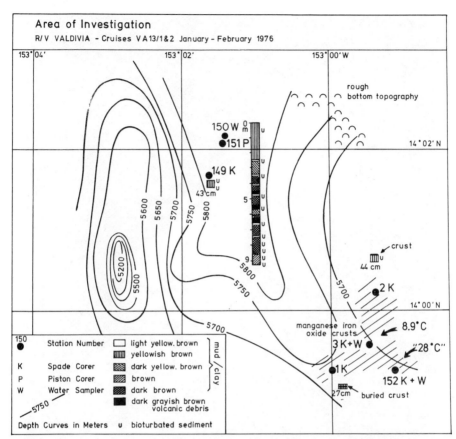

Figure 20-2. Detailed topographic map of the area of investigation showing sample locations, bottom-water temperatures, manganese crust locations, and a description of the piston core for station 151.

In the manganese nodule belt between the Clarion and Clipperton Fracture Zones, manganese crusts are normally restricted to topographic highs such as the tops of the abyssal hills; (Friedrich et al., 1976). However, on cruise VA 13/1 at the "warm water" location of station 152, we found manganese-iron crusts overlying hard, perhaps "baked," yellowish brown sediment (figure 20-3) which came from a topographic depression and not from the top or uppermost slope of an abyssal hill.

A piston corer employed at station 151 (figure 20-2) obtained a core 9.20-m long. But the core barrels were not opened before being shipped to Hannover, so no information about the sediment was immediately available. Nevertheless, the warm water, as well as the manganese crusts, raised so

Figure 20–3. (a and b) Photographs of manganese-iron crust over hard, "baked" sediment, Sample 152.

Figure 20–3b. See 3a.

much interest in the area that the schedule of the following curise of R/V Valdivia was changed to examine the area again.

During the second cruise (VA 13/2), crusts were again found, this time at three stations where spade corers were employed. At station 1, a crust 2-cm thick was observed under a cover of yellowish-brown mud 13 cm thick. Beneath the crust, the sediment was a yellowish-brown, comparatively hard clay. At station 2, two slabs of manganese crust above a yellowish-brown bioturbated clay were found. The slabs showed nodule-like overgrowths. At station 3, only a manganese crust with a few centimeters of "baked" sediment having varve-like bedding was recovered.

Together with the spade corer, a water sampler with a reversing thermometer was used at station 3. Both instruments were attached to the bearing cable 20 m above the spade corer. The temperature measured here was $8.9°C$ and was higher than the normal bottom-water temperature at those depths (about 1 to $4°C$), just as found on cruise VA 13/1.

The piston core from station 151 of cruise VA 13/1—labeled 151 P—showed an unusual sequence when compared to other cores from the central Pacific Ocean (see expansion in figure 20–2 and photograph in figure 20–4). The upper 4.25 m had a yellowish-brown to dark yellowish-brown, partly bioturbated, muddy clay containing some radiolarians. Below that zone were three sequences of layers of fine-bedded dark grayish brown clay with a considerable amount of volcanic material, such as volcanic glass, plagioclase, and magnetite. The sequences began with very sharp upper boundaries and ended more or less diffusely. Each of the grayish brown layers was underlain by a thin layer of light or dark yellowish-brown, heavily bioturbated clay followed by dark brown clay. The thicknesses of the dark brown clay layers varied between 20 and 80 cm. The deepest brown clay layer rested on a light yellowish-brown bioturbated layer (at 7.5 m) similar to the ones just described. Beneath that last yellowish-brown bioturbated layer was another 100 cm of dark brown clay followed by 70 cm of brown clay. The core is still under investigation. Radiolarians were rare, but Quarternary age is established for its uppermost part. Paleomagnetic measurements suggest an age of at least 1 million years for the volcanic layers.

In contrast to piston core 151 P, sediment Sample 149 K, which was the first sample in the area, originally had not caused special attention. It only yielded 43 cm of yellowish-brown, bioturbated, radiolarian-containing muddy clay, as in the top of core 151 P.

Two samples of the manganese crust from station 152, where warm water was found, were analysed (see figure 20–3). Their chemical composition seems to differ from the average composition of the neighboring nodules (table 20–1). All analyses were done on air-dried material. SiO_2 was determined gravimetrically, and the metals were determined by atomic absorption. However, a more meaningful comparison is presented in figure

Figure 20–4. Sample 151 cores obtained by piston corer, with banded sequences of volcanic debris. The core top is the left side of the figure.

Table 20–1
Chemical Analyses of the Crust from Station 152 (fig 20–2) Compared with Average Nodule Analysis from the Neighboring (Siliceous Ooze) Area

	SiO_2 (%)	Mn (%)	Fe (%)	Ni (%)	Cu (%)	Co (%)	Zn (%)
152/I	32.42	7.88	10.97	0.36	0.32	830	455
152/II	32.30	8.45	11.13	0.37	0.32	1000	455
Average nodules VA/04	19.20	22.0	5.91	1.29	1.16	2090	1210

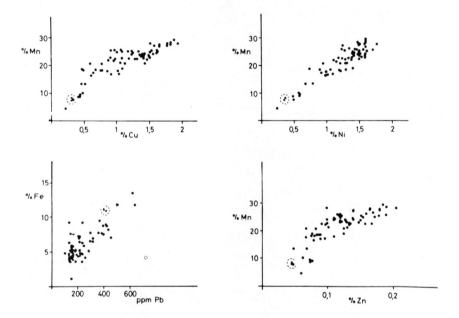

Figure 20–5. Analysis of the crust from station 152 (within dotted circles) compared with analyses of manganese nodules from the neighboring (siliceous ooze) area.

20–5, which shows that the chemical compositions of the crusts fit within the trends in the chemical compositions of the nodules from the neighboring areas of siliceous ooze. The only difference is that the crusts are more diluted by silicate material than the nodules.

The samples of near-bottom seawater on both cruises were obtained with a polyvinyl chloride Niskin bottle or a polyester-coated Nansen bottle. The measurements of the concentrations of dissolved metals were made by

Table 20-2
Preliminary Trace-Element Analysis of the "Heated Bottom Water" (28°C) from Station 152 (Fig. 20-2) Compared with Some Near-Bottom Seawater Samples from the Neighboring Area of Siliceous Ooze Rich in Mn Nodules

Element	"Hot Spot" (152)		Mn Nodule–Rich Area from January 1976
	Not Filtered	Filtered	Cruise (VA 13/1)
Ca (μg/L)	2.6×10^5	3.2×10^5	2.5–3.5×10^5
Br (μg/L)	6.7×10^4	6.7×10^4	—
Fe (μg/L)	32,100	479	14–35,000
Rb (μg/L)	105	100	94–160
Zn (μg/L)	16.10	7.80	6.7–40
Mn (μg/L)	0.37	0.57	Max of 1.9
Cu (μg/L)	16.00	14.00	2.4–25

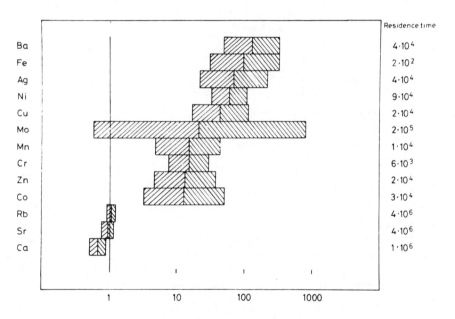

Figure 20-6. Trace-element analyses of Pacific near-bottom seawater from the siliceous ooze area compared with world ocean average seawater. Symbols are: ▨▨▨ Pacific bottom water; ▧▧▧ average seawater. Units are μg/L.

neutron activation analysis. Cu and Mn were first separated and concentrated chemically using radionuclides as tracers. The reported concentrations are only preliminary because matrix corrections have not been applied. However, the data can be used for comparisons between heated bottom water and the surrounding Pacific bottom water. The Rb concentrations, in our opinion, are the least trustworthy.

Table 20-2 shows the preliminary analyses of the sample of hot bottom water. It is compared with some waters from the nodule-rich area of the same cruise. As we can see from this comparison, the hot water is quite similar to cold bottom water, at least in terms of the components measured. The hot bottom water seems to be heated seawater and not a hydrothermal brine. The comparison between the warm near-bottom seawater and cold bottom water from the nearby area seems to us more significant than comparison with the average world ocean water. Moreover, deep water from the Pacific Ocean shows large deviations from the world ocean averages, especially for those elements with short residence times (figure 20-6).

Conclusion

Summarizing our observations, we found the following:

1. Widespread manganese crusts in a topographic depression, where nodules rather than crusts could be expected;
2. The existence of volcanic debris;
3. Heated bottom water.

From our observations, we conclude that volcanic activities are likely in the vicinity of the sampling area and that the volcanic activity could have lasted for at least one million years.

The presence of the crusts suggests, moreover, that volcanic activity can provide material for the formation of manganese deposits. If volcanic activity were responsible for the formation of the crusts we have studied, it could also be responsible for the formation of crusts and nodules in other parts of the ocean.

Reference

Friedrich, G., Plüger, W. L., and Kunzendorf, H. 1976. Geochemisch-lagerstättendundliche Untersuchungen von Manganknollen-Vorkommen in einem Gebiet mit stark unterschiedlicher submariner Topographie (Zentral Pazifik). *Erzmetall* 29:462-468.

21 The Origin of the Volcanic Component in Active Ridge Sediments

Christer Löfgren and
Kurt Boström

Abstract

Statistical analyses of data for pelagic sediments show that the Al, Ti, Fe, and Mn distributions can be explained as simple mixtures of terrigenous matter and a volcanic East Pacific Rise sediment. The small standard errors of estimation obtained in those analyses indicate that the volcanic East-Pacific-Rise component has a very constant composition and is formed at all spreading centers. That constancy suggests that only a very reproducible deep-seated process, such as the leaching of basalt at high temperature and pressure could form the appropriate volcanic emanations. In contrast, shallow leaching of basalt would give rise to active ridge sediments of very *erratic* compositions, and therefore shallow leaching of basalt is probably a negligible source of iron and manganese.

Introduction

The non-carbonate fraction of deep-sea sediments deposited near spreading centers (active-ridge sediments) is rich in Fe, Mn, and several trace elements such as Ba, B, Cu, V, As, Hg, and U, but poor in Si, Al, Ti, Zr, and Th. Generally that fraction is believed to be deposited from hydrothermal solutions formed by processes at spreading centers (Bonatti and Joensuu, 1966; Boström and Peterson, 1966; Boström, 1973; Bertine and Keene, 1975; Piper, 1973; Boström et al, 1976, and Dymond et al., 1976). Seawater leaching of hot basalt may generate such solutions (Boström 1967), and data by Corliss (1971) and Hart (1970) suggest that most of the iron and manganese is delivered by surface alterations of oceanic basalt.

However, the existence of a shallow hydrothermal system implies highly varying residence times for the heated seawater, permitting much unconsumed oxygen and unreduced sulfate to resurface in some instances. Sulfate can then act as a redox buffer and, particularly if some oxygen remains in solution, prevent the formation of iron-rich solutions (Boström, 1973;

Spooner and Fyfe, 1973). Data by Bischoff and Dickson (1975) point in the same direction.

Volcanic deposits should therefore have highly varying compositions, as has been noticed for alteration products in some shallow non-oceanic hydrothermal systems (Barth, 1950; Naboko, 1963). Studies of oceanic basalts by Aumento et al. (1976) confirm that conclusion. They found that the Mn and Fe concentrations of the basalts increased with increased time of alteration, whereas the concentration of Al decreased. Those trends are opposite to the ones required to form active ridge sediments by such processes.

However, volcanic sediments from oceanic spreading centers, are remarkably constant in composition, as their Fe/Ti and Al/(Al + Fe + Mn) ratios show. Shallow hydrothermal alterations are therefore unlikely sources of Fe and Mn for active-ridge sediments (Boström, 1973, 1976; Boström et al., 1976). We will here present further evidence to suggest that the surfacing volcanic solutions are remarkably constant in composition. We will also discuss the genetic implications of those findings.

Data and Methods

The conclusions of this study are based on the average chemical compositions of the following sediments:

1. *Atlantic Ocean*. The samples were recovered during leg 3 of the Deep Sea Drilling Project (DSDP) and are discussed in Boström et al. (1972).
2. *Pacific Ocean*. The samples were recovered during DSDP leg 34 (see Boström et al., 1976). Additional data for areal averages are reported in Boström (1976).
3. *Indian Ocean*. The samples were recovered during several Scripps and Woods Hole cruises and are described in Boström et al., (1969, 1971) and in Boström (1979).

Most data for Fe, Al, and Mn were obtained with atomic absorption (AA); emission spectroscopy (ES) was used to obtain Ti determinations. Additional data for Al, Fe, and Mn were obtained by ES and were used to check the other analytical technique. The analyses for Mn in Indian Ocean sediments were, to a large extent, done with ES and are therefore less reliable than the AA determinations. The error in our AA data is about 5 percent for Al, Fe, and Mn and about 10 percent in the ES determinations, except for Mn, where some ES errors of up to 30 percent may occur. That level of uncertainty is acceptable for this work, as can be seen from comparisons with major element data obtained in other laboratories from similar sediments

Origin of the Volcanic Component

Table 21-1
Average Composition of Terrigenous Matter (TM) and East Pacific Rise Deposits (EPRD).

	TM	EPRD
Al	8.1	0.083
Fe	4.9	7.3
Ti	0.48	0.0085
Mn	0.088	2.8

Source: Data from Boström et al. 1976.

Note: Concentrations given in percent on an absolute basis (not recalculated to carbonate-free basis).

(e.g., Revelle, 1944; El Wakeel and Riley, 1961; and Dymond et al., 1976). In addition, chemical data for the average East Pacific Rise deposit (EPRD) and terrigenous matter (TM) are used (see table 21-1).

Results

All data were used to form the ratios Fe/Ti, Al/(Al + Fe + Mn), Fe/(Al + Fe + Mn), and Mn/(Al + Fe + Mn) and plotted in diagrams (see figures 21-1 through 21-4). Figure 21-1 shows the Pacific data. The solid line connecting EPRD and TM represents the compositions of arbitrary mixtures of EPRD and TM, not a statistically derived best-fit line (see Boström, 1973, 1976; and Boström et al., 1976). The close fit between the sediments and the mixing line suggests that the real sediments are indeed mixtures of components closely resembling TM and EPRD in composition; however, that fitting procedure cannot measure how well such a mixture function describes the real sediments.

In this chapter we have calculated the best-fit regression line for all data points and have subsequently shown how well that line explains the data, using the standard error of estimate of Fe/Ti on Al/(Al + Fe + Mn). To facilitate such calculations, all Al/(Al + Fe + Mn) data were transformed graphically to Al' values so that the mixing curve between TM and EPRD becomes a straight line. The log Fe/Ti data were then plotted against the Al' values, and the linear regression lines and their standard errors of estimate were calculated and plotted (see figures 21-2 and 21-3).

That same statistical method was applied on the Fe-Mn-Al data (see figure 21-4), except that no transform procedure was needed. A Fe-Mn-Al diagram was not made for the Indian Ocean data because the Mn determinations for that ocean were not sufficiently good. However, those errors in Mn are not critical for the function Al/(Al + Fe + Mn) in figures 21-1 through 21-3.

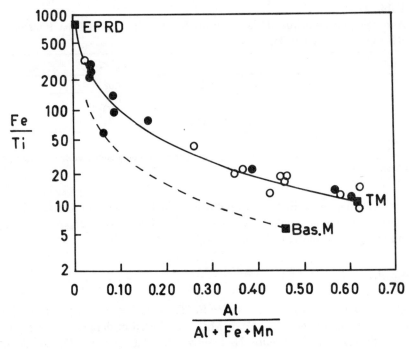

Source: Boström et al., 1976.

Figure 21-1. The relations between Fe/Ti and Al/(Al + Fe + Mn) in Pacific Ocean sediments; based on all data in Boström et al., 1976 and Boström, 1976. EPRD and TM are defined in the text and in table 21-1. BasM represents average basaltic matter.

Discussion

The standard errors of estimate for the regression lines in figures 21-2 through 21-4 are small, and in all graphs, the points for EPRD and TM fall within those standard error limits. That fact supports the hypothesis that the sediments largely are mixtures of two phases, EPRD and TM, insofar as Fe, Al, Ti, and Mn are concerned. Such a conclusion is trivial for points near TM in the graphs, because terrigenous matter is a ubiquitous component in pelagic sediments, but the very small spread in the Pacific Ocean data near EPRD in figures 21-2 and 21-4 is remarkable. Furthermore, the data for the Atlantic and Indian Oceans suggest that TM in those oceans is also being mixed with EPRD of quite constant composition. Because all our data points with high Fe/Ti and low Al' values fall on or near an active ridge, we can

Origin of the Volcanic Component

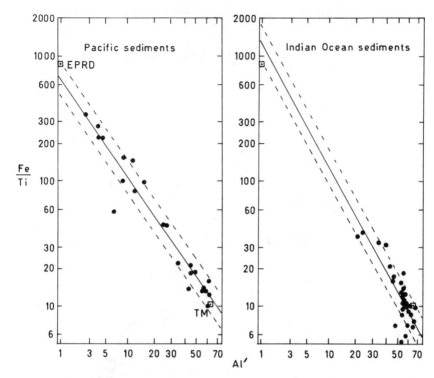

Figure 21–2. The relations between Fe/Ti and Al' in Pacific and Indian Ocean sediments. Al' is a transformation of Al/(Al + Fe + Mn) such that any mixing function between EPRD and TM can become a straight line. The solid lines represent the linear regressions giving the best fits to the data in the graphs, and dashed lines represent the standard errors of estimate for the linear regression. EPRD, TM, and the sources of the data are defined in the text and in table 21–1.

conclude that material identical to EPRD is surfacing at all spreading centers in the World Ocean and, furthermore, that the EPRD component most likely is a volcanic contribution.

The remarkably constant composition of that volcanic contribution strongly argues against the interpretation that seawater leaching of basalt at shallow depths could create the EPRD-forming solutions. Such processes tend instead to form highly varying products, as discussed in the Introduction, and would be revealed as large scatter in the data and very broad error fields in figures 21–2 through 21–4. Moreover, studies by Flower (1975) and

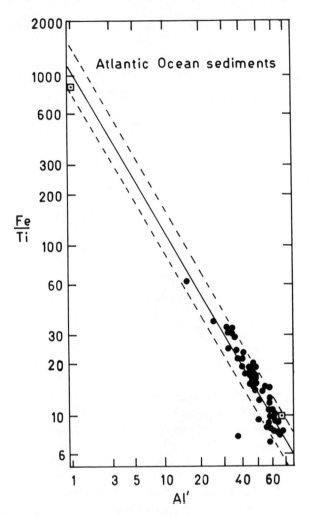

Figure 21-3. The relations between Fe/Ti and Al' in Atlantic Ocean sediments. For further details, see caption to figure 21-2. Sources of data are given in the text.

Aumento et al (1976) in the uppermost few hundred meters of the oceanic crust have failed to demonstrate significant alterations that could be linked to the generation of EPRD-forming solutions. That failure suggests that the likely source of EPRD is located at great depth, probably a kilometer or more below the rock surface.

Experimental studies of basalt-seawater reactions by Bischoff and Dickson (1975), Hajash (1975), and Mottl (1976) show that the amount of

Origin of the Volcanic Component

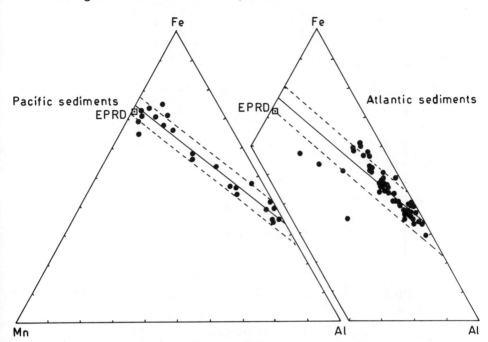

Figure 21-4. The relations between Fe-Al-Mn in Pacific and Atlantic sediments. The solid lines represent the linear regressions, giving the best fit to the data in the graphs, and the dashed lines represent the standard errors of estimate for the linear regressions. Only the position of EPRD is visible; the position of TM is obscured because it is located in the thick cluster of points near the Al-rich end of the regression lines. Sources of data are given in the text.

Fe and Mn dissolved in seawater increases rapidly with increased temperatures and pressures, whereas little or no Al is dissolved. In particular, the experiments by Hajash (1975) and Mottl (1976) suggest that the Fe/Mn ratio in the final solution is a function of temperature. Thus leaching processes at 300°C form Fe/Mn ratios near 1.0, and at 400 and 500°C the corresponding ratios range between 1 to 2 and 5 to 10, respectively, in reactions lasting a few days to a few weeks. With increased time of reaction, the amounts of Fe and Mn in solution decrease rapidly and after 2 years would sink to about 0.04 ppm (Fe + Mn), according to data in Bischoff and Dickson (1975). However, studies by Lister (1975) have shown that the residence time for heated seawater at great depth near spreading centers in the oceanic crust is probably very short, of the order of a few hours or a few

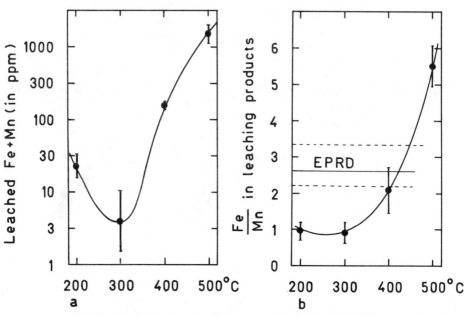

Figure 21-5. Results of experimental leaching of basalt with seawater at 200, 300, 400, and 500°C and pressures of 500, 600, 700, and 1,000 bars, respectively. The graphs are based on data by Mottl (1976) and the errors in the data (one standard deviation) are shown by the extent of the vertical bars. (a) Amount of Fe + Mn (in ppm) that is dissolved into seawater. (b) The ratio of Fe/Mn in the formed solutions. Solid lines show position of EPRD, whereas dashed lines represent positions of error limits for the pure Fe-Mn phase in Pacific sediments (see figure 21-4).

days. That idea suggests that experimental data for prolonged reactions can be ignored. The narrow range of 2.5 to 3.0 in the Fe/Mn ratio of the EPRD phase (see figures 21-4 and 21-5) and the preceding discussion suggest, therefore, that the solutions are formed rapidly at temperatures near 400 to 450°C and at depths of 1 km or more in the crust near the spreading centers.

The compositions of some active ridge deposits are different from those of the sediments just discussed. That difference could be considered to indicate a different mechanism of formation. Thus Bonatti and Joensuu (1966), Scott et al. (1974), Piper et al. (1975), and Moore and Vogt (1976) found hard crusts with Fe/Mn ratios differing from those in unconsolidated active ridge sediments. However, only four locations with such hydrothermal crusts are known out of the many dredging stations that have been occupied on active

Origin of the Volcanic Component

ridges. In contrast, all cored unconsolidated sediments on active ridges are rich in both Fe and Mn. The mass of Fe and Mn in hydrothermal crusts is therefore negligible compared to that in other active ridge deposits.

Furthermore, hydrothermal crusts may form from the same solutions that ordinarily would deposit unconsolidated active ridge sediments. Thus ascending Fe-Mn-rich solutions might react with descending oxygen-rich water in the rock openings. Iron would then deposit because it oxidizes easily, whereas manganese could remain in solution, surface, and deposit as the Mn-oxide slabs observed by Scott et al. (1974) and Moore and Vogt (1976). Iron-rich crusts, however, may deposit where volcanic solutions surface at such a high rate that advected oxygen only suffices to deposit iron, whereas Mn has to migrate further before oxidation and deposition (Bonatti et al., 1972). Because of those quantitative relations and the possibility of the chemical reactions discussed here, we feel that the presence of a few hydrothermal crusts of "unorthodox" composition does not jeopardize our conclusion that volcanic solutions surfacing at spreading centers in the ocean are quite constant in chemical composition.

Mixing of TM with a volcanic phase identical in composition to EPRD can thus explain the range of concentrations of Al, Ti, Fe, and Mn in pelagic sediments. The use of trace-element data has led to the same conclusion, except that a biological source of trace metals is also required. However, biological matter has little importance for the Al, Ti, Fe, and Mn budget in the ocean relative to TM and EPRD (Boström, 1976; Boström et al., 1976), and thus the simplified statistical treatment used in this chapter is successful.

Note added in Proof

In addition to the previously mentioned Fe- and Mn-rich hard crusts at spreading centers, there are reports of two spectacular finds of sulfide deposits on the East Pacific Rise just south of the entrance to the Gulf of California (Francheteau et al., 1979). That location has been suggested by Bischoff in 1969 (personal communication). However, there are no indications that those sediments cover areas much larger than 100 m^2 at each locality, whereas the EPRD-rich sediment mounds on the East Pacific Rise may have areal extents of some 25 km^2 each (Boström et al., 1974).

Acknowledgments

We want to thank Dr. M.J. Mottl, who cordially let us share his experimental data at an early date, and Dr. F.W. Dickson for extensive discussions of some of the problems in this chapter.

References

Aumento, F., Mitchell, W.S., and Fratta, M. 1976. Interaction between sea water and oceanic layer two as a function of time and depth. 1. Field evidence. *Can. Mineral.* 14:269–290.

Barth, T.F.W. 1950. Volcanic geology, hot springs and geysers of Iceland. Carnegie Inst. Washington, Publ. No. 587.

Bertine, K.K., and Keene, J.B. 1975. Submarine barite-opal rocks of hydrothermal origin. *Science* 188:150–152.

Bischoff, J.L., and Dickson, F.W. 1975. Sea water–basalt interaction at 200°C and 500 bars: Implications for origin of sea-floor heavy-metal deposits and regulation of sea water chemistry. *Earth Planet. Sci. Lett.* 25:385–397.

Bonatti, E., and Joensuu, O. 1966. Deep-sea iron deposits from the South Pacific. *Science* 154:643–645.

Bonatti, E., Kraemer, T., and Rydell, H. 1972. Classification and genesis of submarine iron-manganese deposits. In D. Horn (Ed.), *Ferromanganese Deposits on the Ocean Floor*. Washington: Nat. Sci. Found., pp. 149–165.

Boström, K. 1967. The problem of excess manganese in pelagic sediments. In P.H. Abelson (Ed.), *Researches in Geochemistry*, Vol. 2. New York: Wiley & Sons, pp. 421–452.

Boström, K 1973. The origin and fate of ferromanganoan active ridge sediments. *Stockh. Contrib. Geology, Acta Universitatis Stockholmiensis* 27(2):149–243.

Boström, K. 1976. Particulate and dissolved matter as sources for pelagic sediments. *Stockh. Contrib. Geology, Acta Universitatis Stockholmiensis* 30:2,15–79.

Boström, K. 1979. Geochemistry of Indian Ocean sediments (manuscript in preparation).

Boström, K., and Peterson, M.N.A. 1966. Precipitates from hydrothermal exhalations on the East Pacific Rise. *Econ. Geol.* 61:1258–1265.

Boström, K., Peterson, M.N.A., Joensuu, O., and Fisher, D.E. 1969. Aluminum-poor ferromanganoan sediments on active oceanic ridges. *J. Geophys. Res.* 74:3261–3270.

Boström, K., Farquharson, B., and Eyl, W. 1971. Submarine hot springs as a source of active ridge sediments. *Chem Geol.* 10:189–203.

Boström, K., Jeonsuu, O., Valdés, S., and Riera, M. 1972. Geochemical history of South Atlantic Ocean Sediments since late Cretaceous. *Mar. Geol.* 12:85–121.

Boström, K., Joensuu, O., Kraemer, T., Rydell, H., Valdes, S., Gartner, S., and Taylor, G. 1974. New finds of exhalative deposits on the East Pacific Rise. *Geol. För. Stockh. Förh.* 96:53–60.

Boström, K., Joensuu, O., Valdés, S., Charm, W., and Glaccum, R. 1976. Geochemistry and origin of East Pacific sediments, sampled during DSDP leg 34. In *Initial Reports of the Deep Sea Drilling Project*, Vol. 34. Washington: U.S. Govt. Printing Office, pp. 559–574.

Corliss, J.B. 1971. The origin of metal-bearing submarine hydrothermal solutions. *J. Geophys. Res.* 76:8128–8138.

Dymond, J., Corliss, J.B., Heath, G.R., Field, C.W., Dasch, E.J., and Veeh, H.H. 1973. Origin of metalliferous sediments from the Pacific Ocean. *Geol. Soc. Amer. Bull.* 84:3355–3372.

Dymond, J., Corliss, J.B., and Stillinger, R. 1976. Chemical composition and metal accumulation rates of metalliferous sediments from sites 319, 320 B, and 321. In *Initial Reports of the Deep Sea Drilling Project*, Vol. 34. Washington: U.S. Govt. Printin Office, pp. 575–589.

El Wakeel, S.K., and Riley, J.P. 1961. Chemical and mineralogical studies of deep-sea sediments. *Geochim. Cosmochim. Acta* 25:110–146.

Flower, M.F.J. 1975. Low temperature alterations of sea floor basalts, DSDP leg 37 (Abstract). Intern. Conf. on Nature of Oceanic Crust, Dec 4–6, 1975, La Jolla.

Francheteau, J., Needham, H.D., Choukroune, P., Juteau, T., Séguret, M., Ballard, R.D., Fox, P.J., Normark, W., Carranza, A., Cordoba, D., Guerrero, J., Rangin, C., Bougault, H., Cambon, P., and Hekinian, R. 1979. Massive deep-sea sulphide ore deposits discovered on the East Pacific Rise. *Nature* 277:523–528.

Hajash, A., 1975. Hydrothermal processes along mid-ocean ridges: An experimental investigation. *Contrib. Mineral. Petrol.* 53:205–226.

Hart, R., 1970. Chemical exchange between sea water and deep ocean basalts. *Earth Planet. Sci. Lett.* 9:269–279.

Lister, C.R.B. 1975. Rapid evolution of geothermal systems in new oceanic crust predicts mineral output mainly near ridge crests (Abstract). *EOS2, Trans. Am. Geophys. Union* 56:1074.

Moore, W.S., and Vogt, P.R. 1976. Hydrothermal manganese crusts from two sites near the Galapagos spreading axis. *Earth Planet. Sci. Lett.* 29:349–356.

Mottl, M.J. 1976. Chemical exchange between sea water and basalt during hydrothermal alteration of the oceanic crust. Ph.D. thesis, Harvard University.

Naboko, S.I. 1963. *Gidrotermalnii metamorfism porod v vulkanicheskikh oblastyakh*. Moscow: Izd. Akad. Nauk., SSSR.

Piper, D.Z. 1973. Origin of metalliferous sediments from the East Pacific Rise. *Earth Planet. Sci. Lett.* 19:75–82.

Piper, D.Z., Veeh, H.H., Bertrand, W.G., and Chase, R.L. 1975. An iron-rich deposit from the northeast Pacific. *Earth Planet. Sci. Lett.* 26:114–120.

Revelle, R.R. 1944. Marine bottom samples collected in the Pacific by the *Carnegie* on its seventh cruise. Carnegie Inst. Wash., Publ. 556.

Scott, M.R., Scott, R.B., Rona, P.A., Butler, L.W., and Nalwalk, A.J. 1974. Rapidly accumulating manganese deposit from the median valley of the Mid-Atlantic Ridge. *Geophys. Res. Lett.* 1:355–358.

Spooner, E.T.C., and Fyfe, W.S. 1973. Sub-sea-floor metamorphism, heat and mass transfer. *Contrib. Mineral. Petrol.* 42:287–304.

22 The Nature of Hydrothermal Exchange between Oceanic Crust and Seawater at 26°N Latitude, Mid-Atlantic Ridge

Robert B. Scott,
Darcy G. Temple, and
Phillipe R. Peron

Abstract

The hydrothermal exchange between basaltic rocks and seawater on the rift-valley wall on the Mid-Atlantic Ridge at 26°N occurred along closely spaced fractures filled with hydrothermal minerals; also the rocks at vents have been affected only by low-temperature alteration. Yet the nature of hydrothermal exchange between the crust and seawater is predicted by experimental chemical, and theoretical thermal constraints to occur along widely spaced discrete fractures with high discharge rates, high-temperature venting, little mineralization in the fractures, and high water-to-rock ratios (Sleep and Wolery, 1978; Seyfried and Bischoff, 1977).

Hydrothermal deposition fields such as that at 26°N are the exception. The absence of Mg enrichment in both high- and low-temperature hydrothermally altered rocks, the clogging of fractures with quartz-silicate-sulfide assemblages, and the penetrative alteration of hydrothermally affected rocks suggests that low water-to-rock ratios existed in that hydrothermal field. Large volumes of cold oxygenated seawater may advect with hydrothermal fluids in breccias and talus close to the surface. Evidence of episodic faulting, such as a sequence of cross-cutting veins characterized by high-temperature sulfide-bearing veins cut by later low-temperature oxide and clay-bearing veins, presumably reflects the movement of rocks in the fault zones toward progressively cooler and more oxygenated surface conditions.

The low water-to-rock ratios are thought to effectively precipitate most metals as sulfides in the oceanic crust on the Mid-Atlantic Ridge, leaving only Mn to form hydrothermal crusts at the surface and a few metals to be transported from the vent to form metalliferous sediments. More widely spaced faults and higher water-to-rock ratios of hydrothermal systems along the East Pacific Rise allow abundant metals to stay in solution until vented into the water column, where the metals precipitate as oxides or hydroxides and are transported in suspension to form highly metalliferous sediments.

Introduction

The nature of hydrothermal exchange between seawater and basaltic ocean crust has been the objective of numerous direct observations, experimental investigations, and theoretical modeling in the last decade. Specific examples include direct observations by submersibles (Corliss et al., 1977; Ballard et al., 1975; ARCYANA, 1975), deep-tow instrument studies (Lonsdale, 1977; Weiss et al., 1977), numerous studies of hydrothermal specimens collected by dredging and submersible operations (M. Scott et al., 1974; Moore and Vogt, 1976; Bonatti et al., 1976; Thompson et al., 1975; Cann et al., 1977; M. Scott et al., 1978), experiments on hydrothermal exchange under controlled conditions of temperature, pressure, and composition (Mottl, 1975; Seyfried and Bischoff, 1977; Hajash, 1975), and theorectical modeling of seawater circulation and mass exchange (Sleep and Wolery, 1978; Ribando et al., 1976; Lister, 1972, 1974, and 1976; Lowell, 1975; Wolery and Sleep, 1975). However, few studies attempt correlation of sea-floor geologic observations with features of hydrothermally altered basaltic rocks from within the ocean crust using geological, theoretical, and experimental constraints. This chapter is a synthesis of the investigations of the hydrothermal field on the Mid-Atlantic Ridge at 26°N, where a combination of camera traverses across the field, surface samples of hydrothermal deposits, and specimens of hydrothermally altered rocks from fault scarps are available for interpretation.

Between 1970 and 1973 the National Oceanic and Atmospheric Administration (NOAA) conducted a geophysical and geological study of a corridor betweeen Cape Hatteras and Cape Blanc, concentrating on a 2° by 2° region centered at 26°N, 45°W on the Mid-Atlantic Ridge (figure 22-1). Study of this area with cooperating universities was initiated by cruises of the NOAA R/V *Discover* (1972) and *Researcher* (1973); the hydrothermal field was discovered in 1972 and 1973 (R. Scott et al., 1974). In 1975, Soviet, U.S., and Canadian researchers visited the area aboard the R/V *Kurchatov* (R. Scott et al., 1976), and in 1976 and 1977, the R/V *Gyre* of Texas A&M University returned to the hydrothermal site. This chapter utilizes data from all those efforts.

Theoretical, thermal, and chemical constraints on the mechanisms of hydrothermal circulation require widely spaced conduits, high exit temperatures, high fluid-discharge rates, and episodic fluid discharge in the conduits. However, the distribution, dimensions, and evolution of the conduits for hydrothermal fluid circulation are unknown. At what temperatures were the

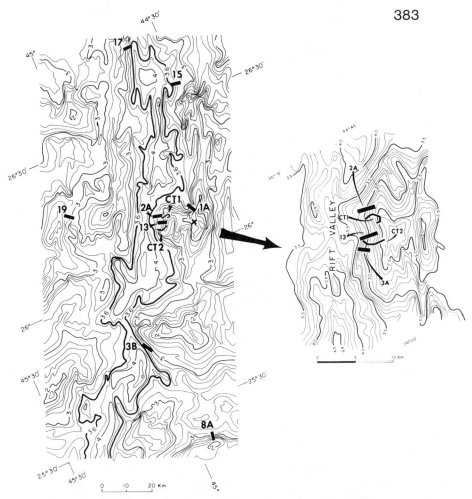

Figure 22–1. (a) (left) Bathymetry and locations of samples and camera traverses used in this study. Contour interval is 0.2 km. The 3.6-km contour is darker to outline the rift valley floor. Note the constricted central regions of segments of rift valley between offsets and broader regions at offsets. Numbered dredge sites are indicated by black bars. CT1 and CT2 mark the two camera traverses on the prominent ridge on the east side of the rift valley where abundant evidence of hydrothermal activity is recorded. X marks the site of trip camera photographs. Note that site numbers (e.g., 19, 2A, 1A) are used as the prefix in sample numbers (e.g., 19-3, 2A8-1, 1A11). (b) (right) Enlargement of hydrothermal field showing details of camera traverses and location of dredge sites where hydrothermal material was recovered.

alteration phases formed at the mouth of conduits? Low temperatures would require a sizable geothermal gradient within the conduit, encouraging precipitation of hydrothermal phases which would clog the circulation system. Are clogged hydrothermal conduits reactivated by subsequent faulting, or are such conduits one-stage processes? Is oxygenated fresh seawater mixed with hydrothermal fluids before they exit from the crust, or are nearly pristine fluids emitted? Can reasonable estimates of the actual water-to-rock ratio in hydrothermal systems be made from the study of samples affected by passage of hydrothermal fluids? How does the Mid-Atlantic Ridge hydrothermal field at 26°N differ from inactive parts of the same ridge system? What similarities does the 26°N field have with other hydrothermal sites? What special structural features of mid-ocean rises are conducive to hydrothermal activity?

We investigated those questions by combining our sea-floor geologic interpretation of two camera traverses with our interpretation of subsurface hydrothermal mechanisms on examination of altered basalts dredged from sites within or close to the hydrothermal field (figure 22–1).

Sea-Floor Observations

Regional Geology

The structure of the ridge crest between 25 and 27°N on the Mid-Atlantic Ridge (MAR) consists of a continuous rift valley having 1.5 to 2.2 km of relief and four minor offsets or fault zones. The northernmost fault zone has a 7-km left-lateral offset and the southern three fault zones have an average of 7-km right-lateral offsets. Distinctive linear valleys and ridges extend from the offsets at high angles to the rift-valley axis; those on the eastern side are perpendicular to the rift axis, but those on the western side are not symmetrical, forming a 60° angle with the ridge crest (figure 22–1). Although those offsets are generally interpreted to be transform faults, the geometric constraint of the lack of symmetry across the rift axis leaves the offset origin difficult to accept using recognized transform-fault mechanisms (Rona et al., 1976). The floor of the rift valley is 10 km wide between the offsets, but at the intersection with the offsets, the rift valley widens to 25 km. Offsets occur every 27 to 50 km along the rift valley. The hydrothermal field is located on the wall of the central eastern valley at 26°06'N and 44°45'W (R. Scott et al., 1974) (figure 22–1). McGregor (1974) shows magnetic evidence that the half-spreading rates are 1.3 cm/yr on the eastern side and 1.1 cm/yr on the western side.

Nature of Hydrothermal Exchange

Bottom Photography

A total of 3,300 stereopair bottom photographs were taken during two camera traverses (figure 22–1) up the eastern rift-valley wall within the hydrothermal field, providing continuous overlapping coverage. Lithology, slope and fault attitudes, and fault frequency were determined systematically. Two geological cross sections were drawn from detailed bottom photography, and from them a geologic map (figure 22–2) was constructed.

Structural Framework

The rift-valley wall consists of a repetitive sequence of narrow, steep, highly faulted zones separated by broad, nearly flat, less faulted terraces; each fault

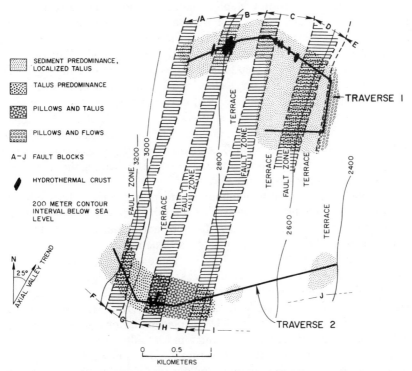

Figure 22–2. Geologic map based on correlation of traverses 1 and 2. Although major terrace and fault-zone structural features can be correleated, lithologic similarities cannot be extended beyond approximately 1 km.

Table 22-1
Large Step-Fault Dimensions. "In" and "out" refer to an inward or outward tilt of the terrace slope relative to the rift valley. See figures 22-3 and 22-4 for location of fault blocks.

Traverse 1 (5 blocks)	Range:	47–128 m	159–760 m	11–42°	0.6° in 0.2°–5° out
	Average:	74 m	408 m	24°	0.6° in 2° out
Traverse 2 (5 complete blocks, incomplete)	Range:	78–265 m	197–702 m	11–69°	5–34° in 3° out
	Average:	170 m	449 m	36°	22° in 3° out
1 and 2 Combined:		47–265 m	159–760 m	11–69°	—
Average of 1 and 2:		105 m	429 m	29°	—

zone and terrace pair form a steplike feature or block. Traverse 1 has five large steps (figure 22-2, blocks A–E); traverse 2 has five visible steps (figure 22-2, blocks F–J) and perhaps others in a section where the camera lost contact with the bottom between 2,700- and 2,500-m depths. The range and average data for the large step-fault features are listed in table 22-1. The steep fault zones have an average slope of 29° but contain closely spaced *en echelon* normal faults. Terraces may slope inward toward the rift valley or outward as much as 5°, indicating rotation of the fault block. The sum of individual displacements within the fault zone averages 105 m, and block width ranges from 159 to 760 m. Traverse 2 fault blocks have steeper slopes and greater displacements than traverse 1 blocks, indicating that fault-block size increases southward in this area.

Individual faults occur within both the steep fault zone of the major large steps and within the broad terraces that resemble treads to major steps. The average fault strike is parallel to the rift-valley axis. Upon approach to visible fault scarps, a lithologic sequence of sediments to talus to breccia is observed; faults may be predicted by the same sequence even if the scarp itself is obscured. Table 22-2 summarizes individual fault-block dimensions within both the fault zones and the terraces. Fault-block tops average 30 m wide perpendicular to strike. Some dip inward toward the rift averaging 7.5°, and others dip outward averaging 13°. Displacements on fault scarps average 10.6 m for faults within fault zones and 5 m for faults within terraces. Faults within terraces may dip inward (average 46°) or outward (average 40°), forming horst-and-graben structures and out-tilted blocks, while faults within fault zones dip more steeply inward or outward (average 53° and 68°, respectively). Average fault spacing is 29.6 m in fault zones and 48.5 m in terraces. The data suggest that the division of fault blocks into fault zones and terraces has a valid structural basis and that the easily observable block-bounding major "faults" seen on seismic records actually are integrations of

Table 22-2
Individual Fault-Block Dimensions. Data have been corrected to a line perpendicular to contours.

	No. of Faults	Vertical Displacement		Fault Spacing		Fault Scarp Dip			Block Top Dip		
		Range	Avg.	Range	Avg.	Range	Avg. In	Avg. Out	Range	Avg. In	Avg. Out
Fault zones	50	1–76m	10.6m	6–113m	29.6m	15–90°	53°	—	0–63°	8.5°	19°
Terraces	21	1–19m	—	6–182m	48.0m	7–88°	46°	40°	0–27°	5.0°	—
Fault zones and terraces	74	1–76m	9.5	5–182m	37.0m	7–90°	52°	47°	0–63°	7.5°	13°

anastomosing *en echelon* faults. The cross sections also support the fault zone–terrace distinction. More detailed discussions can be found in Temple et al. (1979).

Sea-Floor Rocks

The lithology of the rock types observed in bottom photographs is similar to samples dredged in 1972, 1973, 1975, 1976, and 1977 (unpublished cruise reports, R. Scott). Two distinctive sediment types, a talus, three forms of breccias, pillow basalts, and hydrothermal manganese crusts are recognized. A pale brown foraminferal ooze (80 percent $CaCO_3$ and a somewhat metal-enriched carbonate-free fraction) (M. Scott et al., 1978) forms the bulk of terrace deposits. In current-swept parts of the FAMOUS area, darker and coarser lag deposits between ripples are identified as manganese-coated pteropods (W. Bryan, personal communication); in the 26° area, they are identified as manganese-coated bryozoans, pteropods, and other skeletal remains (Figures 22–5a and b). Although sediments commonly occur on gently sloping terraces, they are also found on slopes as steep as 60°. Ripples are common (figures 22–3c and d, 22–4, and 22–5).

Fault-derived talus of basalt and basalt breccia (5 to 50 cm in diameter) is found to increase in concentration and size with proximity to fault scarps (figure 22–3d). Talus is often found scattered on top of sediment, indicating recent faulting. Three origins for the basalt breccia and talus are recognized on a basis of distinctive matrices and physical appearance of the outcrop: (1) unconsolidated talus occurs at the base of fault scarps (figure 22–3d); (2) sedimentary breccia consists of fault-derived basalt talus partially cemented by a matrix of semiconsolidated limestone and finely crushed, weathered basalts. This sedimentary breccia is exposed on numerous small scarps by faulting (figure 22–5c) and in some cases is undercut by current erosion (figure 22–3b); (3) basalt breccia in situ within fault scarps is termed *tectonic breccia* and is exposed in one large scarp and several smaller ones (figure 22–3b).

Where hydrothermal fluids flow from fault zones and deposit hydrothermal manganese-oxide crusts, blocks of basalt talus become cemented by the manganese crusts, forming a hydrothermal breccia (figure 22–3a). Dredge samples show these hydrothermal crusts to be brownish-black to submetallic gray, with a smooth botryoidal surface and a laminated interior. Talus cemented by the manganese crust develops a typical slightly smoothed appearance; the hydrothermal breccia surface often retains the character of the lumpy surface of a talus pile, whereas talus partially smothered by sediment has only low regions filled with sediment, leaving irregular high blocks exposed (compare, upper right portion of figure 22–3a with the right half of figure 22–3d). Numerous specimens have been dredged which contain basalt talus of variable size cemented by layers of manganese deposits that conform to the talus contours. The exposure of laminated manganese-oxide deposits suggests faulting after deposition of the hydrothermal layers because layering in manganese deposits should normally be parallel to the seawater/ deposit interface (figure 22–4b). Dredged samples show broken, brecciated lower layers of manganese crust enveloped by younger layers indicative of faulting followed by renewed hydrothermal-fluid deposition. At drege locality 1A, the breccia apparently forms the wall rocks within the hydrothermal fluid vent. Hydrothermal manganese not only coats the rock/seawater interface but also fills fractures within the breccia (R. Scott et al., 1976). Some hydrothermal manganese crusts are found on sediments without any visible talus or breccia (figure 22–3a).

Several lithologies are associated with pillow-lava flows; the most widely distributed is a talus consisting of wedge-shaped pillow-lava fragments found close to fault scarps or flow fronts. Near the top of traverse 1 are a number of pillow-lava flow fronts which form tiers of pillows (figure 22–4c). Undisturbed pillow-flow tops are exposed with sediments filling the interstices between pillows (figure 22–5d). The abundance of smooth pillow-flow tops undisturbed by faulting found near the top of traverse 1 and at the trip-camera station suggests that off-axis extrusion may have occurred after most of the uplift from rift floors to crest highlands had ceased.

Distribution of Hydrothermal Manganese

In the 3.5-km-long camera traverses, twenty-two discrete hydrothermal manganese deposits occur on an average of 160 m apart, covering 7 percent of the traverses. The deposits are concentrated in patchy zones of continuous crust covering 5- to 40-m segments in between stretches that are 5 to 80 m wide (figure 22–2).

Most of the manganese deposits are related to faults; about 70 percent are either bounding a fault or straddling horsts. Seven deposits are found on heavily sedimented terraces up to 90 m from faults with no evidence of concealed faulting. The faults associated with the manganese deposits are small, averaging less than about 5 m in vertical displacement, and they are reasonably close together, about 37 m. Manganese is commonly found near outward-dipping faults and horst-and-graben structures; average dip of manganese-associated faults is 43°. Hydrothermal manganese occurs 3 times more frequently on faults within terraces than on faults within fault zones. The elongation of the deposits parallel to faults, as depicted in figure 22–2, is speculative and based on analogy with elongate pods or strings of manganese mound deposits at the Galapagos Spreading Center (Lonsdale, 1977).

(a)

(b)
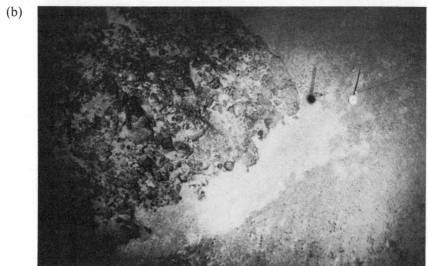

Figure 22–3. Photographs of bottom lithologies encountered on traverses 1 and 2. Field of view is approximately 6 × 4 m; compass with vane is 33.5 cm long. (a) Hydrothermal manganese-oxide-blanketing talus is seen in the upper-right corner forming hydrothermal breccia. Large cup-shaped objects at the bottom of the photographs are sponges with the concave feeding sides facing the right. (b) Tectonic breccia exposed on a steep fault

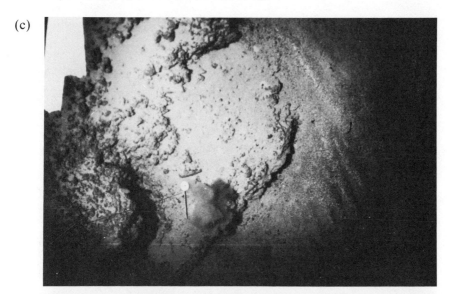

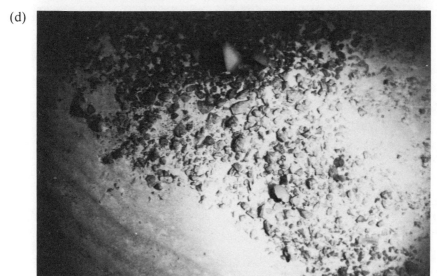

scarp. A few boulders of talus cover the surface and a thin veneer of sediments dusts the indentations of the breccia. Sediments at base show scour. (c) Scarp is semi-consolidated sedimentary breccia. Note the ripples in sediment with light foraminiferal crests and dark lag between. (d) Loose basalt talus covering sediment. Note the uniform orientation of the three sponges growing on the talus.

392 Dynamic Environment of the Ocean Floor

(a)

(b)
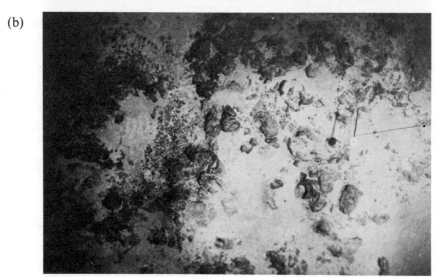

Figure 22–4. Photographs of bottom lithologies encountered on traverses 1 and 2. Field of view is approximately 6 × 4 m; compass with vane is 33.5 cm long. (a) Hydrothermal manganese crust covers sediment to the right. Even ripples of sediment cover the rest of the view. (b) Disrupted hydrothermal manganese

(c)

(d)

crusts showing exposures of interior laminations. (c) Lava flow front formed by three tiers of pillows. Elevation increases 7 m from upper left to lower right corner. (d) Wedge-shaped pillow lava fragments commonly observed close to flow fronts.

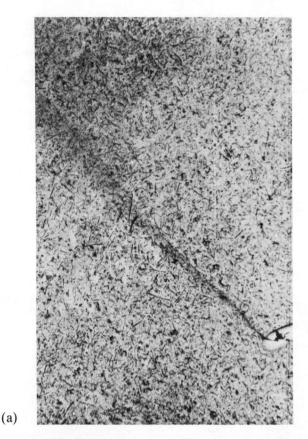

(a)

Figure 22–5. Bottom lithologies at top of rift valley wall 10 km east of traverse 2 and 5 km south of dredge site 1A (marked by X, figure 22–2). Photographs were taken by a trip-weight camera aboard the R/V *Kurchatov*; the field of view is approximately 2.0×1.3 m. (a) Light-colored foraminiferal ooze covered by a lag of twig-like skeletal parts of bryozoans and other organic debris.

Most of the hydrothermal deposits are found on talus ramps and discrete talus as hydrothermal breccia; lesser amounts are found on sediment. Although no manganese deposits were found on tectonic breccia along the camera traverses, samples of tectonic breccia from dredge site 1A are coated and filled with hydrothermal manganese (R. Scott et al., 1976). Of the four fault blocks which contain hydrothermal deposits, three consist of wide sediment expanses with localized talus at faults. The fourth fault block is highly faulted with a steeply dipping fault zone continuously covered with talus, usually consisting of pillow-lava fragments.

(b)

Figure 22–5. (b) Scattered pillow fragments lie on a sediment ooze surface scoured on the current side of basalt clasts with light-colored foraminiferal deposits in the lee of the clasts. Note abundant brittle stars on rocks.

Subinterface Observations

Methods

Dredged rocks which appeared to have been altered by seawater or hydrothermal fluids below the rock/seawater interface were studied in order to characterize hydrothermal mechanisms within the ocean crust. Correlations of fracture characteristics with chemical-mineralogical identification of alteration products provide insight into probable chemical and physical evolution of oceanic crust during hydrothermal activity. Minerals were identified by standard optical petrographic methods, x-ray diffraction

(c)

Figure 22–5. (c) Orthogonal normal fault patterns in sedimented breccia. A small scarp is parallel to the bottom of the photograph, and a small pair trend toward the top of the view. A crinoid grows on the edge of the large scarp.

(powder mounts, powder and single crystal cameras), and scanning electron microscopy. Major elements were analyzed by atomic-absorption spectrophotometry (Al, Fe, Mg, Ca, Na, and Mn), colorimetric spectrophotometry (Si, P), and x-ray fluorescence (Ti); trace elements were analyzed by instrumental neutron-activation analysis (La, Ce, Sm, Eu, Tb, Yb, Lu, Co, Cr, Sc, and Hf) and atomic-absorption spectrophotometry (Cu, Ni, and K). Both H_2O^+ and H_2O^-, along with volatiles such as CO_2, were determined by gravimetric loss of total volatiles. Sulfur content was determined by Leco furnace automatic titration method. USGS standard rocks were used to calibrate the methods.

(d)

Figure 22–5. (d) Top of a pillow basalt flow. This is similar to flat-flow tops seen at the top of traverse 1.

Results

Chemical results are summarized in table 22–2. Two rocks, a relatively unaltered crystalline rock-pillow interior (17–28R) and an unaltered glassy pillow rim (17–20G), provide an unaltered rock to which palagonitized rims can be compared. Figure 22–6 compares the percentage increase or decrease in concentration of each element to unaltered rock. Most rocks, however, do not have unaltered portions and must be compared to ranges of analyses of fresh glasses in the 25 to 27°N segment of the ridge crust; therefore, only

Table 22–2
Chemical Analyses of Altered Rocks: Major Elements

	19–3	17–28–P	17–28–R	17–20–P	17–20–G	1A–11	1A–17	1A–20	1A–24
SiO_2	43.87	36.58	48.60	31.04	49.82	48.18	45.16	48.78	47.78
Al_2O_3	18.32	16.45	16.07	14.11	15.44	14.82	17.59	15.92	14.22
TiO_2	1.87	—	0.91	2.52	1.58	1.71	1.90	1.71	2.00
FeO^a	10.99	13.77	7.56	15.90	9.47	9.75	10.05	9.51	9.88
MgO	5.15	6.26	9.94	4.99	8.36	9.33	5.78	7.46	7.42
CaO	10.20	6.55	22.54	5.51	9.90	6.94	9.13	9.34	7.64
Na_2O	2.84	2.44	4.39	2.28	2.94	3.27	2.75	3.06	3.27
K_2	0.07	0.51	0.21	0.60	0.08	0.09	0.22	0.14	0.22
MnO	0.20	0.45	0.16	0.31	0.18	0.16	0.19	0.46	0.76
P_2O_5	0.30	0.45	0.07	0.45	0.13	0.15	0.19	0.15	0.14
H_2O^+	1.76	4.75	2.02	14.03	1.29	2.11	5.05	2.26	2.91
H_2O^-	2.57	7.92	0.46	7.88	0.37	1.22	2.18	1.01	2.28

Note: Values in weight percent; 19–3 = low-temperature weathered basalt pillow (smectites and oxices); 17–28 P and R, 17–20 P and G = coexisting pairs of fresh rock (R) or glass (G) with palagonitic rims (P) of basalt pillows 1A–11, 1A–17, 1A–20, 1A–24, 1A–31, 1A–28 = low-temperature oxidizing hydrothermal alteration at vent of diabase (oxides, smectites and zeolites); 13–24, 2A8–1 = low-temperature less-oxidizing hydrothermal alteration of basalt pillows (zeolites, oxides, and smectites); 3B–1, 3B–12 =

Table 22–3
Alteration Products

Sample	Description of Alteration Products
19–3	Brown smectite and ferric iron oxides and hydroxides.
17–28–P	Palagonitized glass from rim; no zeolites.
17–28–R	Fresh crystalline interior; no significant alteration.
17–20–P	Palagonitized glass from rim; no zeolites.
17–20–G	Fresh glass rim below zone of palagonitization; no significant alteration.
1A–11, 1A–17, 1A–20, 1A–24, 1A–31	Veins of todorokite; poorly crystalline yellowish smectites.
1A–28	Glass contains veins of analcite with minor phillipsite; minor yellow smectite.
13–24, 2A8–1	Veins of zeolite cut by veins of iron oxides-hydroxides; no chlorite. Small amounts of blue-green smectites.
3B–1, 3B–12	Chlorite (penninite), chalcopyrite, pyrite, quartz in veins; chlorite in groundmass; atacamite, quartz and iron oxide veins cut earlier veins.
8A–3, 15–8	Oldest veins of talc and chlorite, cut by zeolite veins, cut by youngest veins of iron oxides-hydroxides; chlorite in groundmass and replacing olivine.

1A-31	1A-28	13-24	2A8-1	3B-1	3B-12	15-8	8A-3	Percent Standard Deviation
47.58	45.42	47.39	40.37	46.05	49.03	47.27	51.32	1.35
14.37	13.64	15.68	16.46	15.77	15.12	15.48	14.85	3.76
1.69	1.93	1.72	1.91	1.46	1.50	1.46	1.44	5.27
9.81	9.68	8.39	8.58	10.68	9.24	8.57	7.64	1.57
6.98	8.09	8.67	13.19	7.34	7.48	9.45	6.82	1.08
9.58	7.35	8.45	2.53	10.01	9.85	7.74	9.22	2.36
3.37	3.61	2.66	2.74	2.42	2.04	3.43	3.31	1.83
0.21	0.95	0.19	0.82	0.03	0.02	0.09	0.25	1.14
1.05	0.25	0.23	0.23	0.30	0.25	0.13	0.21	1.03
0.13	0.10	0.009	0.08	0.14	0.10	0.09	0.12	0.01
3.47	5.82	3.38	8.76	4.56	4.18	5.05	1.92	2.87
1.23	2.68	3.19	5.38	0.56	0.54	1.76	1.13	1.93

moderate-temperature hydrothermal alteration of diabase (chlorite, quartz, actinolite-tremolite, pyrite, chalcopyrite); 8A-3, 15-8 = moderate-temperature hydrothermal alteration of pillow basalt (chlorite, quartz, pyrite, chacopyrite). Sample numbers (e.g., 19-3, 17-28-P, 1A-11) have the dredge sites as prefixes (e.g. 19, 17, 1A).
[a]Total iron expressed as FeO.

departures from these ranges can be considered to be an alteration. The mineral products of alteration are listed and grouped in table 22-3.

Implications of Chemistry and Mineralogy

The alteration mineral assemblages fall into four classes: (1) low-temperature weathering products, (2) palagonitization that may occur either at elevated temperatures or at ambient seawater temperatures, (3) probable low-temperature hydrothermal mineral deposition and related alteration at or close to vents in the sea floor, and (4) hydrothermal mineral deposition and alteration products resulting from hydrothermal fluids or elevated temperatures. The organization of table 22-2 reflects that classification. Chemical variations in altered rocks are compared to ranges of values found in fresh unaltered samples to correct for variations due to fractionation alone; variations beyond those ranges are assumed to be from alteration.

Class 1. Sample 19-3 is the only example of low-temperature weathering; this rock was dredged at site 19 from 2.8 My-old oceanic crust 30 km from the rift axis. The groundmass has been totally altered to reddish-brown smectites, and plagioclase phenocrysts are coated with iron-oxide hydroxides

400 Dynamic Environment of the Ocean Floor

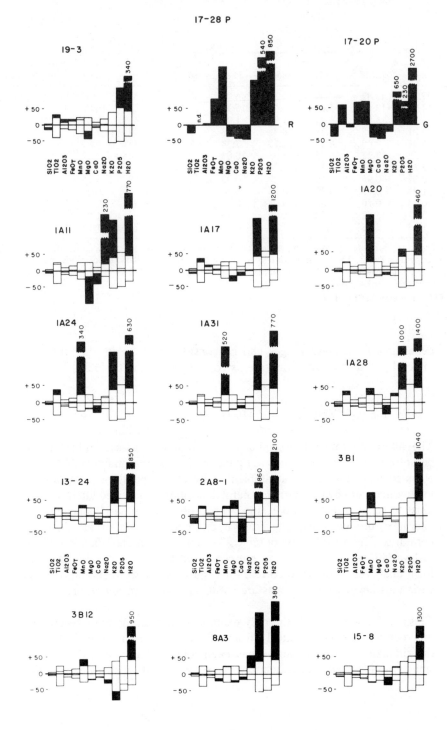

Nature of Hydrothermal Exchange

similar to weathered rocks (Matthews, 1971; Miyashiro et al., 1971). The Mg content from sample 19-3 is significantly lower than the range of Mg contents found in unaltered pillow-glass rims (R. Scott and Hajash, 1976). Such a difference fits the observation that Mg is lost during low-temperature seawater weathering (Thompson, 1973; R. Scott and Hajash, 1976), but the increase in K expected from weathering is not observed. A 15-mm-thick layer of hydrogenous manganese oxide (Fe/Mn = 1.7; Ni = 1,110 ppm, Co = 4,950 ppm, Cu = 405 ppm) encrusts the rock surface.

Class 2. Coexisting pairs of palagonitized glass and unaltered pillow interiors from samples 17-28 and 17-20 were analyzed. Chemically the palagonitized glasses are significantly enriched in K, P, and water and are depleted in Mg, as expected during low-temperature alteration of glasses. However, numerous palagonitized rocks collected from dredge site 17 have palagonite layers several millimeters thick. At measured rates of mid-ocean-ridge palagonitization (Hekinian and Hoffert, 1975), the thickness is proportional to the square root of time (Moore, 1966); such palagonitization would require millions of years to form. Yet the crust at dredge site 17 should not be more than 300,000 years old based on magnetic-anomaly dating. Also, many of the palagonitized rims show evidence of being junctions between two pillows. Thus it is likely that seawater was trapped and heated within a stack of pillows; the retention of higher temperatures would increase palagonitization rates by several orders of magnitude (Friedman and Long, 1976; Lofgren, 1970).

Class 3. The next group, samples 1A–11, –17, –20, –24, –31, and –28 from dredge site 1A within an inactive part of the hydrothermal field, is much more pertinent to hydrothermal-alteration mechanisms. Alteration within those

---

Figure 22-6. Major element variations in altered rocks compared to the range of analyses of fresh basaltic glass from the 25 and 27° segment of the Mid-Atlantic Ridge. Zero marks the average of fresh rock analyses; white bars indicate the range of fresh rock analyses in percent above or below the average; and the black bars indicate alteration that exceeds the fresh range in percent above or below the average. Large enrichments in K_2O, P_2O_5, and H_2O are indicated by discontinuous bars and the percent enrichment value. Palagonitized outer rims of 17-28 and 17-20 are compared to their fresh crystalline rock interior and fresh glassy rim, respectively. Total iron is expressed as FeO_T.

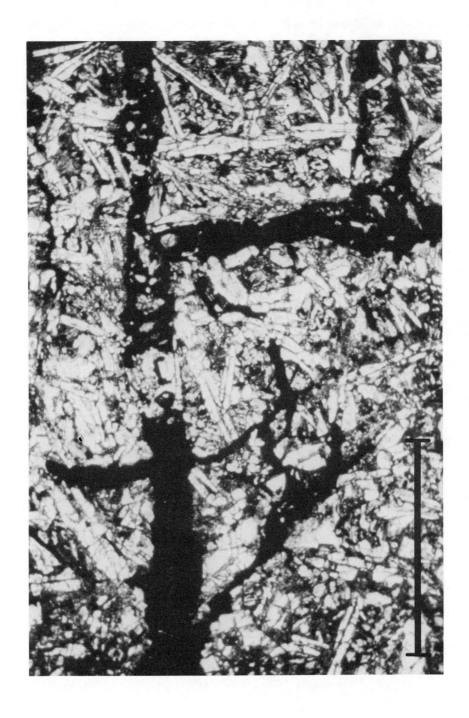

rocks can be attributed, at least in part, to hydrothermal reaction because the rocks were dredged from a tectonic breccia coated with a hydrothermal birnessite and todorokite crust (R. Scott et al., 1976). Chemically the crusts fall into the hydrothermal rather than hydrogenous manganese-oxide field. Fractures below the crust are filled with hydrothermal todorokite (figure 22–7). However, no high-temperature hydrothermal alteration phases are present; only minor yellowish brown smectites are found as alteration products, along with analcite and phillipsite veins. Thus the hydrothermal fluids were apparently vented through fault-fractured rocks at relatively low temperature and were sufficiently oxygenated for todorokite to precipitate in the veins. The mechanical effect of fracturing and hydrothermal alternation has produced a friable rock easily crumbled by hand. A thin section of sample 1A–20 (3.5 × 2.2 cm) exhibits three major interconnecting fractures which have 0.1 mm average width; the largest fracture is 0.9 mm wide. Hand specimens show the density of fractures to create a 1 percent porosity. Some fractures in sample 1A–20 are not completely filled and show a distinct botryoidal growth surface of todorokite apparently terminated by clogging of the more restricted portions of the fractures.

One rock dredged at site 1A does not have abundant manganese oxides filling its fractures; instead, sample 1A–28 has abundant veins of analcite and phillipsite cutting a hydrated basaltic glass. One large vein was only partially filled and contains free-crystal facies of analcite up to 5 mm in diameter. The analcites are low in Ca but somewhat potassic (K/Na = 0.1), suggesting that fluids were fairly low-temperature if analcite aqueous equilibria are similar to alkali-feldspar aqueous equilibria (Orville, 1963).

Samples studied at dredge site 1A show no increase in MgO or decrease in K_2O expected with hydrothermal exchange at elevated temperatures. Instead, the large increase in K_2O and slight decrease in MgO in the solid phases are more suggestive of exchange with aqueous fluids at low temperatures.

Another group of samples, 13–24 and 2A8–1, from within the hydrothermal field have mineralogies similar to what might form during hydrothermal alteration by fluids less oxygenated than those at dredge site 1A, but there is no clear evidence of high-temperature alteration. Even though samples 13–24 and 2A8–1 were collected in a dredge haul containing fragments of hydrothermal manganese-oxide crusts, fresh glass and olivine

Figure 22–7. Veins of todorokite filling fractured tectonic breccia sample 1A–20 in plane light. Bar is 1 mm long. Largest veins in this photograph are about 0.20 mm wide; anastomising veins that follow grain boundaries are typically 0.01 mm wide.

crystals are still preserved. Veins of zeolites in sample 2A8-1 (largely analcite) are cut by iron-oxide veins that form 1.5 percent continuous connecting porosity. Except for an increase in K_2O and H_2O, the rock is not highly altered. Sample 2A8-1 also contains abundant hydrated glass enriched in K_2O with minor devitrification and only a trace of blue-green smectite. However, the MgO content of this rock is considerably higher than that expected by low-temperature alteration. The altered glass does not have a typically palagonitic physical appearance either in hand specimen or in thin section. Although the temperature and composition of aqueous fluids that affected those two rocks cannot be clearly defined, it appears that 13-24 was affected first by less oxidizing fluids which precipitated zeolites and then by a more oxidizing fluid that precipitated iron oxides-hydroxides. Only one stage of minor green-smectite vein precipitation affected sample 2A8-1.

Class 4. Samples 3B-1 and 3B-12, from a steep, west-facing scarp along the eastern rift-valley wall, have textural evidence of a deeper plutonic origin than samples discussed earlier and mineralogical evidence of distinctly high-temperature secondary hydrothermal metamorphism. The rocks display a diabasic intergrowth of clinopyroxene and plagioclase characteristic of shallow plutons or dikes. Much of the original ground-mass mineralogy has been altered; most of the olivine and some of the clinopyroxene has been replaced with chlorite (penninite). However, the plagioclase has not retrograded to albite, and minor orthopyroxene still appears to be fresh. Clots of chlorite in sample 3B-1 contain traces of fibrous actinolite-tremolite, and in sample 3B-12 the same fibrous mineralogy occurs in small pockets of quartz unrelated to visible fracture systems. Veins in both rocks consist of quartz, penninite, patchy or tuftlike growths of fibrous actinolite-tremolite, and the sulfide pair pyrite-chalcopyrite. That association is present in figure 22-8, where the chlorite rims the vein-polygonal quartz and sulfides compete for the center, and actinolite-tremolite fibers grow within the quartz. Two attitudes of veins are present, intersecting at 50 to 80° acute angles; they may represent the filling of sets of conjugate fractures. One set is dominant and forms 3 percent continuous fracture porosity before vein filling, while the other has an average of 1 percent fracture porosity. Thus the rock averages a 2 percent continuous fracture porosity. If the fractures were induced with gravity as the principal stress, then 2 precent is an estimate of vertical porosity within fractured zones on the rift-valley wall.

Figure 22-8. Hydrothermal vein in sample 3B-12 consisting of outer layers of chlorite (penninite) and filled with polygonal quartz and intergrowths of pyrite-chalcopyrite. The vein is about 0.7 mm wide. The bar is 1 mm long. Partially crossed nichols.

Although no obvious cross-cuting relations are clear and relative ages are unknown, veins of zeolites and hematite-like iron oxide are present in 3B–12 in addition to chlorite veins. Therefore, conditions varied from high-grade greenschist facies with actinolite-tremolite-chlorite-sulfide assemblages to zeolite assemblages to nearly ambient-temperature oxide assemblages.

The chemistry of samples 3B–1 and 3B–12 (figure 22–6) is not consistent with the generalization that hydrothermal alteration at elevated temperatures is accompanied by additions of Mg. In fact, both MgO values are close to the lower range of fresh values and may represent a loss of MgO. However, a significant loss of K_2O in both samples does fit trends of hydrothermal metamorphism.

Rocks 8A–3 and 15–8 contain excellent cross-cutting vein relationships that clearly define the age relationships; the mineralogy is similar to that found at dredge site 3B. Dredge site 8A is found on a 1-km-high steep scarp that is roughly orthogonal to the rift valley, and dredge site 15 is parallel to the rift valley and only 0.5 km high (figure 22–1a and b). The sequence of cross-cutting events in sample 15–8, demonstrated in figures 22–11 and 22–12, are (1) chlorite and quartz ± pyrite and chalcopyrite cut by (2) zeolites ± talc cut by (3) iron oxides-hydroxides. Veins make up 3.5 percent of the sample in a continuous subparallel set of filled fractures. Besides vein formation of greenschist-type mineralogy in the oldest veins, the groundmass of sample 15–8 has olivine phenocrysts replaced by chlorite (penninite). Samples 15–8 and 8A–3 both have pillow-basalt textures of phenocrysts set in a fine-grained groundmass. Sample 8A–3 is remarkable primarily because it consists largely of a quartz vein that is more than 2 cm thick. Unfortunately, the entire vein was not recovered in dredging.

Chemically, rocks 8A–3 and 15–8 present an enigma; neither sample shows additions of MgO nor losses of K as predicted for hydrothermal metamorphism at elevated temperatures. In fact, sample 8A–3 shows a slight decrease in MgO and a large increase in K_2O, and sample 15–8 shows no appreciable departure from the range of fresh rocks (figure 22–6).

Mineralogically, the altered rocks can be readily grouped in decreasing grades of hydrothermal metamorphism. Rocks 3B–1 and 3B–12 have the highest greenschist facies and contain actinolite and chlorite. Greenschist-facies rocks 15–8 and 8A–3 contain chlorite but no amphiboles. Zeolite-facies rocks 13–24 and 2A8–1 have no chlorite but contain abundant zeolites. Rocks at site 1A underwent low-temperature hydrothermal metamorphism primarily by alteration to smectite, by precipitation of hydrothermal manganese in fractures, and by minor precipitation of zeolites. The lowest grade of hydrothermal metamorphism may have been the formation of palagonite autometamorphism of trapped seawater and glass within a pillow-lava flow. However, chemically, none of those rocks have the characteristics of both higher MgO and lower K_2O values thought to be related to higher-temperature hydrothermal metamorphism (R. Scott and Hajash, 1976, Thompson, 1973). Those rocks formed by lower-temperature metamorphism do have the character of lower MgO and higher K_2O values expected by low-temperature alteration.

Discussion and Conclusions

During hydrothermal interaction between crustal rocks and seawater circulating in convection cells within the ocean crust, the dynamics of heat exchange and the kinetics of chemical exchange are largely functions of (1) the physical character of the conduits of aqueous transfer, (2) the evolution of those conduits during growth of hydrothermal minerals on conduit walls, (3) the structural evolution of the rift-valley wall, and (4) the evolution of the geothermal gradient as a response to the heat exchange.

Theoretical models have been constructed to describe the hydrothermal systems at mid-ocean ridges using the difference in theoretical heat production associated with sea-floor spreading and observed heat-flow measurements to calculate the rate of hydrothermal heat convection (Wolery and Sleep, 1975). Continued use of both thermal and chemical constraints by Sleep and Wolery (1978) has led to a model that requires the venting of hot water (100 to 300°C) through discrete conduits at high discharge rates [17 to 29 g/(m·s)]. That allows episodic cracks to avoid heat loss by conduction and to avoid sealing off the flow by precipitation of quartz, other silicates, and sulfides. Further, Sleep and Wolery calculated that episodic discharge requires 2-cm-wide cracks distributed several kilometers to tens of kilometers apart.

Comparison of those conclusions with the observations in the 26°N hydrothermal field requires significant changes in concepts of seawater circulation in convection cells in the ocean crust. Low-temperature hydrothermal venting (17°C) has been measured in the Galapagos vents by a submersible (J.B. Corliss, personal communication). At dredge site 1A, the hydrothermal manganese oxides that coat both the rock surface and fractures connecting to that surface and the smectite alteration within the rocks probably require nearly ambient seawater temperatures. The FAMOUS hydrothermal site has no evidence of high-temperature venting (M. Scott et al., 1979). Thus direct observation and indirect evidence suggest that many hydrothermal systems are cool by the time they reach the surface. The cooling may be caused by two mechanisms: (1) conductive heat loss to the surrounding rocks, producing an abnormally high geothermal gradient; and (2) advection of cool seawater into the margins of a brecciated fracture-zone complex or a talus pile, allowing mixing of the hotter waters with seawater. If enough seawater does mix into hydrothermal systems near vent sites below

the surface, the dilution of hydrothermal systems would inhibit silicate precipitation (e.g., a 40:1 seawater to hydrothermal fluid according to Seyfried and Bischoff, 1977), provide a source of oxygen for manganese-oxide precipitation, and lower the temperature of the venting waters (40:1 ratio would lower 300°C hydrothermal fluids to 11°C if the ambient seawater temperature were 4°C).

Clearly the hydrothermal metamorphism at the 26°N. Mid-Atlantic Ridge region is highly episodic. The observation of refaulted cemented talus and hydrothermal crust blocks seen in figures 22-3c, 22-4b, and 22-5c combined with the cross-cutting vein mineralogies seen in thin sections in figures 22-9 and 22-10 requires repeated disruption of fault zones that were previously clogged by minerals. As many as three periods of faulting are recorded before the fourth period of faulting exposed the rocks on the surface. That confirms the episodic aspect of the Sleep-Wolery model already described. But obvious repeated clogging of veins by silicates and sulfides and by hydrothermal fluids closer to the surface invalidates the Sleep and Wolery restriction that a high rate of discharge is required to avoid the loss of conductive heat, which would result in sealing the cracks by precipitation. Because clogging occurs, a high discharge rate is not required.

There is at least qualitative evidence of a stepwise progressive transport of the faulted rocks from a high-grade greenschist metamorphic environment under reducing conditions of sulfide precipitation to an alteration environment under oxidizing conditions that produce precipitation of iron oxides and hydroxides. Pyrite and chalcopyrite formed earlier are unstable under the new low-temperature, oxygen-enriched conditions. Pyrite is oxidized to iron oxides that rim remnant pyrite crystals where oxide-bearing veins are close to older sulfide-bearing veins. In dredge 3B rocks, the copper-chloride hydrate, atacamite [$Cu_2(OH)_3Cl$], is found within fractures with quartz and iron oxides. Bonatti et al. (1976) have also reported the presence of atacamite in oceanic rocks. Prior to those reports, the mineral has been associated with extremely saline and arid environments such as the Atacama Desert of northern Chile (Palache et al., 1963). Apparently the activity of chloride ion in the fluids that secondarily oxidize the sulfide veins is sufficient to create a stable environment in the ocean crust. A possible reaction that may form atacamite from chalcopyrite is

$$2CuFeS_2 + 5H_2O + 17/2\,O_2 + Cl^-$$
$$\rightleftharpoons Cu_2(OH)_3Cl + Fe_2O_3 + 4SO_4^{2-} + 7H^+ + 32e^-$$

Figure 22-9. Cross-cutting veins in sample 15-8. Chlorite vein (C) is cut by a larger zeolite vein (Z), which is in turn cut by a small hematitic vein (H) parallel to the chlorite vein. The zeolite vein is about 0.20 mm wide.

410   Dynamic Environment of the Ocean Floor

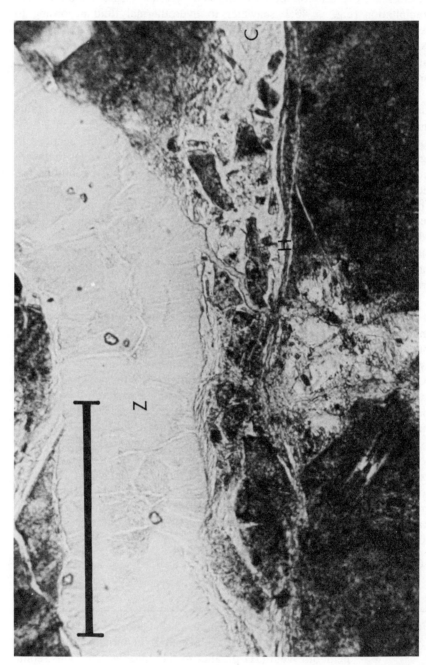

Figure 22-10. Details of cross-cutting veins in 15–8. Chlorite (C), zeolite (Z), and hematite (H). Bar is 0.25 mm long.

This reaction would occur under chlorine-rich oxidizing conditions similar to those of seawater to produce low-pH solutions enriched in sulfate ions. Because Bianchi and Longhi (1973) show atacamite to be stable in seawater, the oxidation of copper sulfides may have occurred by exposure to seawater along fractures near the surface under ambient sea-floor conditions.

Another major problem with the Sleep-Wolery model is the distribution and size of fractures predicted. The model requires 2-cm-wide episodic cracks to be distributed between several kilometers to tens of kilometers apart. Yet in the bottom photographs we observe discrete faults spaced an average of 30 m apart through 3.5 km of observation. It is significant that the hydrothermally affected dredged rocks, which are presumably exposed by faulting and thus represent rocks affected by faulting, contain evidence that continuous fracture porosity may be very high. For example, dredge site 1A rocks have an average of 1 percent fracture porosity prior to mineral clogging, dredge site 2A rocks have 1.5 percent fracture porosity, dredge site 3B rocks have 2.0 percent fracture porosity, and dredge site 15 rocks have 3.5 percent fracture porosity. Dredge site 8A has one specimen with a 2-cm-wide quartz vein in it with an unknown ratio of fracture space to rock. The average fracture width is about 0.5 mm, and the average continuous-crack porosity is 2 percent of the volume. If each fault has a 1-m-wide fracture zone associated with it, it would produce 2 cm of fractures for each fault, giving a cumulative crack width two to three orders of magnitude greater than that predicated in the Sleep-Wolery model. Of course, the fluid dynamics of forty thin 0.5-mm cracks is considerably different from that of one 2-cm crack. Thus a model constructed by using 0.5-mm cracks, a 30-m spacing of faults on the rift valley wall, a low discharge rate, considerable conductive heat loss to the country rock, and consequent frequent clogging of crack conduits may be more realistic model to test for the type of hydrothermal exchange in the 26°N hydrothermal field. Also, thin cracks would create a much lower water-to-rock ratio by increasing the surface area exposed to the water required to convectively cool the new lithosphere. Thus, flow models explaining the patterns of heat-flow anomalies for the entire mid-ocean ridge system do not necessarily account for the hydrothermal exchange requirements, and vice versa. If the model for hydrothermal exchange in the 26°N field was used to explain the worldwide heat-flow patterns, the oceans could become rapidly depleted in Mg. Thus evidence suggests that conditions of flow and exchange vary greatly from place to place.

Based on experimental evidence, Seyfried and Bischoff (1977) consider that water-to-rock ratios play a critical role in the chemical evolution of hydrothermal fluids. They make a case for high effective water-to-rock ratios (50:1), pointing out that the low pH is retained by maintaining a large excess of Mg at equilibrium with alteration phases. If a low water-to-rock ratio (5:1) at low temperatures (200 to 300°C) exists, the fluids will first record a decrease in pH accompanied by release of heavy metals from the rock; but as the Mg in the fluid is depleted and H^+ ions are used in hydrolysis, the pH will rise and concentration of metals will decrease. The assumption that high water-to-rock ratios must exist for effective metal transport in hydrothermal systems does not apply to the 26°N field because (1) excess Mg was not maintained in the fluids, (2) abundant metal-sulfide phases and silicates precipitated in clogged fractures, and (3) penetrative metamorphic alteration of the entire rock far from discrete fractures requires that the bulk of the rock experienced low water-to-rock ratios by diffusion of the aqueous system along narrow-grain boundary conduits with high surface areas. Even if the major fractures contain a high water-to-rock ratio with high Mg values and low pH, that condition will be reversed as the Mg is consumed by reaction within the rock; the pH will rise as H^+ ions are used in hydrolysis reactions with the rock interior. If excess Mg was preserved during reactions, then both high- and low-temperature metamorphic reactions would produce significant Mg enrichment in altered basalts; such is not the case (figure 22–6). Presumably, the water-to-rock ratio was low enough to have effectively removed Mg from the hydrothermal fluids before they affected the rocks studied, and thus no increase in Mg content was possible. As Seyfried and Bischoff (1977) point out, the pH probably increases as OH^- ions are no longer removed by Mg-phase precipitation (and perhaps a Ca-phase precipitation, Andrew Hajash, personal communication), and H^+ ions are used in rock hydrolysis. Those changes cause the precipitation of metals seen as sulfides within veins in rocks from dredge 3B.

Although the model described by Sleep and Wolery may correctly describe the nature of the circulation of hydrothermal fluids for the bulk of the oceanic crust at mid-ocean rises, the circulation at 26°N on the Mid-Atlantic Ridge clearly does not follow such a model. Also, the necessity of a high water-to-rock ratio required for effective metal transport by Seyfried and Bischoff does not seem to operate in the 26°N hydrothermal field.

Two important differences between the Mid-Atlantic Ridge and the East Pacific Rise exist: (1) the Mid-Atlantic Ridge is highly faulted with a distinct rift, whereas the East Pacific Rise has much less pronounced structures; (2) the Mid-Atlantic Ridge has only slightly metalliferous sediments (M. Scott and Salter, 1977; M. Scott et al., 1979), whereas the East Pacific Rise has highly metalliferous sediments (Dymond et al., 1973). Differences in the structure and spreading rates of the two ridges may control the nature of hydrothermal circulation, exchange between crust and seawater, and the degree of hydrothermal metal enrichment of the sediments.

Correlation of all available data from interpretations of bottom photographs along traverses 1 and 2 indicates that a combination of inward-facing normal faulting, outward-facing normal faulting, and outward tilting by rotation of fault blocks accounts for the decay of rift valley relief in the rift mountains (Temple et al., 1979). The fault attitudes are consistent with the model of Needham and Francheteau (1974) and with Osmaston (1971) and Lachenbruch (1973) who favor a mechanism of uplift of the rift walls by viscous drag in the lithospheric conduit. Moreover, the pattern of inward-facing normal faulting in the rift valley, followed by outward-facing normal faulting with outward tilted blocks in the rift mountains, is consistent with the model of Harrison and Stieltjes (1977) and Macdonald and Luyendyk (1977). In their models, the inward-dipping normal faulting creates relief and outward-dipping faulting and outward tilting by block rotation destroys relief. The Mid-Atlantic Ridge at 26°N seems to combine both.

In any case, there is considerable evidence that the episodic faulting must affect the highly faulted rift walls at 26°N. It creates abundant narrow fractures for conduits and effectively lowers the water-to-rock ratio by increasing the surface area. The lower water-to-rock ratio and the closely spaced faults produce a much lower flow rate and discharge rate and considerable conductive heat loss to the country rock. The Mg content of the fluids is depleted, the pH rises, and the temperature drops. The combined effect of those phenomena causes precipitation of metals as sulfides and silicates, clogging the cracks until the next episodic faulting during growth of the rift wall or decay to the highlands (dredge site 3B). As the expanded and cooled fluids reach the highly fractured rocks near the surface (dredge site 1A), they are mixed with cold, oxygenated seawater from lateral mixing in tectonic breccia complexes. Very few metals except Mn are left after removal as sulfides, so hydrothermal Mn-oxides fill fractures and form crusts at the seawater/rock interface. Because relatively few metals are effectively transported to the surface in such an environment, there are few metals which can be incorporated in metalliferous sediments as oxide or hydroxide particulates. It is interesting to note that Lister (1976) predicted an active geothermal area every 20 km along the Mid-Atlantic Ridge, but only every 3 km along the East Pacific Rise. To prevent confusion, it should be made clear that the presence of hydrothermal venting discussed by Lister (1976) and Sleep and Wolery (1978) does not infer the presence of a 26°N or Galapagos-type of hydrothermal *deposit* field. Further special structural environmental conditions, i.e., closely spaced fractures, are necessary for the formation of a hydrothermal deposit field in a hydrothermal venting area.

The East Pacific Rise, with less structural relief, more widely spaced faults, and faster spreading rates opening wider discrete fractures, would more nearly fit the models of Sleep and Wolery (1978) and Seyfried and Bischoff (1977), where high water-to-rock ratios prevail, little metal sulfide precipitation occurs, and large amounts of metals are vented into the seawater. The metals are then transported as oxides to form highly metalliferous sediments.

Acknowledgements

The cooperation of NOAA and Soviet colleagues during surveys and dredging provided the samples and much of the background for this synthesis. Support for this research came from NSF grant OCE76-18567.

References

ARCYANA, 1975. Transform fault and rift valley from bathyscaphe and diving saucer. *Science* 190:108–117.

Ballard, R.D., Bryan, W.B., Heirtzler, J.R., Keller, G., Moore, J.G., and van Andel, Tj. 1975. Manned submersible observations in the FAMOUS area: Mid-Atlantic Ridge. *Science* 190:103–108.

Bianchi, G., and Longhi, P. 1973. Copper in seawater, potential pH diagrams. *Corrosion Science* 13:8–59.

Bonatti, E., Guerstein-Honnorez, B.M., and Honnorez, J. 1976. Copper-iron sulfide mineralizations from equatorial Mid-Atlantic Ridge. *Econ. Geol.* 71:1515–1525.

Cann, J.R., Winter, C.K., and Pritchard, R.G. 1977. A hydrothermal deposit from the floor of the Gulf of Aden. *Min. Mag.* 41:193–199.

Corliss, J.B., Dymond, J., Lyle, M., Cobler, R., Williams, D., Von Herzen, R., and van Andel, Tj. 1977. Observations of the sediment mounds of the Galapagos Rift during the ALVIN diving program. *Geol. Soc. Am. Abstr. with Programs* 9:973.

Dymond, J., Corliss, J.B., Heath, G.R., Field, C.W., Dasch, E.J., and Veeh, H.H. 1973. Origin of metalliferous sediments from the Pacific Ocean. *Geol. Soc. Am. Bull.* 84:3355–3372.

Friedman, I., and Long, W. 1976. Hydration rate of obsidian. *Science* 191:347–352.

Hajash, A. 1975. Hydrothermal processes along mid-ocean ridges: An experimental investigation. *Contrib. Mineral. Petrol.* 53:205–226.

Harrison, C.G.A., and Stieltjes, L. 1977. Faulting within the median valley. *Tectonophysics* 38:137–144.

Hekinian, R., and Hoffert, M. 1975. Rate of palagonitization and manganese coating on basaltic rocks from the rift valley in the Atlantic Ocean near 36°50′N. *Mar. Geol.* 19:91–109.

Lachenbruch, A.H. 1973. A simple model for oceanic spreading centers. *J. Geophys. Res.* 78:3395–3416.

Lister, C.R.B. 1972. On the thermal balance of a mid-ocean ridge: *Geophys. J. R. Astron. Soc.* 26:515–535.

Lister, C.R.B. 1974. On the penetration of water into hot rock. *Geophys. J. R. Astron. Soc.* 39:465–509.

Lister, C.R.B. 1976. Qualitative theory on the deep end of geothermal systems. *Proc. 2nd. U.N. Geothermal Symposium*, Washington: U.S. Govt. Print. Off., pp. 459–464.

Lofgren, G. 1970. Experimental devitrification rate of rhyolite glass. *Geol. Soc. Am. Bull.* 81:553–560.

Lonsdale, P. 1977. Deep tow observations at the Mounds Abyssal Hydrothermal Field, Galapagos Rift. *Earth Planet. Sci. Lett.* 36:92–110.

Lowell, R.P. 1975. Circulation in fractures, hot springs, and convective heat transport on mid-ocean ridge crests. *Geophys. J. R. Astron. Soc.* 40:351–367.

Macdonald, K., and Luyendyk, B. 1977. Deep-tow studies of structure of the Mid-Atlantic Ridge crest near 37°N (FAMOUS). *Geol. Soc. Am. Bull.* 88:621–636.

Matthews, P.H. 1971. Altered basalts from Swallow Bank, an abyssal hill in the NE Atlantic and from a nearby seamount. *Phil. Trans. R. Soc. Lond.* A268:551–571.

McGregor, B.A. 1974. Crest of the Mid-Atlantic Ridge at 26°N: topographic and magnetic patterns. Unpublished Ph.D. dissertation, University of Miami, Coral Gables.

Miyashiro, A., Shido, F., and Ewing, M. 1971. Metamorphism in the Mid-Atlantic Ridge near 24°N and 30°N: *Phil. Trans. R. Soc. London.* A268:589–603.

Moore, J.G. 1966. Rate of palagonitization of submarine basalt adjacent to Hawaii. U.S. Geol. Survey Prof. Paper 550-D, D163-D171.

Moore, W.S., and Vogt, P.R. 1976. Hydrothermal manganese crusts from two sites near the Galapagos spreading axis. *Earth Planet. Sci. Lett.* 29:349–356.

Mottl, M.J. 1975. Chemical exchange between seawater and basalt during hydrothermal alteration of the oceanic crust. Unpublished Ph.D. Thesis, Harvard University, Cambridge, Mass.

Needham, H.D., and Francheteau, J. 1974. Some characteristics of the rift valley in the Atlantic Ocean near 36°48'N. *Earth Planet. Sci. Lett.* 22:29–43.

Orville, P.M. 1963. Alkali-ion exchange between vapor and feldspar phases. *Am. J. Sci.* 264:273–288.

Osmaston, M.F. 1971. Genesis of ocean ridge median valley and continental rift valleys. *Tectonophysics* 11:387–405.

Palache, C., Berman, H., and Frondel, C. 1963. *The System of Mineralogy of James Dwight Dana and Edward Salisbury Dana*, 7th ed. New York: Wiley, pp. 69–73.

Ribando, R.J., Torrance, K.E., and Turcotte, D.L. 1976. Numerical models for hydrothermal circulation in the oceanic crust. *J. Geophys. Res.* 81:3007–3012.

Rona, P.A., Harbison, R.N., Bassinger, B.G., McGregor, B.A., and Scott, R.B. 1976. Tectonic fabric and hydrothermal activity of the Mid-Atlantic Ridge crest (26°N Latitude). *Geol. Soc. Am. Bull.* 87:661–674.

Scott, M.R., Salter, P.S., and Barnard, L.A. 1979. Chemistry of ridge crest sediments from the North Atlantic Ocean. In 2d Maurice Ewing Symposium Volume, submitted.

Scott, M.R., and Salter, P.S. 1977. Metal accumulation rates in sediments from the FAMOUS area on the Mid-Atlantic Ridge. *Trans. Am. Geophys. Union* 58:420.

Scott, M.R., Scott, R.B., Rona, P.A., Butler, L.W., and Nalwalk, A.J. 1974. Rapidly accumulating manganese deposit from the median valley of the Mid-Atlantic Ridge. *Geophys. Res. Lett.* 1:355–358.

Scott, M.R., Morse, J.W., Betzer, P.R., Butler, L.W., and Rona, P.A. 1978. Metal-enriched sediments from the TAG Hydrothermal Field. *Nature,* 276:811–813.

Scott, R.B., and Hajash, A. 1976. Initial submarine alteration of basaltic pillow lavas. *Am. J. Sci.* 276:480–501.

Scott, R.B., Malpas, J., Rona, P.A., and Udintsev, G. 1976. Duration of hydrothermal activity at an oceanic spreading center, Mid-Atlantic Ridge (lat 26°N). Geology 4:233–236.

Scott, R.B., Rona, P.A., McGregor, B.A., and Scott, M.R. 1974. The TAG hydrothermal field. *Nature* 251:301–302.

Seyfried, W., and Bischoff, J.L. 1977. Hydrothermal transport of heavy metals by seawater: The role of seawater/basalt ratio. *Earth Planet. Sci. Lett.* 34:71–77.

Sleep, N.H. and Wolery, T.J. 1978. Egress of hot water from mid-ocean ridge hydrothermal systems: some thermal constraints. *J. Geophys. Res.* 83:5913–5922.

Temple, D.G., Scott, R.B., and Rona, P.A. 1979. Geology of a submarine hydrothermal field, Mid-Atlantic Ridge, 26°N Lat. *J. Geophys. Res.* 84:7453–7466.

Thompson, G. 1973. A geochemical study of the low-temperature interaction of seawater and oceanic igneous rocks. *Trans. Am. Geophys. Union* 54:1015–1019.

Thompson, G., Woo, C.C., and Sung, W. 1975. Metalliferous deposits on the Mid-Atlantic Ridge. *Geol. Soc. Am. Abstr. with Programs* 7:1297–1298.

Weiss, R.F., Lonsdale, P., Lupton, J.E., Bainbridge, A.E., and Craig, H. 1977. Hydrothermal plumes in the Galapagos Rift. *Nature* 267:600–603.

Wolery, T.J., and Sleep, N.H. 1975. Hydrothermal circulation and geochemical flux at mid-ocean ridges. *J. Geol.* 84:249–275.

23 On Mantle Helium, Argon, and Methane Discharge in Thermal Spring Waters of Ocean Margin and Ridge Areas

L. K. Gutsalo

Abstract

Outgassing from the mantle and atmospheric gases are the major sources of He, Ar, and methane in thermal springs along the ocean margin at such places as the Kurile Islands and Kamchatka and in ridge areas like Iceland. The contribution of gases from the Earth's crust is relatively unimportant.

A direct relationship between ^{3}He/^{4}He and ^{40}Ar/^{36}Ar isotope ratios in gases of thermal springs is demonstrated and reflects the isotopic mixing of helium and argon from the upper mantle with gases from the atmosphere. Migration of helium, argon, and methane from the upper mantle takes place mainly in a free-gas phase.

Based on experimental and calculated data, the most probable values for the isotopic ratios of He and Ar in the upper mantle are established at ^{3}He/^{4}He $= 1.1 \pm 0.2 \times 10^{-5}$ and ^{40}Ar/^{36}Ar $= 309 \pm 5$ for the Kurile Islands and Kamchatka; and ^{3}He/^{4}He $= 2.2 \pm 0.2 \times 10^{-5}$ and ^{40}Ar/^{36}Ar $= 323 \pm 2$ for Iceland. An ^{40}Ar/^{36}Ar ratio of 334 ± 13 is calculated for the Earth's mantle as a whole.

Introduction

At present there are few reliable criteria for establishing crustal or mantle origin for different substances. Isotopic composition of a number of elements and noble gases in particular is believed to be one of the most valid criteria for revealing the primary source of a substance. According to Shukolyukov and Levskii (1972), "Helium and argon are unique elements in the Earth because of the changes in their isotopic composition by addition of radiogenic isotopes in the course of Earth's history." The mantle, crust, and atmosphere are supposed to have differing isotopic composition (Tolstikhin et al., 1975; Ozima, 1975). On that basis, Tolstikhin et al. (1975) suggested that the helium-isotope ratio could serve as a criterion for the "juvenility" of some rocks and the natural gases from some rocks. Such juvenility may indicate large tectonic dislocations that provide transfer of substances from the mantle.

Anomalously high ^{3}He/^{4}He ratios ($0.5 - 3.6 \times 10^{-5}$) were discovered for the first time in thermal waters of the Kurile Islands, Kamchatka, and Iceland and in rocks of presumed mantle origin (Mamyrin et al., 1969, 1972, 1974; Kamenskii et al., 1971, 1974, 1976; Devirts et al., 1971; Gerling et al., 1971, 1972; Shukolyukov and Levskii, 1972; Tolstikhin et al., 1972a, 1972b, 1974, 1975; Baskov et al., 1973; Baskov and Surikov, 1975; Kononov et al., 1974; Krylov et al., 1974). According to those workers, the anomalously high ratios formed as a result of mixing of upper-mantle helium with helium from the Earth's crust. ^{3}He/^{4}He isotopic ratios of $3 \pm 1 \times 10^{-5}$, 3×10^{-8}, and $1.40 \pm 0.01 \times 10^{-6}$ have been proposed by Mamyrin et al. (1974) to characterize the mantle, crust, and atmosphere, respectively. Kamenskii et al. (1976) recommended a ^{3}He/^{4}He ratio of 1.4×10^{-5} as more characteristic of Kamchatka, the Kurile Islands, and perhaps the entire Pacific area. They also considered that the crust, mantle, and atmosphere all contributed He to the thermal-spring gases and calculated proportional contributions from each source as 17 to 50, 35 to 70, and 0.5 to 30 percent, respectively.

Opinions on the isotopic composition of mantle argon are much more diverse than for helium. Assuming a cold accretion model for Earth and gradual degassing, Tolstikhin et al. (1975) interpreted observations on inert gases as yielding a mantle argon isotopic composition in the ratio ^{40}Ar/^{36}Ar = 500 to 1,200. Based on theoretical and experimental data, Ozima (1975) postulated a mantle ^{40}Ar/^{36}Ar ratio greater than 2,000 and argued that gradual degassing of the Earth could not explain the evolution of rare gases earlier than 4.35 billion years ago.

Experimental determination of argon isotopic composition in thermal waters and rocks of presumed mantle origin from Italy, the Kurile-Kamchatka volcanic region, the United States, Japan, Iceland, Hawaii, and the Atlantic and Pacific Oceans generally vary (Cherdyntsev et al., 1966, 1967; Dalyrymple and Moore, 1968; Fisher, 1971; Baskov et al., 1973; Mazor and Fournier, 1973; Smelov et al., 1975; Ozima, 1975; Baskov and Surikov, 1975; Kamenskii et al., 1976, and others). Usually, however, ^{40}Ar/^{36}Ar ratios in thermal fluids do not differ appreciably from atmospheric ratios. Furthermore, from the preceding considerations, Smelov et al. (1975) concluded that, although mantle ^{40}Ar/^{36}Ar ratios may vary because of primary gas concentration, potassium content, thermal history, and other factors, in general ^{40}Ar/^{36}Ar in the mantle should not differ greatly from atmospheric ratios.

According to Galimov (1973) and others, mantle methane, which forms from the reaction of mantle carbon with water, has a carbon-isotope ratio of

($\delta^{13}C \approx -7‰$), roughly similar to that of mantle carbon. However, during migration toward lower temperatures, primary methane begins to become enriched in ^{13}C by exchanging with CO_2. $\delta^{13}C$ values of -20 to $-30‰$ in the methane of thermal springs in Yellowstone National Park, New Zealand, and Italy are explained by Craig (1953), Hulston and McCabe (1962), and Ferrara and Gonfiantini (1965) in terms of exchange processes in the CH_4-CO_2 system.

The purpose of this chapter is to draw on the available information in order to delineate better the modern degassing of He, Ar, and CH_4 from the mantle in thermal waters of the Kurile-Kamchatka region and Iceland and thereby derive better values for isotopic ratios of He and Ar in the Earth's mantle.

Discussion

Helium Isotopic Composition of the Mantle

Sheppard et al. (1971) and White (1974) pointed out that not only helium ratios but also D/H and $^{18}O/^{16}O$ ratios in juvenile water of the upper mantle differ from those in meteoric and ocean waters. If migration of helium from the mantle occurs in aqueous solution, then solvent waters of thermal springs enriched in ^3He should also contain juvenile components, and one should expect correlation between $^3He/^4He$ and D/H and $^{18}O/^{16}O$ ratios in mixtures of aqueous phases. However, with rare exceptions, analyses of the latter isotopic ratios in the world's ^3He-enriched thermal areas shows an absence of juvenile water components (Baskov et al., 1973), suggesting that migration of mantle helium from magma chambers probably occurs chiefly in a free-gas phase.

Baskov et al. (1973) and Gutsalo and Vetshtein (1974, 1976) have established that D/H and $^{18}O/^{16}O$ ratios in thermal waters of the Kurile-Kamchatka volcanic region mainly correspond to those in local meteoric waters, and similar observations are typical for other geochemical systems of the world (Craig, 1963). Since meteoric waters are naturally saturated with atmospheric gases, one can conclude that they must play an important role in the gas balance of thermal springs and serve as the main source of the nitrogen, argon, krypton, and xenon dissolved in these waters, as observed in different parts of the world (Mazor and Fournier, 1973; Gunter, 1973; Mazor, 1972; Smelov et al., 1975). In the helium balance of the thermal springs the atmosphere also takes part, but to a lesser extent. Proof of that may be seen in table 23-1 and figure 23-1, which show the linear relationship between ^3He and ^4He in thermal waters of the Kurile-

Kamchatka area in the atmosphere (Gutsalo, 1975). That relationship is

$$^3\text{He}_{therm} = K(^4\text{He}_{therm} - {^4\text{He}_{air}}) + {^3\text{He}_{air}} \qquad (23.1)$$

where $^3\text{He}_{therm}$ and $^4\text{He}_{therm}$ are concentrations in thermal waters in volume percent, and $^3\text{He}_{air}$ and $^4\text{He}_{air}$ are amounts (3.2×10^{-10} and 2.3×10^{-4} in volume percent, respectively) dissolved in the freshwater under equilibrium with the atmosphere at $T°C = 15$ and 760 bars. K is the slope of line I in figure 23–1:

$$K = {^3\text{He}_{mantle\text{-}cr}}/{^4\text{He}_{mantle\text{-}cr}} = \frac{{^3\text{He}_{therm}} - {^3\text{He}_{air}}}{{^4\text{He}_{therm}} - {^4\text{He}_{air}}}$$

$$= 1.1 \pm 0.2 \times 10^{-5}$$

The behavior of He isotopes of atmospheric origin in thermal waters will depend on the temperature and the salt concentration. However, investigations reveal that saturation of thermal waters with atmospheric helium in the Kurile-Kamchatka volcanic region takes place mainly in zones of infiltration of meteoric waters under temperatures of 10 to 20°C and concentrations of from 2.1×10^{-4} to 2.4×10^{-4} volume percent. The dissolution of atmospheric helium in meteoric waters is accompanied by fractionation of He isotopes of less than 1.2 percent (Weiss, 1970).

Accordingly, the relationship shown by line I of figure 23–1 permits the following conclusions: (1) atmospheric helium is present in all thermal springs; (2) as a consequence of (1), it is necessary to subtract atmospheric helium to determine the "true" ratio of helium isotopes supplied from the Earth's crust and mantle, i.e., K; (3) the "true" value of K ($1.1 \pm 0.2 \times 10^{-5}$) and its relative constancy over large areas suggest that the mantle may function as a single source of helium with a constant isotopic ratio; and (4) as will be discussed, the process of helium migration from the upper mantle through the Earth's crust does not significantly affect the $^3\text{He}/^4\text{He}$ ratio.

Deviation of points in line I in figure 23–1 serves to measure helium supply from the Earth's crust. On the whole, those deviations are small and within limits of the range of the measurements of $^3\text{He}/^4\text{He}$ (Gerling et al., 1972), thus indicating a relatively insignificant contribution by the crust to total helium balance in thermal springs. A linear dependence could be maintained in principle for line I if the amount and isotopic ratio of helium supplied from the crust were constant. Although the $^3\text{He}/^4\text{He}$ ratio in mantle He is three orders of magnitude larger than that for crustal He, the total concentration of He is two to three orders of magnitude larger in the crust than in the mantle, if the value of 0.2×10^{-6} cm^3/g (Gerling et al., 1972) is accepted for initial concentration of He in the mantle. If a significant

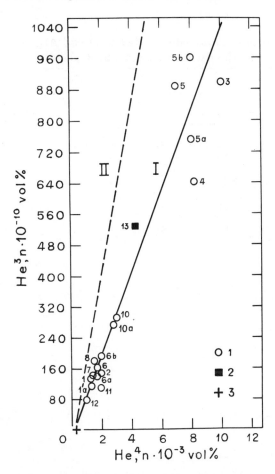

Figure 23-1. Relations of ^{3}He and ^{4}He contents in some thermal waters and in the atmosphere (table 23-1). (1) ^{3}He and ^{4}He contents in thermal waters of the Kuril Islands and Kamchatka. (2) ^{3}He and ^{4}He contents in thermal waters of the Iceland (Krabla) volcano. (3) Saturation concentrations of ^{3}He (3.2 10^{-10} vol %) and ^{4}He (2.3 10^{-4} vol %) obtained when water is equilibrated with the atmosphere at $t = 15°C$ and $p = 760$ torr. The small numbers next to the data points indicate the specimen number in table 23-1. I = regression line representing the ^{3}He and ^{4}He contents of thermal waters of the Kuril Islands, Kamchatka, and the atmosphere; II = regression line representing the ^{3}He and ^{4}He contents of the thermal waters of Iceland and the atmosphere using the data of Kononov et al. (1974).

Table 23-1
Contents and Isotopic Composition of Helium and Argon in Some Thermal Waters of the Kuril Islands, Kamchatka, and Iceland

No.	Place of Sampling	Date	Temp. (°C)	pH	Total Dissolved Solids. (g/kg)	$^3He/^4He \cdot 10^{-6}$	$^3He \cdot 10^{-10}$ (vol %)
	Kunashir Island:						
1	Upper Mendeleyev springs	8/7/68 (about)	99	1.85	3.5	10.3	134
1a	Upper Mendeleyev springs	8/17/68	82	2.50	1.0	8.8	114
2	Mendeleyev volcano, southern part of eastern fumarole field	8/25/67	95	3.1	0.7	7.2	145[a]
3	Stolbovsky springs	8/3/67	78	7.75	2.7	8.9	896[a]
3a	Stolbovsky springs		67	—	—	—	7.6	760
4	Tretyakovsky springs	8/13/68	80	7.60	3.8	7.8	640
	Iturup Island:						
5	Hot source springs	8/18/67	55	7.50	2.3	14.0	885[a]
5a	Hot source springs	9/21/68	—	—	—	9.4	752
5b	Hot source springs	9/21/68	—	—	—	12	—
6	Springs at the headwater of the Sernaya River on the Southwestern slope of Baransky volcano	8/28/68	40	2.20	2.0	9.5	162
6a	Spring at the headwater of the Sernaya River on the southwestern slope of Baransky volcano	8/28/68	40	2.20	2.0	8.1	138
6b	Spring at the headwater of the Sernaya River on the southwestern slope of Baransky volcano	—	—	—	—	9.5	190

^4He (vol %)	$He + Ne$ (vol %)	$Ar + Kr + Xe$ (vol %)	$^{40}Ar/^{36}Ar$	Ar_{air} (vol %)	$^3He/Ar_{air} \cdot 10^{-8}$	Reference
0.0013	0.0013	0.0910	—	—	—	Tolstikhin et al., 1972a
0.0013	0.0013	0.084	295.6	—	—	Baskov et al., 1973 Baskov and Surikov, 1975
0.002	0.002[b]	0.303	295.6	0.303	4.8	Kamenskii et al., 1971
0.01	0.01[b]	0.66	305.9	0.64	14	Kamenskii et al., 1971
0.01	0.01	0.66	304.0	—	—	Mamyrin et al., 1969; Baskov et al., 1973
0.0082	0.0082	0.5350	—	—	—	Tolstikhin et al., 1972a
0.007	0.007	0.334	311.9	0.316	28	Kamenskii et al., 1971
0.008	0.008[b]	0.321				Tolstikhin et al., 1972a
—	—	—	310.0	—	—	Baskov et al., 1973
0.0017	0.0017	0.5190	—	—	—	Tolstikhin et al., 1972a
0.0017	0.0017	—	—	—	—	Baskov et al., 1973
0.002	0.002	—	—	—	—	Tolstikhin et al., 1972a

Table 23-1 *(continued)*

No.	Place of Sampling	Date	Temp. (°C)	pH	Total Dissolved Solids. (g/kg)	$^3He/^4He \cdot 10^{-6}$	$^3He \cdot 10^{-10}$ (vol %)
	Paramushir Island, Ebeco Volcano:						
7	Spring in the southwestern solfatara field	7/20/68	87	2.50	0.9	9.9	139
7a	Spring in the southwestern solfatara field	7/25/68	97	2.50	0.9	8.5	—
8	Zelezisty spring in the eastern solfatara field	8/30/68	97	2.55	2.0	12.0	180
	Kamchatka Peninsula, Uzon kaldera:						
9	Site of sulphur hillocks	10/14/68	28	5.1	0.2	9.4	190[a]
9a	Site of sulphur hillocks	—	—	5.1	0.2	8	—
10	Central fumarole	—	—	—	—	9.8	290
10a	Central fumarole field	10/14/68	—	—	—	9.8	274
11	Middle solfatara field	10/14/68	23	2	2	5.4	108
12	Northern part of Center Lake	10/14/68	37	5.9	1.1	7.8	78
	Iceland:						
13	Krabla volcano, Tekstareykir region	—	—	—	—	12.3	529

[a]The value of ^3He was determined from the ratio value $^3He/Ar_{air}$ and the value Ar_{air}.
[b]The mixture of neon and helium makes up no more than 20% (Devirts et al., 1971). In other samples, the separation of ^3He and ^4He was calculated on the assumption that $^3He + ^4He \sim He + Ne$.

contribution from the crust existed, we would expect a marked shift of K to lower values, and the irregularity of contributions regionally and with depth should disrupt the linearity of line I. Crustal helium may be less than 15 to 20 percent of total helium and therefore has a minor influence on He isotope ratios.

^4He (vol %)	$He + Ne$ (vol %)	$Ar + Kr + Xe$ (vol %)	$^{40}Ar/^{36}Ar$	Ar_{air} (vol %)	$^3He/Ar_{air} \cdot 10^{-8}$	Reference
0.0014	0.0014	0.0215	—	—	—	Tolstikhin et al., 1972a; Baskov et al., 1973
—	—	0.028	302.2	—	—	Baskov and Surikov, 1975
0.0015	0.0015	0.0260	—	—	—	Tolstikhin et al., 1972a
0.002	0.002	0.226	295.6	0.226	8.4	Kamenskii et al., 1971
—	—	—	—	—	—	Baskov et al., 1973
0.003	0.003	0.372	—	—	—	Kamenskii et al., 1971
0.0028	0.0028	0.372	—	—	—	Tolstikhin et al., 1972
0.002	0.002	0.170	—	—	—	Tolstikhin et al., 1972a
0.001	0.001	0.044	—	—	—	Tolstikhin et al., 1972
0.0043	0.0043	—	—	—	—	Mamyrin et al., 1972

Line II in figure 23–1 is based on $^3He/^4He$ ratios for Icelandic thermal waters obtained by Kononov et al. (1974). A K value derived for these data would be larger ($2.2 \pm 0.2 \times 10^{-5}$) than that discussed above. Mamyrin et al. (1974) explained those values as potentially arising through admixture of crustal helium, but I feel that this question requires further investigation.

Although the constancy of ^{3}He/^{4}He ratios establishes criteria for the supply of mantle helium, one cannot necessarily use the presence of mantle helium as an indicator of the supply of other gases or water, since it was shown that helium probably migrates in a free-gas phase. Relationships between the proportion of mantle helium and isotopic relationships for other elements have been used to define mantle constituents (Kamenskii et al., 1976), but such comparisons should be considered more qualitative than quantitative (Gutsalo, 1976). Since gas migration from the mantle takes place primarily via zones of tectonic disruption, fractionation of isotopes during migration is believed to be absent. Consequently, correlations between helium isotopic composition and other elements must reflect the mixing of mantle material with material from other layers enroute.

Argon Isotopic Composition of the Mantle

Smelov et al. (1975) already noted the "coincidence of areas enriched in ^{40}Ar with those enriched in ^{3}He in thermal fluids of Iceland." The subsequent discovery of correlations between ^{3}He/^{4}He and ^{40}Ar/^{36}Ar ratios in thermal waters and in the atmosphere (Gutsalo, 1976) is important in elucidating the problem of origin and distribution of helium and argon isotopes. As shown in figure 23-2, thermal spring waters enriched in ^{3}He may either be enriched in ^{40}Ar or contain argon with isotopic ratios typical for the atmosphere (295.6). In the second case, there is no relationship (line II in figure 23-2), whereas in the first there is a direct proportion (line I) between helium and argon isotopic composition:

$$^{40}\text{Ar}/^{36}\text{Ar}_{\text{therm}} = C(^3\text{He}/^4\text{He}_{\text{therm}} - {}^3\text{He}/^4\text{He}_{\text{air}}) + {}^{40}\text{Ar}/^{36}\text{Ar}_{\text{air}}$$

(23.2)

where ^{40}Ar/^{36}Ar and ^{3}He/^{4}He are isotopic ratios in thermal waters and in atmospheric air, and C is a constant equal to 1.35×10^6 for the investigated areas.

As seen from figure 23-2, line I reflects the process of mixing helium and argon of thermal springs with those of the atmosphere. The correlation between helium and argon isotopic composition, together with the dependence of helium isotopic composition on degassing from the upper mantle as discussed previously, leads to the assumption that helium and argon migrated from the same source, i.e., upper mantle. I also conclude that changes of argon isotopic composition are ultimately controlled by argon discharge from the mantle and subsequent mixing with atmospheric argon. Assuming that the ^{3}He/^{4}He ratio for mantle helium under the Kurile and

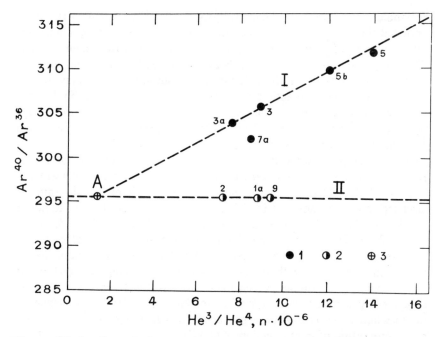

Figure 23-2. Correlation between isotopic compositions of argon ($^{40}Ar/^{36}Ar$) and helium ($^{3}He/^{4}He$) in some thermal waters from the Kuril Islands, Kamchatka, and the atmosphere. I = regression line of waters enriched by ^{40}Ar (closed circles); II = regression line of waters containing argon with the atmospheric air ratio $^{40}Ar/^{36}Ar$ of 295.6 (half-filled circles); A = the point indicating the isotopic ratios of $^{40}Ar/^{36}Ar$ (295.6) and of $^{3}He/^{4}He$ (1.4 10^{-6}) in atmospheric air and air dissolved in water (marked by ⊕). For description of small numbers, see figure 23–1.

Kamchatka area is $1.1 \pm 0.2 \times 10^{-5}$, then according to equation 23.2 the corresponding value of $^{40}Ar/^{36}Ar$ for the mantle in this area must be 309 ± 5. Likewise, based on a $^{3}He/^{4}He$ value of $2.2 \pm 0.2 \times 10^{-5}$, the $^{40}Ar/^{36}Ar$ value for the upper mantle in Iceland will be 323 ± 2. If we further assume a fixed relationship for argon and helium isotopes for the entire upper mantle of the Earth, then accepting a $^{3}He/^{4}He$ ratio of $3 \pm 1 \times 10^{-5}$ (Mamyrin et al., 1974), the corresponding $^{40}Ar/^{36}Ar$ ratio will be 334 ± 13.

It is important to emphasize that calculated $^{40}Ar/^{36}Ar$ ratios for the mantle under Iceland exceed that for atmospheric air by 8 percent. This conclusion agrees roughly with data of Smelov et al. (1975) from a number of

Icelandic thermal springs. Their work found that ^{40}Ar was 4 to 10 percent enriched with respect to air, and occasionally 12 percent enriched. They attributed those enrichments to a "partial gaseous flow from mantle rocks." The increased ^{40}Ar/^{36}Ar ratios (about 305 to 314) in thermal waters of Yellowstone National Park (Mazor and Fournier, 1973) may also be explained by input of mantle argon. ^{40}Ar/^{36}Ar ratios considerably less than 500 were also found in four out of five submarine basalts from the Atlantic and Pacific Oceans (Ozima, 1975). Ozima explained these data by contamination with atmospheric argon.

Relationships between helium and argon concentrations in the Kurile-Kamchatka area reflect simultaneous enrichment of thermal spring waters in ^{3}He and ^{40}Ar. That process appears to occur where Ar/He ratios in discharging gas average 65 (line I in figure 23-3). When the Ar/He ratio in gas approaches 180, the enrichment of thermal waters by ^{40}Ar is not observed (line II in figure 23-3). Similar relationships are typical for Iceland (Smelov et al., 1975), where maximum enrichment of ^{40}Ar (more than 12 percent in comparison with air) is common for gas having very high helium concentration (0.3 volume percent) and "with isotopic composition corresponding to probable mantle origin" (^{3}He/^{4}He $= 2 \times 10^{-5}$). Consequently, the enrichment of thermal-spring gases in ^{40}Ar is accompanied by an increase in the proportion of mantle helium.

For argon and helium of exclusively atmospheric origin in thermal waters, correlations between their concentrations should approximate straight lines, and those lines should pass through the origin and through points giving the concentrations of argon and helium dissolved in the water when it is in equilibrium with the atmosphere under the given conditions of temperature, pressure, and ionic strength. However, line II of figure 23-3, which shows data for the same specimens as line II of figure 23-2, does not pass through the origin and is displaced to the right, i.e., toward enrichment in He. Such deviations may be partly explained by input of mantle He, as determined by He isotopic composition (figure 23-2, line II). However, the displacement may also be affected by Rayleigh distillation while sampling gases (Gunter, 1973), which leads to an overestimation of He and underestimation of Ar. In general, however, the isotopic correlations tend to be confirmed by comparing the isotopic ratios to concentration (lines I and II in figures 23-2 and 23-3).

The ratio of radiogenic ^{40}Ar to He strongly confirms the existence of argon discharge from the mantle to thermal springs in the Kurile-Kamchatka region. According to Cherdyntsev et al. (1966), an "excess of ^{40}Ar$_{rad}$/He above 0.16 may be regarded as a criterion of radiogenic discharge of argon from the earth's interior, probably from subcrustal material." All thermal springs identified as being related to mantle discharge by their isotopic ratios of argon and He (line I in figure 23-2) have ^{40}Ar$_{rad}$/He ratios (0.30 to 2.63) greater than that for the earth's crust. For example, ratios of 2.63 and 2.31

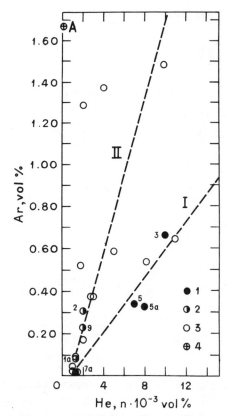

Figure 23-3. Correlation between concentrations of helium and argon in some thermal waters of the Kuril Islands and Kamchatka including the data of Baskov and Surikov (1975) (open circles 3). For I and II, see figure 23-2. A = ratio of concentrations of argon to helium dissolved in freshwater equilibrated with the atmosphere at $t = 15°$ C, $p = 750$ torr (marked by ⊕). For other symbols, see figure 23-1.

were observed in Hot Springs and Stolbovsky Springs, respectively. Ratios of 0.30 to 0.35 are found in springs within the southwestern solfatera field of Ebeco volcano.

Isotopic Composition of Mantle Methane

The tendency for increase in the $^{13}C/^{12}C$ ratio with increasing mantle helium in gas from thermal springs on Kamchatka was first observed by Kamenskii

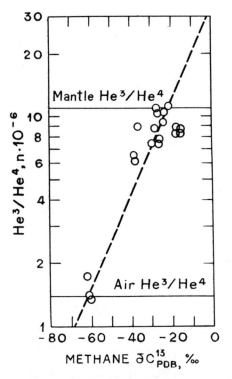

Figure 23-4. Correlation between isotopic compositions of helium (^{3}He/^{4}He) and carbon in methane ($\delta^{13}C_{PDB}$) in some thermal and cold waters from Kamchatka springs.

et al. (1974, 1976), who suggested that the methane was also of mantle origin. Figure 23-4, based on data of Kamenskii et al. (1976) shows the relations between ^{3}He/^{4}He and the isotopic ratio of carbon ($\delta^{13}C_{PDB}$) in methane. The proportional relationship suggests that mantle-derived helium and methane may be mixing with He and methane from other zones within the earth. Since helium from the mantle seems to mix with atmospheric helium, one can anticipate a similar process for methane.

Figure 23-4 shows us that a $\delta^{13}C$ ratio for methane of $-63‰$ corresponds to an atmospheric ^{3}He/^{4}He ratio of 1.40×10^{-6} in gases of nitrogen-methane composition from cold springs of Kamchatka (table 23-2). This $\delta^{13}C$ value is typical for carbon of biochemically formed methane in swamps at near-surface conditions and in contact with the atmosphere (Galimov, 1973).

Galimov (1973) has shown that isotopic transformations in the system $CO_2 = CH_4$ begin to become significant at temperatures above 150° to

Table 23-2
Chemical and Isotopic Gases Composition Out of the Waters of Some Cold Kamchatka Springs

| Place of Sampling | Temp. (°C) | Gas Composition (vol %) | | | $^3He/^4He \cdot 10^{-6}$ | $\delta^{13}C_{PDB}$ (CH_4) (‰) | $\delta^{13}C_{PDB}$ (CO_2) (‰) | Equilibrium Constants in the System CO_2-CH_4 | |
		$H_2S + CO_2$	CH_4	N_2				K_{theor}	K_{calc}
Opala river, the left bank lake I (large)	5	2.8	73.9	23.1	1.75	−62.1	−3.8	1.068	1.062
Opala river, the left bank lake II	18	3.2	73.2	23.0	1.43	−61.1	−7.7	1.063	1.057
Shikova river, the interstitial gas in the riverbed at the right bank	7	1.4	54.6	44.0	1.35	−59.9	−12.4	1.067	1.050

(Reference: Kamenskii et al., 1976).

200°C. Isotopic equilibrium under 25°C is not reached even in the course of 400 million years. However, data in table 23–2 show that the $\delta^{13}C$ for the CO_2–CH_4 system in cold springs at 5 to 18°C is close to the thermodynamic equilibrium values for those conditions.

According to Galimov (1973), the preceding relationships must be due to primary thermodynamic isotopic equilibrium, modified by biochemical or chemical transformation. Experimental exchange constants for CO_2 and CH_4 were calculated from equation 23.3 and related to theoretical equilibrium constants at temperatures used by Craig (1953) (table 23–2):

$$K = \frac{1{,}000 + \delta CO_2}{1{,}000 + \delta CH_4} \qquad (23.3)$$

For $^3He/^4He = 1.1 \pm 0.2 \times 10^{-5}$ established for mantle helium beneath the Kurile Islands and Kamchatka, we obtain a calculated value of $\delta^{13}C = -22‰$ for mantle carbon. However, this value differs from the $\delta^{13}C$ value for mantle methane according to Galimov (1973). Investigations in other hydrothermal fields (Craig, 1953; Hulston and McCabe, 1962; Ferrara and Gonfiantini, 1965) suggest that $\delta^{13}C$ depletion of the calculated "mantle methane" discharged in thermal-spring waters is not inherited from primary mantle carbon. It may be a secondary depletion due to exchange with CO_2 during migration. Under the high-temperature conditions in the mantle and lower zones of the earth's crust, possibilities for isotopic fractionation of carbon in the CO_2-CH_4 system are small ($K = 1.003$ under 1,000°C). However, as the juvenile methane and carbon dioxide reach temperature zones of 300 to 500°C, better conditions for exchange occur ($K = 1.019$ under 400°C). Since in the thermal springs of Kamchatka, CO_2 concentrations exceed those of CH_4 by factors of dozens or hundreds and equilibrium is not achieved in geologic time under temperatures of 150 to 200°C (Galimov, 1973), the $\delta^{13}C$ of endogenic methane will not be lowered more than -25 to $-30‰$.

Observed values of $\delta^{13}C$ for methane of thermal springs in Kamchatka vary mainly from -15 to $-30‰$ (figure 23–4), suggesting that they are of endogenic origin. Again according to Galimov (1973), exogenic sedimentary) methane has mean values of $\delta^{13}C$ varying from -32 to $-72‰$.

$\delta^{13}C$ in methane and CO_2 show not only the endogenic character of methane but also indicate that the methane has been depleted in ^{13}C by isotopic exchange with carbon dioxide (figure 23–5). From the figure (line I), one can observe the tendency for depletion of mantle methane and enrichment of mantle CO_2 in $\delta^{13}C$ during migration to the crust. An analogous tendency has been found in Yellowstone National Park and New Zealand (figure 23–5). It is also important to note that isotopic exchange ratios depend largely on the relative concentrations of CO_2 and methane. Where

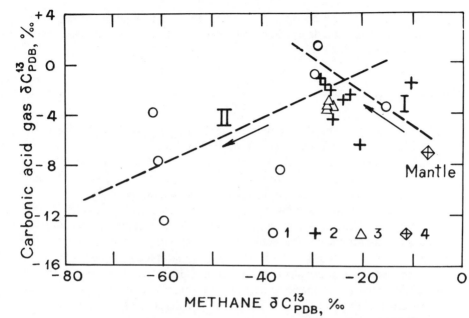

Figure 23-5. Correlation between methane $\delta^{13}C$ and carbon dioxide $\delta^{13}C$ values in some waters from: (1) thermal and cold springs in Kamchatka (Kamenskii et al., 1976); (2) thermal springs in Yellowstone National Park (Craig, 1953); (3) thermal springs in New Zealand (field Wairakei) (Hulston and McCabe, 1962); (4) the isotopic ratio of carbon dioxide and methane $\delta^{13}C$ values in the mantle (Galimov, 1973). I = trace line for hypothetical carbon isotopic exchange between CO_2 and CH_4 while co-migrating from the mantle to the Earth's crust; II = trace line for hypothetical isotopic mixing process between both mantle-derived methane and carbon dioxide and both subsurface methane and carbon dioxide of biochemical origin. The arrows indicate the direction in isotopic processes.

CO_2 predominates in the gas phase, mantle-methane values values decrease during exchange with only minor enrichment in ^{13}C. Craig (1953) was also right in pointing out that the relative proportions of CO_2 and CH_4 could vary in thermal-spring gases after full isotopic equilibrium had been reached.

Upon completion of isotopic exchange, methane of mantle origin ($\delta^{13}C = -24‰$) and CO_2 ($\delta^{13}C = -1.4‰$) mix with biochemical methane ($\delta^{13}C = -61‰$) and CO_2 ($\delta^{13}C = -8‰$), as shown by line II in figure 23-5. The proportion of endogenic methane in total methane is usually 80 to 90 percent or more. Notwithstanding, it is conceivable that, when the $\delta^{13}C$ of methane varies from -15 to -26 or -30, all the methane in thermal springs may merely represent mantle methane depleted in ^{13}C through exchange with CO_2.

Conclusions

1. Evidence of discharge of helium, argon, and methane from the mantle is observed in waters and gases of thermal springs.
2. The contribution of helium, argon, and methane from the crust is relatively insignificant compared to the contribution from mantle and atmospheric sources to the balance of these gases.
3. The isotopic composition of helium in thermal springs of the Kurile Islands and Kamchatka shows a linear relationship between ^3He and ^4He. Similar relationships are noted in Iceland. The constancy of $^3He/^4He$ ratios for pure mantle helium over large geographic or widely separated areas was first established by the author (Gutsalo, 1975).
4. Mantle and atmospheric argon form the main active sources for argon balance in thermal springs. Supply of mantle argon is accompanied by helium supply.
5. A direct relationship between $^3He/^4He$ and $^{40}Ar/^{36}Ar$ of apparent mantle origin is established here. The relationship demonstrates isotopic mixing of mantle argon and helium with that of the atmosphere.
6. Methane from thermal springs of Kamchatka is proposed to have formed as a result of mixing of mantle methane (having ^{13}C of about $-7‰$, reduced by exchange with CO_2 to $-22‰$) with biochemical methane (having ^{13}C values of approximately $-63‰$) in the upper earth's crust. Mantle methane forms at least 80 to 90 percent of the total methane.
7. Helium, argon, and methane migrate from the upper mantle chiefly in a free-gas phase.
8. Based on experimental and calculated data, most probable values for helium and argon isotopic ratios for the upper mantle are $^3He/^4He = 1.1 \pm 0.2 \times 10^{-5}$ and $^{40}Ar/^{36}Ar = 309 \pm 5$ for the Kurile Islands and Kamchatka. $^3He/^4He = 2.2 \pm 0.2 \times 10^{-5}$ and $^{40}Ar/^{36}Ar = 323 \pm 2$ for Iceland. A calculated value of $^{40}Ar/^{36}Ar$ ratio of 334 ± 13 is proposed for the earth as a whole.
9. $^{40}Ar_{rad}/He$ ratios exceeding 2.0 in gas extracted from thermal springs may serve as an index of argon and helium supply from the earth's mantle.

References

Baskov, E.A., Vetshtein, V.E., Surikov, S.N., Tolstikhin, I.N., Malyuk, G.A., and Mishina, T.A. 1973. H, O, C, Ar and He isotopic composition

of thermal waters and gases of the Kurilie-Kamchatka volcanic area as indicators of their origin (Russ.). *Geokhimiya* 2:180-189.
Baskov, E.A., and Surikov, S.N. 1975. *Gidrotermy Tikhookeanskogo segmenta zemli.* (Russ.) Moscow: "Nedra".
Cherdyntsev, V.V., Kolesnikov, E.M., and Lizarskaya, I.V. 1966. Isotopic composition of argon in natural gases (Russ.). *Geokhimiya* 5:604-606.
Cherdyntsev, V.V., Shitov, I.V., and Lizaiskaya, I.V. 1967. The isotopic composition of argon in volcanic gases of the USSR (Russ.). *Doklady Akad. Nauk* USSR 172:1180-1182.
Craig, H. 1953. The geochemistry of the stable carbon isotopes. *Geochim. Cosmochim. Acta* 3:53-92.
Craig, H. 1963. The isotopic geochemistry of water and carbon in geothermal areas. In E. Tongiorgi (Ed.), *Nuclear Geology in Geothermal Areas.* Pisa: Consiglio Nazionale delle Richerche Lab. de Geologia Nucleare, pp. 17-53.
Devirts, A.L., Kamenskii, I.L., and Tolstikhin, I.N. 1971. Helium and tritium isotopies in igneous springs. (Russ.) *Doklady Akad. Nauk, USSR* 197:450-452.
Dalrymple, G.B., and Moore, J.G. 1968. Argon-40: Excess in submarine pillow basalts from Kilauea Volcano, Hawaii. *Science* 161:1132-1135.
Ferrara, G., and Gonfiantini, R. 1965. Isotopic composition of argon and carbon from Larderello (Tuscany) steam jets. *Trans. Am. Geophys. Union* 46(1):170-181.
Fisher, D.E. 1971. Incorporation of Ar in East Pacific basalts. *Earth Planet. Sci. Lett..* 12:321-324.
Galimov, E.M. 1973. Isotopy ugleroda v neflegazovoi geologii. (Russ.). Moscow: "Nedra".
Gerling, E.K., Mamyrin, B.A., Tolstikhin, I.N., and Yakovleva, S.L. 1971. Isotopic helium composition in some rocks. (Russ.). *Geokhimia* 10:1209-1217.
Gerling, E.K., Tolstikhin, I.N., Mamyrin, B.A., Anufriev, S.G., Kamenskii, I.L., and Prasolov, E.M. 1972. New investigation of isotopic geochemistry of helium (Russ.). In First International Geochemical Congress, Moscow 20-25 July 1971. *Magmaticheskie protessy,* Vol. 1. Moscow: Akademiya Nauk, SSSR, pp. 200-216.
Gunter, B.D. 1973. Aqueous phase-gaseous phase material balance studies of argon and nitrogen in hydrothermal features at Yellowstone National Park. *Geochim. Cosmochim. Acta* 37:495-513.
Gutsalo, L.K., and Vetshtein, V.E. 1974. Hydrogen and oxygen isotopes as criteria for the origin of natural waters (Abstract). In *International Symposium "Water-Rock Interaction."* Czechoslovakia, Prague, pp. 13-14.
Gutsalo, L.K. 1975. Helium isotopic geochemistry in thermal waters of Kuril Islands and Kamchatka (Abstract). In *Abstracts of Second United*

Nations Symposium on the Development and Use of Geothermal Resourses. Abstract No. III-36, Lawrence Berkeley Laboratory, University of California, Berkeley, California.

Gutsalo, L.K. 1976. On the nature and regularities of the distribution of helium and argon isotopes in thermal waters of the Kuril Islands and Kamchatka (Russ.). *Geokhimiya* 6:886–895.

Gutsalo, L.K., and Vetshtein, V.E. 1976. Isotopes of hydrogen and oxygen as criteria of the origin of natural waters. In *Proceedings of the International Symposium on Water-Rock Interaction.* Geological Survey, Czechoslovakia, Prague, pp. 323–330.

Hulston, J., and McCabe, W. 1962. Mass spectrometer measurements in thermal areas of New Zealand. Part 2. Carbon isotopic ratios. *Geochim. Cosmochim. Acta* 26:399–410.

Kamenskii, I.L., Yakutsen, V.P., Mamyrin, B.A., Anufriev, S.G., and Tolstikhin, I.N. 1971. Helium isotopes in nature (Russ.). *Geokhimiya* 8:914–931.

Kamenskii, I.L., Lobkov, V.A., Prasolov, E.M., Beskrovnyi, N.S., Kudryavtseva, E.I., Anufriev, G.S., and Pavlov, V.P. 1974. Helium and methane of the upper mantle in gases of Kamchatka (Abstract, Russ.). In *5th Vsesoyznyi simposium po geokhimii stabil'nykh izotopov. 30 Sept.– 30 Okt. 1974.* Moscow: Tesisy dokladov., pp. 148–149.

Kamenskii, I.L., Lobkov, V.I., Prasolov, E.M., Beskrovnyi, N.S., Kudryavtseva, E.I., Anufriev, G.S., and Pavlov, V.P. 1976. Components of the upper mantle of the Earth in gases from Kamchatka, based on He, Ne, Ar, and C isotopes. (Russ.). *Geokhimiya* 5:682–695.

Kononov, V.I., Mamyrin, B.A., Poliak, B.C., and Khabarin, L.V. 1974. Helium isotopes in gases of Iceland thermal springs. (Russ.). *Doklady, Akad. Nauk, SSSR* 217:172-175.

Krylov, A.Ya., Mamyrin, B.A., Khabarin, L.V., Mazina, T.I., and Silin, Yu.I. 1974. Helium isotopes in rocks of the bottom of oceans. (Russ.). *Geokhimiya* 8:1220–1225.

Mamyrin, B.A., Tolstikhin, I.N., Anufriev, G.S., and Kamenskii, I.L. 1969. Anomalous isotopic composition of helium in volcanic gases. (Russ.). *Doklady Akad. Nauk, SSSR* 184:1197–1199.

Mamyrin, B.A., Tolstikhin, I.N., Anufriev, G.S., and Kamenskii, I.L. 1972. Isotopic composition of helium in thermal springs of Iceland. (Russ.). *Geokhimiya* 11:1396.

Mamyrin, B.A., Tolstikhin, I.N., Anufriev, G.S., Kamenskii, I.L., Khabarin, L.V. 1974. Distribution of helium isotopes in upper Earth's crust (Abstract). (Russ.). In *5th Vsesoyuznyi simposium po geokhimii stabil'nykh isotopov.*, 30 Sept.–30 Okt. 1974. Moscow: Tesisy dokladov, pp. 149–151.

Mazor, E. 1972. Paleotemperatures and other hydrological parameters deduced from rare gases dissolved in groundwaters, Jordan Rift Valley, Israel. Geochim. Cosmochim. Acta 36:1321–1336.
Mazor, E., and Fournier, R.O. 1973. More on noble gases in Yellowstone National Park hot waters. *Geochim. Cosmochim. Acta* 37:515–525.
Ozima, M. 1975. Ar isotopes and Earth-atmosphere evolution models. *Geochim. Cosmochim. Acta* 39:1127–1134.
Sheppard, S.M.F., Nielsen, R.L., and Taylor, H.P., Jr. 1971. Hydrogen and oxygen isotope ratios in minerals from porphyry copper deposits. *Econ. Geol.* 66:515–542.
Shukolyukov, Yu.A., and Levskii, L.K. 1972. *Geokhimiya i kosmokhimiya isotopov blagorodnykh gazov.* (Russ.). Moscow: "Atomizdat."
Smelov, S.B., Vinogradov, V.I., Kononov, V.I., Polyak, B.G. 1975. Isotopic composition of argon in thermal fluids of Iceland (Russ.). *Dokl Akad. Nauk* 222:429–432.
Tolstikhin, I.N., Mamyrin, B.A., Baskov, E.A., Kamenskii, I.L., Anufriev, G.S., and Surikov, S.N. 1972a. Isotopes of helium in gases of thermal springs of the Kurile-Kamchatka volcanic region (Russ.). In *Ocherki sovremennoi geokhimiya i analiticheskoi khimiya.* Moscow: "Nauka." pp. 405–414.
Tolstikhin, I.N., Mamyrin, B.A., and Khabarin, L.V. 1972b. Anomalous isotopic composition of helium in some xenoliths (Russ.). *Geokhimiya* 5:629–631.
Tolstikhin, I.N., Mamyrin, B.A., Khabarin, L.V., and Erlich, E.N. 1974. Isotopic composition of helium in ultrabasic xenoliths from volcanic rocks of Kamchatka. In *Geokhimiya radiogennykh i radioaktivnykh isotopov.* (Russ.). Leningrad: "Nauka," pp. 90–104.
Tolstikhin, I.N., Azbel, I.Ya., and Khabarin, L.V. 1975. Isotopes of light inert gases in the mantle, the crust, and the atmosphere of the Earth (Russ.). *Geokhimiya* 5:653–666.
Weiss, R.F. 1970. Helium isotope effect in solution in water and seawater. *Science* 168:247–248.
White, D.E. 1974. Diverse origin of hydrothermal ore fluids. *Econ. Geol.* 69:954–973.

Part IX
Geophysics of Hydrothermal Processes

24 "Active" and "Passive" Hydrothermal Systems in the Oceanic Crust: Predicted Physical Conditions

C.R.B. Lister

Abstract

Oceanic crustal rock is delivered to new lithospheric plate as a magma. The surface material cools rapidly in direct or indirect contact with the seawater, but the bulk of the crustal layer crystallizes without water contact. Nevertheless, the entire crustal section is cooled more rapidly than is possible by conduction alone and is substantially permeable to circulating waters by the time enough sediment has collected for heat-flow measurements to be feasible. The process of water penetration into cooling, shrinking, and cracking rock is probably rapid (~ 10 m/yr), and the thermal output per unit area is high. That is the "active" stage of hydrothermal circulation, during which the heat stored in the fresh rock is extracted rapidly. Once penetration has ceased, due to failure or alteration of the rock at depth, the remaining permeable region continues to convect the heat conducted up from below. This "passive" stage of hydrothermal activity may continue even in the oldest ocean crust if permeability is retained. "Active" systems are characterised by high water temperature and fast flow: water of 200 to 350°C passes up the hot column of the convection cells in a few hours. The system itself has a life of the order of hundreds of years. The "passive" system, on the other hand, has typical water temperatures of less than 100°C and hot water residence times of years to hundreds of years. As the lithosphere ages and the mantle heat flow decreases, temperatures drop, but water residence times increase. If the surface becomes covered by relatively impermeable sediment, the circulating water may be trapped for long periods of time. Initial crack spacings of 1 to 2 m are expected from a revised penetration theory, and such cracks are still visible in ophiolite suites. The diffusion of chemical species in rock is probably slow enough for most bulk chemical exchange to occur during the low-temperature, long-time "passive" phase of hydrothermal circulation.

Introduction

The idea that seawater circulates through the rocks of the oceanic crust arose from two principal lines of geochemical evidence. Metal-rich sediments have been found near the crests of mid-ocean ridges where sea-floor spreading exposes fresh rock to interaction with seawater (Bostrom and Peterson, 1969; Bender et al., 1971; Dymond et al., 1973). Laboratory experiments have shown that hot water can react with rock and extract the trace-metal species often observed in hydrothermal fluids (Ellis and Mahon, 1964; Ellis, 1968), and two ideas were easily put together (for example, Bostrom et al., 1971; Corliss, 1970). Further data on the chemical details of seawater/basalt interaction are being obtained by experiment (for example, Hajash, 1974; Bischoff and Dickson, 1975), but subsequent improvement in the interpretation of submarine processes is hindered by inadequate knowledge of physical conditions expected in hydrothermal systems.

The geophysical approach to the problem has been largely independent of the geochemical work and perforce highly indirect. Suggestions that hydrothermal heat transport should be significant at mid-ocean ridges were made years ago, but they were hard to substantiate (Palmason, 1967; Talwani et al., 1971). Measurements of heat flow through the ocean floor were highly variable and averaged to unexpectedly low values near the oceanic spreading centers. However, nothing systematic was discernible until some measurements were fortunately placed on a sedimented spreading center and demonstrated a distribution that could be compatible only with large-scale hydrothermal circulation (Lister, 1972). Similar coherent large-scale fluctuations have been found in other sedimented areas (Williams et al., 1974; Davis and Lister, 1977), and the scale of the surface fluctuations suggests that the circulatory systems penetrate to depths of several kilometers in the crust (see, for example, Elder, 1965; Hartline and Lister, 1977). Water interaction with the rocks thus seems to be no haphazard surface effect, but instead a systematic bulk process.

If a hard rock is to be permeable to fluid flow, it must be cracked, and the question at once arises: on what scale is the cracking? At one extreme, water could circulate in widely separated fault zones or thermal contraction cracks and produce the observed thermal effects (Bodvarsson and Lowell, 1972). Major chemical interaction with the bulk rock is unlikely in such a model, because diffusive processes can affect only relatively shallow thicknesses of rock, even over long time spans (Sales and Meyer, 1948; Nigrini, 1969). The quasi-regular fabric structures sometimes observed in exposed oceanic rocks (Coleman and Keith, 1971) are on the scale of a few millimeters. Such an extremely small scale could not be consistent with kilometer-scale convection because rock alteration proceeds rapidly, and the expansion associated with hydration would soon reduce the permeability to low values (see Page, 1967). The circumstantial evidence favors an intermediate scale of cracking, perhaps like that of columnar basalts (Beard, 1960), but probably not as regular in morphology. An analysis of water penetration into uniform hot rock, based on the concept of a sharp "cracking front" between the hydrothermal system and undisturbed rock, predicts crack spacings from

Hydrothermal Systems

Figure 24-1. Photograph of a horizontal erosion surface of the sheeted complex in the Smartville (California) Ophiolite (river bed). The thin knobbly dike crossing the center of the picture is about 7 cm wide. Note the cracking.

several centimeters to meters (Lister, 1974; appendix 24A). The treatment of physics of cracking is rudimentary, but cracks at a similar spacing are common in the field (figure 24–1).

A rock matrix, cracked and thermally contracted on a scale of a meter or two, is consistent with reasonable bulk permeabilities and a medium rate of chemical interchange. The latter is a significant point because the metalliferous sediments are found only on very young oceanic crust (Piper, 1973), but the thermal-transport effects of circulation persist to substantial ages (Sclater et al., 1976). If the sub-meter cracking scale can be accepted through its general agreement with the circumstantial evidence, then predictions can be made about the physical conditions in the geothermal systems. This chapter is an attempt to extend the basic theory (Lister, 1974) to such predictions based on a qualitative structural model of ridge-crest processes that takes hydrothermal cooling effects into account (Lister, 1977a).

The Semiquantitative Framework

When penetrative cooling of hot rock by water was analysed physically, there was one big surprise: the rapidity of the process. The speed of the advance of the cracking front is controlled by the thermal-transport capability of the permeable-convective system. This capability is a function of the permeability of the cracked rock, the effective bottom boundary temperature, and the intrinsic properties of convection in porous media. The faster the rate of penetration, the thinner must be the conductively cooled boundary layer ahead of the front, and the thinner the layer stressed by thermal contraction. The average spacing between cracks is related to the thickness of the stressed layer, and, since the permeability is a strong function of the crack spacing (Bear, 1972; Lister, 1974), the system is stabilized by negative feedback. The details of the cracking process are difficult to quantify, but the predictions of the simplified theory are within the order of magnitude that includes the field data. Hence the permeability predictions should also be of the right order of magnitude.

The effective bottom-boundary temperature is related to the temperature at which the rock is sufficiently stressed to crack, and that is not predictable from current knowledge of the mechanical behavior of rocks. Results have been computed for the full plausible range of cracking temperatures (Lister, 1974; appendix 24A) and predict considerable differences in the physical regime for water/rock interaction. The variation will be retained in the numerical presentations (figures 24–3 and 24–4), but there remains one other significant problem concerning heat transport and Rayleigh number (figure 24–2). The intrinsic properties of convection in porous media are not well known for high Rayleigh numbers. The heat transport is usually expressed nondimensionally as a ratio to the conductive heat flux that would be driven

Hydrothermal Systems

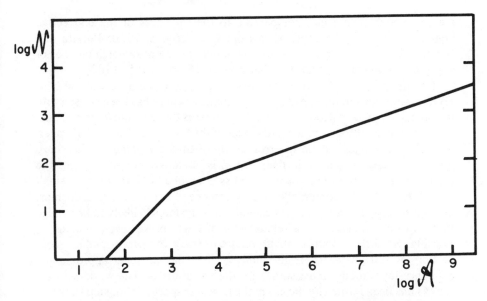

Figure 24–2. The heat transport ($\mathcal{N}$) versus Rayleigh number ($\mathcal{A}$) relationship used in the text. Published data confirm the steeper initial part of the relation (Buretta and Berman, 1976; Combarnous, 1970), but the breakover to ⅓ power dependence is less secure. At Rayleigh numbers over 1,000, most experimental convection cells contain beads comparable in diameter to the boundary-layer thicknesses and are no longer true porous media.

by the same temperature difference in the absence of convection: this is the Nusselt number $\mathcal{N}$. The Rayleigh number $\mathcal{A}$ is a nondimensional expression for the level of instability in the system, and the onset of convection in a porous slab with impermeable conductive boundaries occurs at $\mathcal{A} = 40$ (Lapwood, 1948).

The Nusselt number should be some function of Rayleigh number, and early experiments suggested that it might be simply $\mathcal{N} = \mathcal{A}/40$ (Elder, 1965). A first-order boundary-layer analysis has suggested $\mathcal{N} \propto \mathcal{A}^{\frac{1}{3}}$ should be more applicable at medium to high Rayleigh numbers (J. Booker, personal communication). The validity of this analysis is supported by experimental measurements of the flow velocity (Hartline and Lister, 1977), but published data on heat transport in laboratory convection cells cannot be used to establish where the initial $\mathcal{A}/40$ variation changes to the form $\theta' \mathcal{A}^{\frac{1}{3}}$. The apparent breakover in laboratory experiments occurs under conditions where the particle size of the artificial porous medium is larger than the thickness of the convective plume and comparable to the thickness of the convective boundary layer (for example, data from Combarnous, 1970; Buretta and Berman, 1976). Data not showing a breakover below a Rayleigh number of 2000 are, however, suspect for other reasons (Elder, 1965; 1967).

For the purpose of making order-of-magnitude calculations of the convective conditions in the ocean crust, an arbitrary curve has been chosen with the breakover at $\mathscr{A} = 1,000$ (figure 24–2). It fixes the constant θ' as 2.5, and, because it depends only on the ⅔rd power of the breakover Rayleigh number, the value is unlikely to be in error by more than 50 percent. The sharp arbitrary breakover shown in figure 24–2 is obviously not realistic for the transition region, but the error in the Nusselt number so introduced is small enough to justify the computational convenience of two simple relationships.

The Rayleigh number in active geothermal systems is likely to be so high that a further transition to a relationship of lower power even than $\mathscr{A}^{\frac{1}{3}}$ is possible. So little is known about what to expect in porous-medium convection at ultra-high Rayleigh numbers that the present, approximate, boundary-layer theory must suffice for the calculations. The work of Lister (1974) has been refined by the use of the heat-transport relationship shown in figure 24–2 and an improved analysis of the mechanism that controls the crack spacing. The calculations are presented in appendix 24A and show that somewhat slower water penetration, larger crack spacing, and higher water temperatures are predicted than were expected in the earlier work.

Qualitative Analysis of a Ridge Crest: "Active" and "Passive" Hydrothermal Systems

The basic order-of-magnitude rates of water penetration are meters per year, a conclusion not likely to be modified by further corrections to the theory or by the largely unquantifiable effects of wall-rock alteration on the permeability. The rate is much higher than the maximum rate of sea-floor spreading, and therefore the penetration of hot rock must be an episodic phenomenon at a ridge crest. The implications for the thermotectonics at a spreading center are profound and have been discussed qualitatively by Lister (1977a). The principal result is that the oceanic crustal section is cooled by "active" water penetration very close to the injection zone for new oceanic rock. The penetration is halted at depth, either by a change to ultramafic rock that is highly susceptible to serpentinization or by a process of static fatigue in rock subjected to both high temperature and heavy overburden pressure. In either case, a substantial permeable region remains above the level of crack closure and continues to be heated by conduction from below. Moderate hydrothermal activity continues as long as the heat flow and permeability remain high enough to sustain convection; such a condition is the "passive" phase of circulation.

Hydrothermal Systems

It is important to remember that the "active" and "passive" hydrothermal systems occur in the same permeable rock matrix. The differences between them arise from a great difference in the vigor of convection, and from the fact that fresh rock is continually exposed during the active phase but probably not during the passive phase. Although the chemical implications will be discussed briefly, the aim of this chapter is to present rough estimates of the range of physical conditions in the circulation and the effects on them of the principal variable parameters. The equations used for handling the two classes of system are rather different, but results will be presented on similar diagrams for comparison.

General Relationships in Porous-Medium Convection

The two parameters most useful for specifying the conditions of hydrothermal convection are Q (the total heat transport per unit area) and D (the permeability). If the properties of the fluid are known, the thermal conductivity of the rock is not a major variable, and the permeable layer thickness can be assumed. The system is fully specified. It might be argued that specifying the layer thickness is inappropriate for a region as little known as the oceanic crust, and especially so when an actively penetrating hydrothermal system is being considered. However, the results do not vary much more rapidly than linearly with layer thickness and are in any case only good to an order of magnitude, so the simple assumption of a plausible layer thickness simplifies the presentations without losing essential information. A similar argument can be made for the choice of a median rock conductivity somewhere within the not inconsiderable range between that of rapidly cooled basalts and that of high-olivine gabbroic cumulates. The effects of all these parameter changes can be followed best if a fully algebraic analysis is performed before inserting numerical data. The analysis follows.

The basic relationships are

$$\mathcal{A} = \alpha_w g \Delta T h D / \kappa v_w \quad \text{Definition} \qquad (24.1)$$

$$\mathcal{N} = hQ/\kappa\rho c\Delta T \quad \text{Definition} \qquad (24.2)$$

$$\mathcal{N} \simeq \mathcal{A}/40 \quad (\mathcal{A} < 1{,}000) \text{ (Elder, 1965)} \qquad (24.3a)$$

$$\mathcal{N} \simeq 2.4\, \theta \mathcal{A}^{\frac{1}{3}} \quad (\mathcal{A} > 1{,}000) \qquad (24.3b)$$

$$\alpha_w/v_w \simeq w(T_w - 273)^2 \quad (T_w \geq 313°\text{K; Elder, 1965}) \qquad (24.4a)$$

$$\alpha_w/v_w \simeq w \cdot 40^2 \quad (T_w < 313°\text{K; room temperature values}) \qquad (24.4b)$$

$$v \simeq 0.17\,(\mathscr{A}^2 - 40^2)^{1/2} \quad \text{(Hartline and Lister, 1977)} \quad (24.5)$$

$$\tau_{\text{hot}} \simeq h/v = h^2\beta/\kappa v \quad \text{Definition} \quad (24.6)$$

where the symbols are as listed in appendix 24B. There are three principal cases: (1) when equations 24.3a and 24.4a are operative, as in a low-permeability altered oceanic crust, (2) when equations 24.3b and 24.4b apply, as in high-permeability young ocean crust, and (3) when equations 24.3b and 24.4a must be used, as in any "active" penetrating system. The results of combining and solving the relevant equations can be summarized for the three cases as follows:

1. Low-permeability "passive" convection:

$$\mathscr{A} \simeq (40^3 g h^4 Q^3 Dw/\kappa^4 \rho^3 c^3)^{1/4} \simeq 1.33 \times 10^6 (Q^3 D)^{1/4} \quad (24.7a)$$

$$T_w - 273 \simeq (40Q/\rho c g Dw)^{1/4} \simeq 0.0573\,(Q/D)^{1/4}\,°K \quad (24.7b)$$

$$\tau_{\text{hot}} \simeq 5.9 h^2 \beta/\kappa(\mathscr{A}^2 - 40^2)^{1/2} \simeq \quad (24.7c)$$
$$1.15 \times 10^{12}(\mathscr{A}^2 - 40^2)^{-1/2} \quad \text{s} \quad (\beta = 0.007)$$

where β is assumed to have about one-half the value derived from the formula in appendix 24B (see case 2).

$$F \simeq (\mathscr{N} - 1)Q/\mathscr{N}\,(T_w - T_0)\rho_w c_w \simeq \quad (24.7d)$$
$$2.39 \times 10^{-7}(\mathscr{N} - 1)Q/\mathscr{N}\,(T_w - T_0) \quad \text{m/s}$$

2. High-permeability "passive" convection:

$$\mathscr{A} \simeq (0.4\alpha_w g h^2 QD/v_w \kappa^2 \theta \rho c)^{3/4} \quad (24.8a)$$
$$\simeq 8.92 \times 10^{11}(QD/\theta)^{3/4}$$

$$T_w - 273 \simeq (0.064 Q^3 h^2 v_w/\alpha_w \theta^3 \kappa^2 \rho^3 c^3 g D)^{1/4} \quad (24.8b)$$
$$\simeq 0.079(Q^3/\theta^3 D)^{1/4}\,°K$$

$$\tau_{\text{hot}} \simeq 5.9 h^2 \beta/\kappa\mathscr{A} \simeq 2.29 \times 10^{12}/\mathscr{A} \quad \text{s} \quad (\beta = 0.014) \quad (24.8c)$$

where β is given a suitable "median" value of 0.014 based on the plausible cracking temperature range of 0.007 to 0.022, using the formula in appendix 24B for thermal contraction.

$$F \simeq (\mathcal{N} - 1)Q/\mathcal{N}\,(T_w - T_0)\rho_w c_w \qquad (24.8d)$$
$$\simeq 2.39 \times 10^{-7} Q/(T_w - T_0) \text{ m/s}$$

3. "Active" penetrating system:

$$\mathcal{A} \simeq (0.064\, wgh^4 Q^3 D/\theta^3 \kappa^4 \rho^3 c^3)^{1/2} \qquad (24.9a)$$
$$\simeq 1.76 \times 10^9 (Q^3 D/\theta^3)^{1/2}$$

$$T_w - 273 \simeq (0.064 Q^3 h^2 / wg\theta^3 \kappa^2 \rho^3 c^3 D)^{1/6} \qquad (24.9b)$$
$$\simeq 0.63\, (Q^3/\theta^3 D)^{1/6} \quad °K$$

$$\tau_{\text{hot}} \simeq 5.9 h^2 \beta/\kappa \mathcal{A} \simeq 9.31 \times 10^4 \beta (\theta^3/Q^3 D)^{1/2} \text{ s} \qquad (24.9c)$$

$$F \simeq (\mathcal{N} - 1)Q/\mathcal{N}\,(T_w - T_0)\rho_w c_w$$
$$\simeq 2.39 \times 10^{-7} Q/(T_w - T_0) \quad \text{m/s} \qquad (24.9d)$$

where β should be determined individually for systems based on different cracking temperatures because of substantially varying T_w (see table 24A-1).

The results should have order-of-magnitude validity for 5-km-thick permeable layers in typical ocean crust, provided that convective conditions permit the basic equations (24.1–24.6) to represent the physics. There are several obvious differences between ocean crust and an idealized convection cell. First, flow may be semi-open to the ocean through surface faulting or outcrops of permeable rock, or there may be a semi-insulating sediment blanket over the permeable layer. Also, the basal boundary is underlain by rock of conductivity only somewhat greater than that of the porous matrix, rather than the perfect conductor assumed in theory and approached in laboratory convection cells. The most serious difference is probably the existence of a sediment blanket, since all temperatures are now increased by the thermal drop across it. However, if the convection cells are sufficiently isolated from the open ocean to have a substantially higher temperature, there is relatively little chemical interchange between the ocean and the buried rocks. Thus the effects may be important in the consideration of the history of rock alteration, but are less so to the chemical balance of the oceans.

The largest errors are likely to occur in the active systems, where the Rayleigh numbers are so high that the simple formulae are unlikely to apply. If porous-medium convection shows any analogy to free convection, there are no stable patterns (at least near the boundary), and the heat transport is somewhat lowered, both by the unsteadiness (or thermal turbulence) and by the inherent failure of the boundary-layer theory itself when the plumes become too narrow (see appendix 24A). Temperatures may be slightly higher near the bottom boundary, but the biggest error is in the residence time of the hot water in the ascending plume, since the thermal spreading of the narrow plume will reduce the hot fluid temperature and its rate of ascent through the matrix. Nevertheless, the times are short enough overall for the chemical implications of any errors to be small; that is, the water is unlikely to come to equilibrium with the wall rock.

This section has presented formulae for the estimation of the chemically important parameters, temperature and time, given a basal heat flow input and a permeability. Once some alteration has occurred, the choice of the permeability is mainly a matter of taste, and the material and diagrams that follow are intended to show plausible ranges rather than to make specific predictions. Only in the active systems are conditions well enough constrained by their youth and the processes of cracking and penetration to allow moderately specific predictions of temperatures, water flux, and perhaps water residence time.

Conditions in "Passive" Convective Systems

The "passive" phase of hydrothermal activity is the convection of fluid in the surface or near-surface permeable layer of the oceanic crust, driven by conductive heat flow from below. In that case, the heat flow Q is relatively well quantified as a function of age. It has been shown (Davis and Lister, 1974) that ocean-ridge topography follows a square-root dependence on age from 1 to 80 million years (My) and that the dependence is consistent with the isostatic sinking of rock which is cooling and contracting in a thickening boundary layer. The corresponding conductive heat flux through the surface follows the relation

$$Q = 0.5 t^{-1/2} w m^{-2} \qquad (24.10)$$

with reasonable rock parameters (Lister, 1977b), and the relation has been verified by analysis of field data (Sclater et al., 1976). Beyond 80 My, the topography seems to flatten out over most old oceanic areas, presumably because of some kind of heat or magma supply from below, but it is possible that equation 24.10 is still sufficiently valid due to the slow thermal response of a thick lithosphere. Field data in the old ocean basins are sparse and poorly controlled relative to environmental effects, but such values that exist do not differ from equation 24.10 by more than 20 percent.

With the heat input Q determined as a function of age, it is only necessary

Hydrothermal Systems

to specify some value of permeability D to compute the probable conditions in the convective circulation. When the oceanic crust is young, it should still contain the thermal-contraction cracks left by the "active" penetrating systems (appendix 24A). The permeability calculated for near-ocean temperatures in the cracked matrix, although obviously an upper bound, may not be too unrealistic. As the crust ages, however, the cracks will begin to fill up with mineral deposits, and hydration of the rock matrix will cause expansion and general closure of all the cracks. How much the permeability is imagined to decrease is largely a matter of taste, but, for the purpose of providing a reasonable lower bound for the graphs, I have chosen the value quoted by Palmason (1967) as representative of rocks in Iceland. Values much lower than his (10^{-14} m^2) would cause rapid cessation of water circulation and would be in conflict with the heat-flow data that point to continued open circulation through outcropping rocks for long periods of time (Sclater et al., 1976).

Variations of the main circulation parameters as a function of age can be computed readily for the two extreme cases. The forms of the results differ, mainly because variable fluid properties must be included when the permeability is low and the temperature differences are relatively high, but low-temperature values suffice when the permeability is high.

1. Low-permeability (10^{-14} m^2) passive convection (symbols defined in appendix 24B):

$$\mathcal{A} = 250 t^{-3/8} \tag{24.11a}$$

$$(T_w - T_0) = 152 t^{-1/8} \ ^\circ K \tag{24.11b}$$

$$\tau_{hot} \simeq 148 t^{1/8} \text{ yr} \quad t < 10 \text{ My} \tag{24.11c}$$

$$F \simeq 24.7 t^{-3/8} \text{ km/My} \quad t < 10 \text{ My} \tag{24.11d}$$

where the last two relations apply only approximately to the larger ages because of the approach of the Rayleigh number to the critical cutoff: 40 for sealed circulation and 27 for open circulation.

2. High-permeability (1.2×10^{-7} m^2) freshly cracked rock:

$$\mathcal{A} = 3.42 \times 10^6 t^{-3/8} \tag{24.12a}$$

$$(T_w - T_0) = 2.52 t^{-3/8} \ ^\circ K \tag{24.12b}$$

$$\tau_{hot} = 0.021 t^{3/8} \text{ yrs} \tag{24.12c}$$

$$F = 1490 t^{-1/8} \text{ km/My} \tag{24.12d}$$

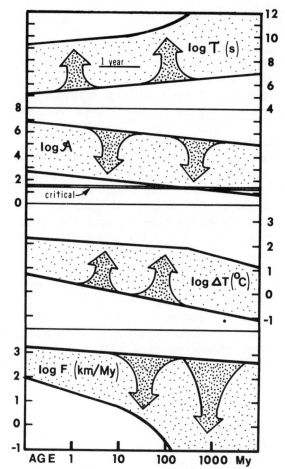

Figure 24–3. A generalized plot of the main parameters in a convecting porous slab 5 km thick against basal heat-source age in millions of years. τ is water residence time in the hot upwelling plume (s), $\mathscr{A}$ is Rayleigh number, ΔT is temperature difference across the convecting medium (°C), and F is the circulating fluid flux per unit area (km My^{-1}). The temperature difference is expressed as ΔT rather than $(T_w - T_0)$ because the circulation may be insulated from the ocean by beds of sediment. Alteration arrows mark the transition from the permeability (1.2×10^{-7} m^2) predicted by the water-penetration model (appendix 24A; Lister, 1974) to a plausible limiting low value (10^{-14} m^2) quoted by Palmason (1967). Real conditions in the ocean crust might be expected to trend from the former regions at young ages to the latter in older material, subject to a large degree of geologic "noise." The circulation flux may or may not be open to the ocean, depending on the thickness, universality, and permeability of surface sediment layers.

Hydrothermal Systems

The results of those computations are plotted in logarithmic form in figure 24–3, where the large stippled arrows show the trend in directions for rock alteration and the consequent reduction in permeability. The real history of a parameter with age might be expected to begin near the high-permeability curve and trend toward, or perhaps beyond, the low-permeability curve. The exact path followed may depend considerably on local conditions: the crustal content of ultramafic minerals, the initial cracking temperature during water penetration (appendix 24A), and the rapidity of sediment blanketing. In the case of circulation underneath a thick and impermeable blanket of sediments, the temperature at the permeable surface (T_0) is no longer the ocean temperature but that at the base of the relatively poorly conducting sediments. Such modified systems may have profound implications for the nature and rate of rock alteration beneath sediment-filled basins but little for the chemical balance of the oceans because of the restricted, or non-existent, exchange between the circulating fluid and the ocean above.

Conditions in "Active" Systems

The calculations given in appendix 24A for active systems are highly dependent on the simplifying assumptions of the theory. Prime among them is the one-dimensional approximation, where the rock is treated as uniform infinite layers that cannot shrink without cracking (Lister, 1974). In fact, even in the absence of environmental variations, a geometric instability may be expected to occur, and each active convection cell would become a separate penetrating system. The qualitative aspects of the problem have been discussed by Lister (1976), and some reduction is expected in penetration rate, permeability, and cracking temperature.

The estimation of a cracking temperature is the key to the whole problem, since every result depends on it directly. An application of the sketchy data available on both rock creep and rock cracking resulted in the nominal choice of 800°K as a plausible cracking temperature (Lister, 1974). Some readers may prefer other values; so the results are presented as if that were a free parameter (figure 24–4). Localized one-dimensional-model conditions may be approached at very high Rayleigh numbers because of the instability in the position of boundary-layer separation into upwelling plumes; thus the heat transfer may be statistically uniform over a substantial area.

Another serious potential problem with the theory is that it assumes free cold-water access to the region of hot rock being actively cracked. The surface of the ocean crust is frequently sealed by a layer of sediment or a mixture of baked sediment and pillow lavas, with permeability confined to tectonic fault zones. Since the cracking temperature is determined by the thermal shrinkage rate as well as intrinsic rock-creep properties, it is higher for higher penetration rates. Thus the response of the system to restricted water access is the reduction of actively cracking area rather than a general

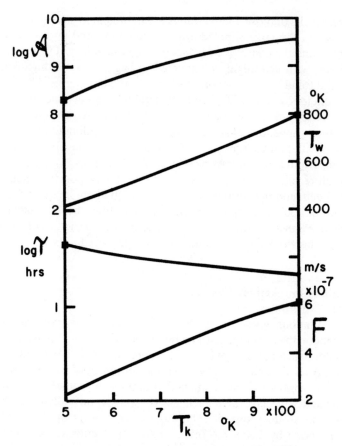

Figure 24-4. Conditions in an "active" penetrating geothermal system, as predicted by the revised theory given in appendix 24-A. The abscissa is cracking temperature; it is not really a free parameter, but the creep and cracking properties of representative rocks are simply not well enough known. The water temperature T_w is given in absolute units because the system is assumed to open either to the ocean or to a large cold recharge volume. The residence time τ should be considered as no more than an order-of-magnitude lower limit. The flux F is given in different units from figure 24-3 because of the relatively small size and short lifetime expected for active areas (see Lister, 1976).

Hydrothermal Systems   455

reduction in cracking temperature. The instability arises because a region of faster penetration produces cracks more resistant to closure as the general temperature rises and also produces the hottest upwelling water that must dominate the upwelling plume. That mechanism is, in fact, the key to the geometric instability referred to earlier.

Values calculated in appendix 24A are presented in figure 24-4, having been abstracted from the more complete table 24A-1. The power dependencies on the recognized fudge factor θ are relatively high, but all the properties are relatively insensitive to the poorly known fudge factor ϕ, related to the cracking properties of the rocks. The biggest problem with the results is the failure of the boundary-layer flow approximation for plumes as narrow as would be predicted by the theory (flow in a single crack!). The hot-column residence time has been corrected for the effect of turbulent flow in the crack(s) but not for the thermal diffusive spreading of the plume. It must be recognized that all the properties graphed except $\mathscr{A}$ are interconnected. Thus, if the hot-column temperature rise above that of the ocean is halved, the residence time rises by $\sqrt{2}$, and the total volume flux is doubled. In practice, the temperature, flow velocity, and flux must change gradually up the upwelling plume, and no one set of conditions could describe the chemical interchange in the system. The characteristics of an active system are, however, clear in a broad sense: high water temperature, high flux, and short water-residence time.

Conclusions

The principal result of the investigation into how circulating water may penetrate into cooling hot rock is that the process is rapid. Convective transport of heat is so much more efficient than conduction on a large scale that the permeable region should be separated from essentially uncooled hot rock by a thin conductive boundary layer (Lister, 1974). It is this boundary layer that contains the tensile stresses which cause the rock to crack and become permeable, and it is the local temperature gradient which controls the crack spacing (appendix 24A). That local temperature gradient, maintained only by the continual advance of the cracks into the rock, is far higher than any purely conductive heat supply could achieve, save for the unlikely situation where magma exists a few meters away from circulating water. The energy output of the convective system is therefore high even for modest areal extent and leads to the designation of the rock-penetrating phase of convection as an "active" geothermal area.

Once active cracking of fresh rock has ceased due to some combination of physical (static fatigue) and chemical (alteration) effects, heat is supplied by simple conduction to the base of the permeable region. If the permeability is high enough, convective fluid circulation will occur, either open to the ocean or perhaps sealed by sediment accumulations on the surface. That can be termed the "passive" phase of geothermal activity, where the water temperatures may be high enough to warrant the description "hot springs" where they appear but are not comparable to the (surface) boiling temperatures observed in geothermal areas.

The geochemical conditions in the two cases are very different. During "active" penetration, fresh rock is continually exposed to very hot water, but the lifetime of the whole system is short by geological standards. Profoundly different alteration histories apply to material that goes from the hot dry state into the cold recharge column of the convective cycle compared to material that happens to underlie the hot upwelling column. The former undergoes a brief exposure to hot water in the convective boundary layer and is then chilled to essentially ocean temperature. Alteration should be mostly of the low-temperature type. The material in the hot column, however, remains exposed to hot fluid throughout the lifetime of the "active" geothermal phase, and labile elements may be extracted efficiently, possibly even under reducing conditions maintained by the constant exposure of fresh rock at the base of the system. Here may be the source of the heavy metals found as metalliferous sediments near the ridge crests.

The "passive" circulation is much slower and occurs at relatively low temperatures, a maximum of 100°C being possible in very young but relatively impermeable ocean crust (figure 24–3). Although the fluxes are smaller than in "active" systems, circulation may continue for many millions of years and involve a considerable fraction of the world's oceans (Wolery and Sleep, 1976). For example, if a median flux like 20 km/My is applied to a substantial portion of the square kilometers forming the ocean floor, the entire volume of the ocean could be cycled through in a fraction of a million years! In practice, much of the ocean floor is thickly sedimented basin, so cycling every few millions of years would be a more realistic estimate. The implications for the chemical balance of the oceans are certainly profound— no longer do the surface fluxes of all the chemical species have to balance to an equilibrium. Those that can interact significantly with cracked basalt or ultramafic rock can have large sources or sinks in the oceanic crust.

Acknowledgments

This work was supported in part by National Science Foundation Grant DES73-06593-AO1. I wish to thank F.T. Manheim for stimulating the project and B.K. Hartline for invaluable assistance in developing the properties of porous-medium convection.

References

Bear, J. 1972. *Dynamics of Fluid in a Porous Medium.* New York: American Elsevier.

Beard, C.N. 1960. Quantitative study of columnar jointing. *Geol. Soc. Am. Bull.* 70:379–381.

Bender, M., Dymond, J.R., and Heath, G.R. 1971. Isotopic analyses of metalliferous sediments from the East Pacific Rise. *Geol. Soc. Am. Abstr.* 3:537.

Bischoff, J.L., and Dixon, F.W. 1975. Sea water–basalt interaction at 200°C and 500 bars: Implications for origin of sea-floor heavy metal deposits and regulation of sea water chemistry. *Earth Plan. Sci. Lett.* 25:385–397.

Bodvarsson, G., and Lowell, R.P. 1972. Ocean-floor heat-flow and the circulation of interstitial waters. *J. Geophys. Res.* 77:4472–4475.

Bostrom, K., and Peterson, M.N.A. 1969. The origin of aluminum ferromanganoan sediments in areas of high heat flow on the East Pacific Rise. *Mar. Geol.* 7:427–447.

Bostrom, K., Farquharson, B., and Eyl, W. 1971. Submarine hot springs as a source of active ridge sediments. *Chem. Geol.* 10:189–203.

Buretta, R.J. and Berman, A.S. 1976. Convective heat transfer in a liquid saturated porous layer. *J. Appl. Mech.* 43:249–253.

Clark, S.P. (Ed.). 1966. Handbook of physical constants. *Geol. Soc. Am. Mem.* 97.

Coleman, R.G., and Keith, T.E. 1971. A chemical study of serpentinization—Burro Mountain, California. *J. Petrol.* 12:311–328.

Combarnous, M. 1970. Convection naturelle et convection mixte daus une couché poreuse horizontale. *Rev. Gen. Therm.* 108:1355–1375.

Corliss, J.G. 1970. "Mid-ocean Ridge Basalts. 1. The Origin of Submarine Hydrothermal Solutions. 2. Regional Diversity along the Mid-Atlantic Ridge. Ph.D. dissertation, Univ. of California, San Diego.

Davis, E.E., and Lister, C.R.B. 1974. Fundamentals of ridge crest topography. *Earth Plan. Sci. Lett.* 21:405–413.

Davis, E.E., and Lister, C.R.B. 1977. Heat flow measured over the Juan de Fuca Ridge on a quasi-regular grid. *J. Geophys. Res.* 82:4845–4860.

Dymond, J., Corliss, J.G., Heath, G.R., Field, G.W., Dasch, E.J., and Veeh, H.H. 1973. Origin of metalliferous sediments from the Pacific Ocean. *Geol. Soc. Am. Bull.* 84:3355–3372.

Elder, J.W. 1965. Physical processes in geothermal areas. *Am. Geophys. U. Mon.* 8:211–239.

Elder, J.W. 1967. Steady free convection in a porous medium heated from below. *J. Fluid Mech.* 27:29–48.

Ellis, A.J. 1968. Natural hydrothermal systems and experimental hot-water/rock interaction: Reactions with NaCl solutions and trace metal extraction. *Geochim. Cosmochim. Acta.* 32:1356–1363.

Ellis, A.J., and Mahon, W.A.F. 1964. Natural hydrothermal systems and experimental hot-water/rock interactions. *Geochim. Cosmochim. Acta.* 28:519–538.

Hajash, A. 1974. An experimental investigation of high temperature sea water-basalt interactions (Abstract). *Geol. Soc. Am. Ann. Meeting* 771.

Hartline, B.K., and Lister, C.R.B. 1977. Thermal convection in a Hele-Shaw cell. *J. Fluid Mech.* 79:379–389.

Hoerner, S.F. 1958. *Fluid Dynamic Drag*, 2d ed. Midland Park, N.J.: Hoerner, Chap. 2, p. 6.

Lapwood, E.R. 1948. Convection of a fluid in a porous medium. *Proc. Camb. Phil. Soc.* 44:508–521.

Lister, C.R.B. 1972. On the thermal balance of a mid-ocean ridge. *Geophys. J. R. Astr. Soc.* 26:515–535.

Lister, C.R.B. 1974. On the penetration of water into hot rock. *Geophys. J. R. Astr. Soc.* 39:465–509.

Lister, C.R.B. 1976. Qualitative theory on the deep end of geothermal systems. *Proc. 2nd U.N. Geothermal Symposium.* U.S. Government Printing Office, Washington, pp. 459–464.

Lister, C.R.B. 1977a. Qualitative models of spreading center processes, including hydrothermal penetration. *Tectonophysics* 37:203–218.

Lister, C.R.B. 1977b. Estimators for heat flow and deep rock properties based on boundary layer theory. *Tectonophysics* 41:157–771.

Nigrini, A. 1969. "Prediction of Ionic Fluxes in Rock Alteration Processes at Elevated Temperatures." Ph.D. dissertation, Northwestern Univ., Evanston, Illinois.

Page, N.J. 1967. Serpentinisation considered as a constant volume metasomatic process: A discussion. *Am. Mineralogist* 52:545–549.

Palmason, G. 1967. On heat flow in Iceland in relation to the Mid-Atlantic Ridge, in *Iceland and Mid-Ocean Ridges*. S. Bjornsson (Ed.), *Soc. Sci. Islandica Publ.* 38:11–127.

Piper, D.Z. 1973. Origin of metalliferous sediments from the East Pacific Rise. *Earth Plan. Sci. Lett.* 19:75–82.

Sales, R.H., and Meyer, C. 1948. Wall rock alteration at Butte, Montana. *Am. Hist. Min. Met. Eng. Trans.* 178:9–35.

Sclater, J.G., Crowe, J., and Anderson, R.N. 1976. On the reliability of oceanic heat flow averages. *J. Geophys. Res.* 81:2997–3006.

Talwani, M., Windish, C.C., and Langseth, M.G., Jr. 1971. Reykjanes Ridge crest: A detailed geophysical study. *J. Geophys. Res.* 76:473–517.

Williams, D.L., Von Herzen, R.P., Sclater, J.G., and Anderson, R.N. 1974. The Galapagos spreading center: Lithospheric cooling and hydrothermal circulation. *Geophys. J. R. Astr. Soc.* 38:587–608.

Wolery, T.J., and Sleep, N.H. 1976. Hydrothermal circulation and geochemical flux at mid-ocean ridges. *J. Geol.* 84:249–275.

Appendix 24A: Revision of Water-Penetration Theory

Formulae to estimate the speed of water penetration into cracking hot rocks have been published by Lister (1974). The derivation used the old convective heat-transfer relationship $\mathcal{N} = \mathcal{A}/40$, and it is now clear that, at high Rayleigh numbers, a form $\mathcal{N} = 2.5\, \theta \mathcal{A}^{1/3}$ is more appropriate. Moreover, the original derivation used a condition to control the crack spacing that is not properly representative of the heat transfer near the edge of a large-scale boundary layer. The basic idea was that the rock columns between the cracks are cooled mainly by the fluid circulation once cracks have opened. If the columns try to become too large, by the process of "obdivision" (Lister, 1974), a radial temperature gradient is set up that places the periphery of the columns under tension, and they crack and subdivide again. An order-of-magnitude estimate of that radial temperature difference (Lister, 1974, eq. 24) was combined with the best available data to yield a relationship between crack spacing and the rate of advance of the cracking "front" (Lister, 1974, eq. 31). The crack spacings calculated are considerably smaller than those found in nature (5 versus 20 cm; see figure 24–1) and substantially smaller than the thickness of the conductive thermal boundary layer x_o (Lister, 1974, eq. 1) that forms a kind of upper limit to plausible crack spacings.

The idea that water circulating through individual cracks would do all the cooling near the boundary of the permeable region is false. Convective advection is primarily lateral near a planar boundary unless a local convective instability causes turbulence; due to inertia forces, the usual form of turbulence is impossible in a truly porous medium. Lateral temperature gradients near a boundary are small and vanish as the boundary itself is approached, so that fluid advection should *not* account for more than a small proportion of column cooling. The thermal gradient is primarily perpendicular to the boundary, and most of the heat transfer is still by vertical conduction of heat through the columns (figure 24A–1). There is *some* vertical advection, however, because the cracking front is advancing downward, and the new cracks must fill with fluid. The downward fluid flow is against a substantial temperature gradient, and therefore the peripheries of the rock columns are cooled, although not by as much as under the old assumption of full advective cooling.

The advection per unit surface of crack is given by

$$\tfrac{1}{2} du \rho_w c_w \partial T/\partial z \qquad (24\text{A}.1)$$

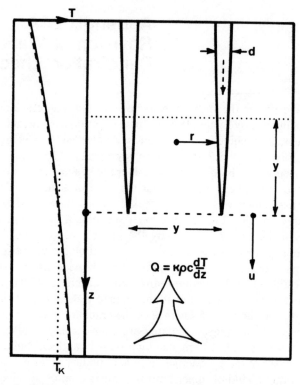

Figure 24A-1. Diagram of the principal features and dimensions of cracks near a cracking front (not to scale). The cracking front is the dashed line moving downward at velocity u. Behind it for a distance y, equal to the mean crack separation, extends the region where column shrinkage strongly influences the propagation of the crack tips. A typical column radius is shown as r; d is the actual crack width after column shrinkage below T_κ, the cracking temperature. A schematic temperature curve is shown above the large basal arrow. Q may be statistically steady due to convective instability, or it may vary with position in the cell, as suggested by Lister (1976).

and, if this caused the steady cooling of a cylinder, the radial temperature difference would be

$$\Delta T_r = y^2 \dot{T}/16\kappa \quad \text{(Lister, 1974, eq. 24)} \quad (24\text{A}.2)$$

with a heat output per unit area of

Hydrothermal Systems

$$\tfrac{1}{2} r \dot{T} \rho c \qquad (24A.3)$$

where $\dot{T}$ is the time rate-of-change of temperature (due to advection *alone*) at a point on the cylinder surface. The steady-state assumption, while obviously not strictly applicable, is sufficiently accurate with the predicted column size and crack advance rates, considering that the real geometry is more complex than the cylindrical abstraction being used (for example, see Lister, 1974, figure 3). The relations combine to give

$$\Delta T_r = u d y \rho_w c_w (\partial T / \partial z) / 8 \kappa \rho c \qquad (24A.4)$$

and the local vertical temperature gradient is (Lister, 1974)

$$\partial T / \partial z = u \kappa^{-1} (T_1 - T_\kappa) \exp(-uz/\kappa) \simeq u(T_1 - T_\kappa)/\kappa \qquad (24A.5)$$

as long as it is being evaluated near the cracking boundary ($T = T_\kappa$), compared to a scale $x_o = \kappa/u$. Important effects of the column shrinkage on the propagating crack tips are confined to one column diameter from the front (figure 24A–1). Therefore, using equation 24A.6, d can be evaluated a distance y above the cracking front:

$$d = \alpha y (T_p - T) \simeq \alpha y (T_\kappa - T) \simeq \alpha y^2 \partial T / \partial z \qquad (24A.6)$$

In this situation T_p and T_κ are close enough to be substituted, since

$$T_p - T_\kappa \simeq \phi' y^{-1/2} \qquad \text{(Lister, 1974, eq. 30)} \qquad (24A.7)$$

is the result of analyzing the best available creep and cracking data. The final step is to guess that a reasonable equilibrium value for ΔT_r is $T_p - T_\kappa$ because, if the column is stressed by a greater differential thermal contraction than is needed to propagate the main existing cracks, it should subdivide. In case the reader does not like such a sweeping assumption, the numerical fudge factor ϕ, besides containing all the creep and cracking problems, is retained in the calculations to the very end and appears at a low enough power in most results to inspire some confidence. Hence if

$$\Delta T_r = T_p - T_\kappa \qquad (24A.8)$$

all the equations can be combined into

$$y = [8 \phi' \kappa^3 \rho c / \alpha u^3 \rho_w c_w (T_1 - T_p)^2]^{2/7} \qquad (24A.9)$$

which embodies the crucial relationship between y and u needed for the analysis to proceed.

The effective boundary temperature is not that at which the cracks open (T_κ) because the permeability at the boundary is zero and there is no circulation. The real problem is again extremely difficult, but results accurate to a few precent can be obtained simply by choosing the convective boundary temperature as that at which the permeability of the expanding cracks has reached about two-thirds that in the hot plume itself:

$$T_2 = \tfrac{1}{9}(T_p + 8T_w) \tag{24A.10}$$

Following Lister (1974) for the derivation of an effective permeability and the local heat balance of an advancing front

$$D = 0.118 y^2 \alpha^3 (T_p - T_w)^3 \tag{24A.11}$$

$$Q = \mathcal{N} \kappa \rho c (T_2 - T_0)/h = u\rho c(T_1 - T_w) \tag{24A.12}$$

the equations can be solved for u to yield

$$u^{11/7} = 0.09730 \left[\frac{\kappa^{33/7} \alpha^{17/7} \alpha_w g}{\kappa' v_w h^2} \right]^{1/3} \left[\frac{\phi' \rho c}{\rho_w c_w} \right]^{4/21}$$

$$\times \frac{(8T_w + T_p - 9T_0)^{4/3}(T_p - T_w)}{(T_1 - T_w)(T_1 - T_p)^{8/21}} \tag{24A.13}$$

and if α_w/v_w is assumed to have the temperature dependence quoted by Elder (1965), a quasi-numerical result is

$$u = \theta^{7/11} \phi^{4/33} \times 3.01 \times 10^{-10} (T_1 - T_p)^{-8/33}$$

$$\times \left[\frac{(T_w - T_0)^2 (8T_w + T_p - 9T_0)^4 (T_p - T_w)^3}{(T_1 - T_w)^3} \right]^{7/33} \tag{24A.14}$$

Equation 24A.14 can be evaluated for any given T_p by maximizing with respect to T_w, the hot fluid temperature, since that is a free parameter able to adjust itself for the fastest water penetration. The chief difference in philosophy behind this result and behind that of Lister (1974) is the inclusion of the *water properties* term in the maximization for T_w. The effect is small. For example, if $T_p = 800°K$, T_w is now $630°K$ and would have been $585°K$, while the term *inside* the brackets changes by 20 percent, changing u by only 4 percent. Furthermore, the powers of dependence on the fudge factors θ and ϕ are small, especially on the relatively poorly known ϕ'. Thus it is hard to

Hydrothermal Systems

see how these alone could introduce more than an order of magnitude of error. It should perhaps be noted that the values now computed for u are about one-seventh of those in Lister (1974) after a really substantial change in Nusselt number relationship.

Once u is known, the other parameters can be calculated readily by means of the following relations:

$$x_0 = \kappa/u = 9 \times 10^{-7}/u \quad \text{m} \tag{24A.15}$$

$$y = [0.8\kappa^3 \rho c/\alpha u^3 \rho_w c_w (T_1 - T_p)^2]^{2/7}$$

$$= 1.368 \times 10^{-4}[u^3(T_1 - T_p)^2]^{-2/7} \quad \text{m} \tag{24A.16}$$

$$\dot{\varepsilon}_p = 2\alpha u^2(T_1 - T_p)/\kappa = 33.33 u^2(T_1 - T_p) \quad \text{s}^{-1} \tag{24A.17}$$

$$D = 0.118 y^2 \alpha^3 (T_p - T_w)^3$$

$$= 3.977 \times 10^{-16} y^2 (T_p - T_w)^3 \quad \text{m}^2 \tag{24A.18}$$

$$Q = u\rho c(T_1 - T_w) = 3.22 \times 10^6 u(T_1 - T_w) \quad \text{wm}^{-2} \tag{24A.19}$$

$$\mathscr{A} = (0.064 wgh^4 Q^3 D/\theta^3 \kappa^4 \rho^3 c^3)^{1/2}$$

$$= 1.755 \times 10^{11} y u^{3/2} (T_p - T_w)^{3/2} (T_1 - T_w)^{3/2} \tag{24A.20}$$

$$\tau_{\text{hot}} = 2\alpha h^2(T_p - T_w)/0.17 \kappa \mathscr{A}$$

$$= 4.902 \times 10^9 (T_p - T_w)/\mathscr{A} \quad \text{s} \tag{24A.21}$$

$$D_0 = 3.977 \times 10^{-16} y^2 (T_p - T_0)^2 \quad \text{m}^2 \tag{24A.22}$$

$$d = \alpha y (T_p - T_w) = 1.5 \times 10^{-5} y (T_p - T_w) \quad \text{m} \tag{24A.23}$$

$$B_H \simeq (40/\mathscr{A})^{1/3} h = 1.71 \times 10^4 \mathscr{A}^{-1/3} \quad \text{m} \tag{24A.24}$$

$$F = Q/\rho_w c_w (T_w - T_o) = 0.77 u(T_1 - T_w)/(T_w - T_o) \quad \text{m/s} \tag{24A.25}$$

$$\tau_{\text{hot}}^{\text{corr}} = \tau_{\text{hot}} (R_p/1000)^{1/2}$$

$$= 1.51[\tau_{\text{hot}} y(T_p - T_w)(T_w - T_0)]^{1/2} \quad \text{s} \tag{24A.26}$$

$$B_P = (40/\mathscr{A})^{2/3} h \tau_{\text{hot}}^{\text{corr}}/\tau_{\text{hot}} = 5.85 \times 10^4 \tau_{\text{hot}}^{\text{corr}}/\tau_{\text{hot}} \mathscr{A}^{2/3} \quad \text{m} \tag{24A.27}$$

The calculated data are presented in table 24A-1.

A few words of explanation follow about the equations for $\tau_{\text{hot}}^{\text{corr}}$ and B_p. The simple linear flow relations predict a plume velocity so fast that it is well into a regime of turbulent flow inside an individual crack. An estimate of the

Table 24-1
Flow Conditions in Active Systems

Parameter	Factor Multiplier	Values							Units
T_p		500	600	700	800	900	1000		°K
T_w		422	490	560	630	712	795		°K
u	$\theta^{0.64}\phi^{0.12}$	4.2	9.0	16	26	40	59 $\times 10^{-8}$		ms^{-1}
u	$\theta^{0.64}\phi^{0.12}$	1.3	2.8	5.0	8.2	12.5	18.5		myr^{-1}
x_0	$\theta^{-0.64}\phi^{-0.12}$	21	10	5.6	3.4	2.3	1.5		m
y	$\theta^{-0.55}\phi^{0.18}$	5.6	3.1	2.0	1.4	1.1	0.9		m
ε_p	$\theta^{1.27}\phi^{0.24}$	0.58	2.4	6.9	16	32	57 $\times 10^{-10}$		s^{-1}
D	$\theta^{-1.09}\phi^{0.36}$	5.9	5.0	4.4	4.0	3.1	2.6 $\times 10^{-9}$		m^2
Q	$\theta^{0.64}\phi^{0.12}$	0.15	0.29	0.49	0.73	1.0	1.3		k w m^{-2}
$\mathscr{A}$	$\theta^{-1.09}\phi^{-0.36}$	0.20	0.54	1.1	1.9	1.7	3.7 $\times 10^9$		
τ_{hot}	$\theta^{1.09}\phi^{-0.36}$	1900	1000	640	440	340	270		s
D_o	$\theta^{-1.09}\phi^{0.36}$	1.4	1.3	1.2	1.2	1.2	1.0 $\times 10^{-7}$		m^2
d	$\theta^{-0.55}\phi^{0.18}$	6.6	5.1	4.2	3.6	3.1	2.7		mm
B_H	$\theta^{0.36}\phi^{-0.12}$	29	21	17	14	12	11		m
F	$\theta^{0.64}\phi^{0.12}$	2.3	3.2	4.1	4.9	5.5	6.1 $\times 10^{-7}$		ms^{-1}
τ_{hot}^{corr}	$\theta^{0.55}\phi^{-0.18}$	4.6	3.6	3.0	2.7	2.3	2.1		hr
B_p	$\theta^{0.18}\phi^{-0.06}$	1.5	1.1	0.9	0.8	0.7	0.7		m

Table interrelationships were calculated before the final rounding to two significant figures. The factor multiplier should be applied to the values in the table if the fudge factors θ and ϕ are felt to differ from unity. Values of τ_{hot}, τ_{hot}^{corr} and B_p are not expected to be representative of real conditions and are included mainly for completeness.

real flow velocity can be made by considering that the local pressure-drag coefficient of thin plates in parallel flow is a very slow function of the Reynolds number (calculated on the basis of distance from the leading edge), that is, $\propto R^{-1/7}$ (Hoerner, 1958). It is also known that the transition from laminar flow to turbulent flow occurs in pipes at $R = 1,000$ or on free surfaces at similar Reynolds numbers when they are calculated on the basis of boundary-layer thickness. Thus the real flow velocity may be expected to rise only proportionally to the square root of pressure gradient once the transition has taken place, and since R_p was calculated on the basis of the linear (laminar flow) relations, it is in fact a measure of the driving pressure gradient. Hence the correction factor for the real reduced flow velocity is simply the square root of the ratio of R_p and the transition value of 1,000. The plume thickness is increased in the same proportion as the flow velocity is decreased, since the fluid flux is determined by conditions in the horizontal boundary layer and is not affected by what happens to the plume.

The plume thickness B_p, even after correction for turbulent flow in the cracks, is in all cases less than the crack spacing y. Thus the upwelling would occur in a single crack, and indeed this may occur near the lower boundary. The boundary-layer theory (J. Booker, personal communication, 1977) assumes that lateral heat leakage from the plume is negligible, and this is clearly not so for such a narrow plume. Thermal conduction of heat to neighboring cracks causes a broadening of the plume with increasing height, increasing the effective fluid flux but decreasing the temperature of the hottest water. It should be noted that the extra fluid entrained from the porous matrix was never a part of the hot basal boundary layer and does not mix efficiently with the original plume fluid. The temperature history of fluid in the upwelling plume is thus not uniform or easily determined; there is even some upwelling fluid that simply carries off the heat left in the rock matrix at the water temperature of the hot boundary layer. The last fluid is not considered by the theory or included in the flux estimates F; its chemical effects are not large during the active penetration phase because of low temperatures and low flow rates, which are more comparable to conditions in the "passive" circulation that follows penetration.

Appendix 24B: List of Symbols used in the Analysis

$\mathcal{A}$ = Rayleigh stability number for porous-medium convection

α = Coefficient of linear expansion of rock $\sim 1.5 \times 10^{-5}$ °K^{-1} (Clark, 1966)

α_w = Coefficient of cubical expansion of water $\sim 4.2 \times 10^{-6} \, (T-273)$ °K^{-1}, for $370 < T < 570$° K (Elder, 1965)

B_H = Horizontal boundary-layer thickness (estimate by J. Booker, personal communication)

B_P = Vertical plume thickness (similar estimate, neglecting thermal diffusion)

β = Porosity of the permeable matrix = $2\alpha(T_p - T_w)$ (Lister, 1974)

c = Specific heat of rock $\sim 1{,}000$ j kg^{-1} °K^{-1}

c_w = Specific heat of circulating fluid 4184 j kg^{-1} °K^{-1}

d = Crack width = $y\alpha(T_p - T_w)$ m   (Lister, 1974)

D = Permeability at $T_w = (\sqrt{2}/12)y^2\alpha^3(T_p - T_w)^2$ m^2 *or a variable* (Lister, 1974)

D_o = Permeability at sea water temperature = $(\sqrt{2}/12)y^2\alpha^3(T_p - T_o)$ m^2 *or a variable*

$\dot{\varepsilon}_p$ = Rate of rock creep (equivalent uniaxial) just prior to cracking
= $2\alpha u^2(T_1 - T_p/\kappa \text{ s}^{-1}$ (Lister, 1974)

F = Circulation flux, volume per unit area and time, ms^{-1} or kmMy^{-1}

g = Acceleration of gravity $\simeq 10$ m/s^2

h = Depth of convecting slab ~ 5000 m

θ = Thermal-transport parameter correction factor ~ 1

$\kappa\rho c$ = Rock thermal conductivity = 2.63 w m^{-1} °K^{-1}

κ = Thermal diffusivity of rock $\sim 9 \times 10^{-7}$ m^2 s^{-1}

κ' = Effective thermal diffusivity of matrix for porous convection
$\simeq 7 \times 10^{-7}$ m^{2}s^{-1} (conductivity divided by fluid heat capacity)

$\mathcal{N}$ = Nusselt number (thermal-transport number)

v_w = Kinematic viscosity of fluid $\sim 3.3 \times 10^{-5}(T - 273)^{-1}$ m²/s ($T > 313°$K) (Elder, 1965)

ϕ' = Dimensional and numerical correction factor between cracking of quartz and mafic rock in a polygonal system = $0.1\,\phi\,°K\,m^{1/2}$, where ϕ is purely numerical

Q = Heat transport per unit area = $u\rho c(T_1 - T_w)$ wm^{-2} *or a variable*

R_p = Reynolds number in plume calculated on a laminar flow basis = vd/v_w.

ρ = Density of rock $\simeq$ 3,200 kg/m³ (Clark, 1966)

ρ_w = Density of circulating fluid $\simeq$ 1,000 kg/m³

t = Time expressed in millions of years

T = Temperature in degrees Kelvin (°K)

T_κ = Cracking temperature

T_p = Temperature at which overburden pressure is supported by creep stress

T_w = Water temperature in the hot (upwelling) convection column

T_o = Reservoir temperature (ocean) $\sim 273°$K

T_1 = Initial rock temperature = $1,500 \pm 50°$K (magma eruption; Clark 1966)

T_2 = Hot boundary effective temperature $\simeq 1/9(8T_w + T_p - 9T_o)$

ΔT_r = Radial temperature difference in a rock column

τ_{hot} = Residence time of hot water in the upwelling column, s

τ_{hot}^{corr} = Residence time of hot water in the upwelling plume, corrected for turbulent flow in the cracks, s

u = Cracking front velocity, ms^{-1} or myr^{-1}

v = Real mean fluid velocity in the cracks, ms^{-1}

$\mathcal{v}$ = Nondimensional real fluid velocity in the porosity = $h\beta v/\kappa$ $\simeq 0.17\,(\mathcal{A}^2 - 40^2)^{1/2}$ (Hartline and Lister, 1977)

w = Water properties parameter = $\alpha_w/v_w \sim 0.127$ s°K^{-3}m^{-2} (Elder, 1965)

x_o = $1/e$, thermal boundary-layer thickness, m

y = Mean crack spacing, m

25 Numerical Models of Hydrothermal Circulation for the Intrusion Zone at an Ocean Ridge Axis

Patricia L. Patterson and *Robert P. Lowell*

Abstract

Hydrothermal circulation in the oceanic crust for the intrusion zone at a ridge axis is modeled numerically using finite differences. Episodic emplacement of new crustal material at the axis is assumed. The circulation is driven by transient, horizontal temperature gradients arising after one episodic emplacement and by an assumed uniform heat flux from below. The oceanic crust is treated as a porous medium for various cases of permeability conditions. Although the ocean bottom is considered to be a permeable boundary, convective heat transfer across that boundary is precluded by other assumptions.

Conductive heat transfer through the ocean bottom resulting from fluid circulation within the crust is found to be directly proportional to crustal permeability for cases of homogeneous, isotropic permeability. Conductive heat transfer through the ocean bottom becomes more concentrated at the ridge axis with increasing permeability and with decreasing ratios of horizontal to vertical permeability. Changes in permeability conditions from homogeneous, isotropic distributions do not significantly alter cooling effects.

Introduction

Measurements of ocean-floor heat flow in the vicinity of ridge crests show several noteworthy features. The distribution of values exhibits a high degree of scatter and includes values close to zero. In some cases, the scatter is opposite to that expected from the effects of topographic refraction (Lister, 1972). Talwani et al. (1971) have noted that the average heat flux measured on the Reykjanes Ridge crest is less than the average flux that would be generated by a cooling lithospheric plate originating at and moving with constant speed from the ridge axis (e.g., McKenzie, 1967; Sclater and

Francheteau, 1970; Parker and Oldenburg, 1973; Sleep, 1974). A similar observation has been made for the Galapagos Spreading Center (Williams et al., 1974) where, in fact, the heat flow exhibits a roughly sinusoidal pattern with a wavelength of approximately 6 km. Compilation of heat-flow data from the world ocean shows that mean heat fluxes to considerable distances from ridge axes are less than predicted by models of a conductively cooling crust (Anderson et al., 1977).

To account for those features of oceanic heat flow, circulation of seawater in the oceanic crust with convective heat transfer across the ocean bottom has been proposed (e.g., Palmason, 1967; Talwani et al., 1971; Williams et al., 1974; Lister, 1972). The existence of hydrothermally altered rock (Aumento et al., 1971) and hydrothermal deposits (Corliss, 1971; Scott et al., 1974), near-bottom photographs of cracks on the Mid-Atlantic Ridge (Ballard, 1975), and measurements of water-temperature anomalies near the seafloor (Rona et al., 1975) give evidence for the occurrence of such hydrothermal circulation. Observations of active hydrothermal vents at the axis of the Galapagos Spreading Center (Von Herzen et al., 1977; Corliss et al., 1979) clearly confirm its occurrence.

Theoretical models for hydrothermal circulation in the oceanic crust have been based on convection in narrow, deep, widely spaced vertical fractures (Bodvarsson and Lowell, 1972; Lowell, 1975), as well as on steady-state convection in a porous layer with constant temperature or constant heat flux at the lower boundary (Lapwood, 1948; Ribando et al., 1976). The models have given widely divergent results with regard to convective heat losses at the sea floor. Based on his fracture models, Lowell (1975) concluded that a significant portion of lithospheric heat is lost by convection, the heat being carried by hot springs discharging at the sea floor. However, Ribando et al. (1976) suggested that little heat is lost by convection and that the heat-flow measurements reported by Williams et al. (1974) on the Galapagos Spreading Center are indicative of the true heat flux. None of the preceding models, however, has considered conditions at the intrusion zone at a ridge axis, where transient, horizontal temperature gradients must exist. More recent work by Fehn et al. (1977) has included the intrusion zone.

The purpose of this chapter is to examine the effects of some of the conditions not included in the earlier works of others. The oceanic crust is treated as a porous layer, and the hydrothermal environment of a single convection cell at a ridge axis is modeled following the emplacement of a high-temperature intrusion into the layer. Several conditions of crustal permeability are considered: (1) homogeneous, isotropic permeabilities; (2) nonhomogeneous, isotropic permeabilities; and (3) homogeneous, anistropic permeabilities. The results provide insight into possible cooling histories of the oceanic crust and are of interest not only for the explanation of measured heat fluxes, but also for the prediction of environments for hydrothermal reactions with crustal rocks.

The Basic Model

The basic model consists of a block of oceanic crust of linear extent $h = 4$ km in the x and y directions and infinite extent in the z direction (see figure 25-1.) The xy plane of the model corresponds to a cross section of oceanic crust perpendicular to a ridge axis; the xz plane, the ridge axis; and the yz plane, the ocean bottom. There is a uniform heat flux $q_o = 10$ μcal/(cm$^2 \cdot$ s) across the lower boundary. The block is overlain by seawater at temperature $T = 0°$C, assumed to act as a perfect conductor, and the vertical sides of the block are insulated. The block is permeable to depth d, where d may be less than or equal to h; the vertical boundaries are impermeable; and the upper boundary is permeable, so that seawater may enter and exit, circulating through the permeable region. For time $t < 0$, no circulation is assumed to occur, and a steady-state conductive temperature field exists: $T(x, y, t < 0) = (q_0/\lambda)x$, where $\lambda = 0.01$ cal/(°C $\cdot$ cm $\cdot$ s) is the thermal conductivity of the rock. At time $t = 0$, a rectangular intrusion of width 0.2 km and temperature $T_0 = 1,200°$C is emplaced at a depth of 0.2 km beneath the spreading center of the oceanic block. The event represents the sudden emplacement of a dike or

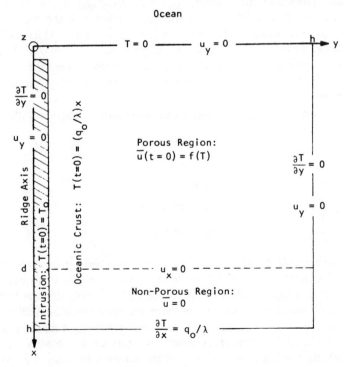

Figure 25-1. Basic model with initial and boundary conditions.

dike swarm. The crust is impermeable for $T \geq T_c$, where $T_c = 690°C$ is the cracking temperature, and permeable for $T < T_c$. (This is a much simplified treatment of the thermally induced rock fracturing process given by Lister (1974).) The depth of the permeable layer d and the permeability K vary according to table 25–1 for different models. (Case 1 is a conductive model, involving no hydrothermal circulation.)

Assuming the rock and fluid to be in thermal equilibrium and the fluid to be incompressible, the pertinent equations of conservation of fluid mass, fluid momentum, and thermal energy and the equation of state for the fluid are (Lapwood, 1948):

$$\vec{\nabla} \cdot \vec{u} = 0 \tag{25.1}$$

$$-\vec{\nabla} P - \rho_f \nu \underset{\sim}{K}^{-1} \cdot \vec{u} + \rho_f \vec{g} = 0 \tag{25.2}$$

$$\rho c \partial T/\partial t + \rho_f s \vec{u} \cdot \vec{\nabla} T = \lambda \nabla^2 T \tag{25.3}$$

$$\rho_f = \rho_{f_0}(1 - \alpha T) \tag{25.4}$$

where $\vec{u}$ is the Darcy velocity of the fluid, P its pressure, ρ_f its density, $\nu = 0.01$ cm^2/s its kinematic viscosity, $\underset{\sim}{K}$ the permeability, $\vec{g}$ the acceleration of gravity, $\rho c = 0.75$ cal/(cm$^3 \cdot °C$) the heat capacity of the rock, $s = 1.00$ cal/(°C·g) the specific heat of the fluid, $\alpha = 1.60 \times 10^{-4}/°C$ its coefficient of thermal expansion, and $\rho_{f_0} = 1.00$ g/cm^3 its density at $T = 0°C$. In equation 25.2 the inertial terms are omitted, being much smaller than the viscous term (Straus, 1974). To allow for anisotropies in permeability, permeability is written as a tensor.

Conditions at $t \geq 0$ are treated as perturbations on the conditions existing at $t < 0$, so that

$$\begin{aligned} T &= T^0 + T' \\ \vec{u} &= \vec{u}' \\ P &= P^0 + P' \\ \rho_f &= \rho_f^0 + \rho_f' \end{aligned} \tag{25.5}$$

where the zero superscripts designate the unperturbed values at $t < 0$, and the primes designate the perturbations to these values at $t \geq 0$. Equations 25.5 are substituted into equations 25.1 through 25.4, employing the Boussinesq approximation; the pressure term of the momentum equation is eliminated by taking the curl of this equation; and the resulting equations are non-dimensionalized by measuring length in units of h, velocity in units of $\lambda/(\rho_{f_0} s h)$, time in units of $\rho c h^2/\lambda$, and temperature in units of T_0. A stream function is introduced, given by:

Numerical Models

Table 25-1
Cases Modeled

Case	d	K_0	$\dfrac{\partial K_y}{\partial x}$	$\dfrac{K_x}{K_y}$
1	0			
2	4		0	1
a		0.1		
b		0.0275		
c		0.01		
d		0.0075		
e		0.005		
f		0.001		
3	4	0.01	$-\dfrac{0.01}{4}$	1
4		0.01	0	1
a	2			
b	3			
5	4	0.01	0	
a				2
b				3
c				10

Explanation:
d = depth of porous layer in kilometers.
K_0 = permeability to vertical flow at top boundary in Darcies (1 Darcy = 0.987×10^{-8} cm^2).

$\dfrac{\partial K_y}{\partial x}$ = change in permeability with depth in Darcies per kilometer.

$\dfrac{K_x}{K_y}$ = ratio of permeability to vertical flow to permeability to horizonal flow.

$$v_x = -\partial \psi/\partial Y$$
$$v_y = \partial \psi/\partial X \tag{25.6}$$

where v_x and v_y are the dimensionless Darcy velocity components, and X and Y are dimensionless coordinates. The resulting dimensionless equations are

$$\partial^2 \psi/\partial Y^2 - 1/K_y \, (\partial K_x/\partial X)(\partial \psi/\partial X) + K_x/K_y \, \partial^2 \psi/\partial X^2 - RK_x/K_0 \, \partial \theta/\partial Y = 0 \tag{25.7}$$

$$\partial \theta/\partial \tau = \nabla^2 \theta - \vec{v} \cdot \vec{\nabla} \theta - v_x R_q/R \tag{25.8}$$

where θ and τ are dimensionless perturbation temperature and time, respectively, K_x/K_y (permeability to vertical flow/permeability to horizontal flow) equals a constant, K_x is a function of depth only, K_0 is $K_x(x=0)$, and R and R_q are Rayleigh parameters given by

$$R = \rho_{f0} s \alpha g T_0 K_0 h / (\nu \lambda)$$

$$R_q = \rho_{f0} s \alpha g q_0 K_0 h^2 / (\nu \lambda^2)$$

The initial temperature conditions are

$$\theta = \begin{cases} 1 + (R_q/R)X & \text{at and inside the intrusion} \\ 0 & \text{outside the intrusion} \end{cases} \quad (25.9)$$

and the boundary conditions are

$$\theta(0,Y) = \partial \theta(X,0)/\partial Y = \partial \theta(X,1)/\partial Y = \partial \theta(1,Y)/\partial X = 0 \quad (25.10)$$

$$\partial \psi(0,Y)/\partial X = \psi(X,0) = \psi(X,1) = \psi(d/h,Y) = 0 \quad (25.11)$$

with modification to allow for displacement of the convection cell boundary inward, out of impermeable regions where $T \geq T_c$.

Numerical Solution

Solutions were obtained to equations 25.6, 25.7, and 25.8 with conditions 25.9 through 25.11 computationally by standard finite-difference methods. A network of 20 × 20 grid points was superimposed on the XY plane of the model such that the grid-point spacing $\Delta X = \Delta Y$ equaled 1/20. All equations were approximated by algebraic equations derived from truncated Taylor's expansions about individual grid points. The form of equation 25.8 is given by method 1 of Torrance (1968).

Solutions were obtained at each grid point at discrete times τ^l, where τ^l is the time after the $(l-1)^{th}$ time step, $\Delta \tau^{l-1}$. The magnitude of $\Delta \tau$ was varied for each time step to equal 90 percent of its maximum allowable value for stability, ensuring numerically that heat would flow from higher to lower temperature points:

$$\Delta \tau^l = 0.9 \min_{x_{i,j}} \left[\frac{|v^l_{x_{i,j}}| + |v^l_{y_{i,j}}|}{\Delta X} + \frac{4}{(\Delta X)^2} \right]^{-1} \quad (25.12)$$

where $v^l_{x_{i,j}}$ is velocity at the (i,j)th grid point at time τ^l (see Torrance, 1968).

Numerical Models

The procedure for the solution was as follows: (1) by the method of successive over-relaxation, solve the finite-difference representation of equation 25.7 for the stream-function field using the known (initial) temperature field; (2) use the stream-function solution to solve the finite-difference representation of equations 25.6 for the velocity field; (3) use the velocity solution to solve equation 25.12 for the length of the time step $\Delta\tau$; (4) use the velocity solution and the known temperature field to solve the finite-difference representation of equation 25.8 for the temperature field at the new time $\Delta\tau$ greater than the old; and (5) return to step 1 using the new temperature field as the known temperature field and thus advance through time, obtaining solutions for all variables at discrete times τ^l.

From those variables, the surface heat conduction through the top boundary, the total heat conducted out of the system, and the volume of water circulated through the system as functions of time were also calculated. Solutions were obtained from time $t = 0$ to $t = 15,000$ years. The average time step was about 190 years for the cases considered.

Accuracy was limited by truncation errors introduced when the differential equations were converted to algebraic representations. Excluding the contribution of the nonlinear convection term of equation 25.8, inaccuracies would have been on the order of ΔX or 5 percent (Torrance, 1968). The truncation error involved in the representation of equation 25.8, however, included the quantity

$$\frac{\Delta X |v_{x_{i,j}}|}{2} \frac{\partial^2 \theta_{i,j}}{\partial X^2} + \frac{\Delta Y |v_{y_{i,j}}|}{2} \frac{\partial^2 \theta_{i,j}}{\partial Y^2}$$

so that the temperature at every grid point was reduced by at least that amount every computation step. When velocities were large, the magnitude of the error therefore became large and heat energy effectively disappeared from the system. Cases 2a and 2b (see table 25-1) involved the highest velocities of all models, and by the end of 15,000 years, 153 percent of the excess heat introduced by the intrusion had been removed spuriously in case 2a and 31 percent in case 2b. All other cases were at least 95 percent conservative.

Further details of the numerical methods are given in Patterson (1976).

Results

The principal results are shown in figures 25-2 through 25-8. The cases are identified in table 25-1. The dashed lines with arrows in figures 25-2 through 25-4 represent streamlines, lines of constant ψ. Darcy velocities parallel streamlines and are inversely proportional to streamline spacing in accord

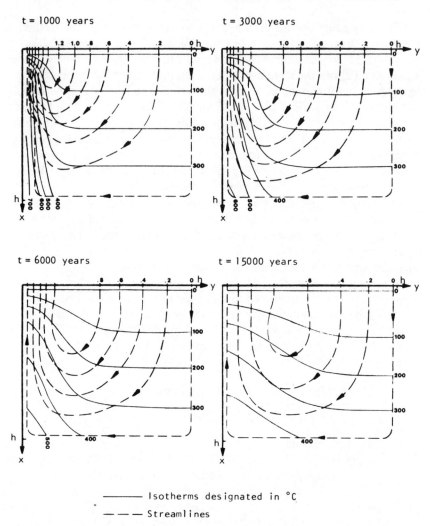

Figure 25-2. Typical time development of hydrothermal field (case 2c).

with equations 25-6. Thus the average vertical and horizontal Darcy velocities between streamlines may be determined from the figures by the formulas

$$u_x = (2.5 \times 10^{-8})(-\partial\psi/\partial Y) \text{ cm/s}$$

$$u_y = (2.5 \times 10^{-8})(\partial\psi/\partial X) \text{ cm/s}$$

Figure 25-2 (for case 2c) shows features typical of the time development of a hydrothermal system. The convection cell is asymmetric, with the center of the cell displaced toward the intrusion. As cooling progresses, the cell becomes more symmetric. Velocities are greatest at shallow depths near the axis and decrease with depth and with distance from the axis. At a given point in space the velocity increases to a maximum and then decreases with time. Note that after 1,000 years about one-half the intrusion is still at temperatures greater than $T_c = 690°C$. The approximate rate at which water penetrates the intrusion is therefore 2 m/yr, which is somewhat slower than Lister's (1974) estimate, partly because of the low permeability assumed here. For the case where $K_0 = 0.1D$, the entire intrusion has cooled below T_c in less than 1,000 years.

Figure 25-3 shows how the temperature field in the presence of hydrothermal circulation differs from that in its absence. The isotherms for convective cases are displaced relative to those for the conductive case in the direction of fluid flow so that, at the surface of the convective models, higher temperature gradients exist near the intrusion, but lower gradients away from it.

Figure 25-4 (a, b, c, and d) shows the thermal and flow regimes at 15,000 years for most of the models. For the cases of constant, isotropic permeability (figure 25-4a) a comparison of the streamline spacing shows the Darcy velocity varies approximately as $1.5K_0$. For the case of linear decrease in permeability with depth (figure 25-4b), flow is more concentrated near the surface. For the cases of anisotropic permeability (figure 25-4d), flow is more concentrated near the intrusion.

Figure 25-5 shows the perturbation surface heat flux [excess over the uniform heat flux, $q_0 = 10\mu\text{cal}/(\text{cm}^2 \cdot \text{s})$] at 15,000 years for most of the models. Note how hydrothermal circulation redistributes the heat flux conducted through the upper surface relative to that of the nonconvective model. Near the axis the heat flux is increased, while at distances greater than 1 to 1.5 km, the heat flux is decreased.

Figure 25-6 shows the total volume of fluid that has entered or left the system (total volume of fluid exchange) as a function of time for the models. The slopes of the curves indicate that for all cases the rate of fluid exchange reaches a maximum early in the cooling history and then gradually decreases with time. The average rates of fluid exchange for all the models fall within the range estimated by Wolery and Sleep (1976).

Figure 25-7 shows the excess heat conducted out in 15,000 years over that for the conductive case and the total volume of fluid exchange by the end of 15,000 years as functions of permeability for cases 2b through 2f, which have homogeneous, isotropic permeability distributions. (The values for case 2a are excluded because of nonconservation in the numerical solution.) For this range of permeabilities (0.0275 to $0.001D$), both functions are roughly

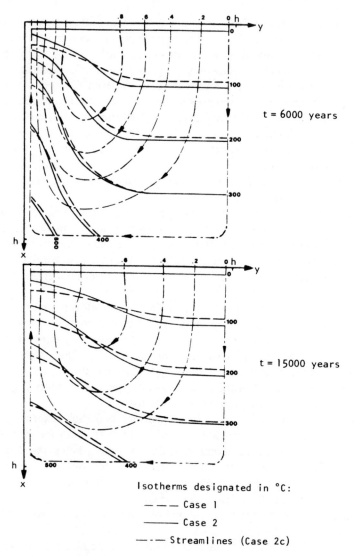

Figure 25-3. Comparison of temperature fields for conductive case (case 1) with those for typical convective case (case 2c).

linear. The values of excess heat output and volume of fluid exchange (dependent variables) for the other cases with no single-number permeability designation (independent variable) are superimposed on those functions in alignment with the magnitude of the dependent variables.

Figure 25-8 shows the total perturbation heat (excess over that resulting from the uniform heat flux q_0) conducted out of the models as functions of

Numerical Models

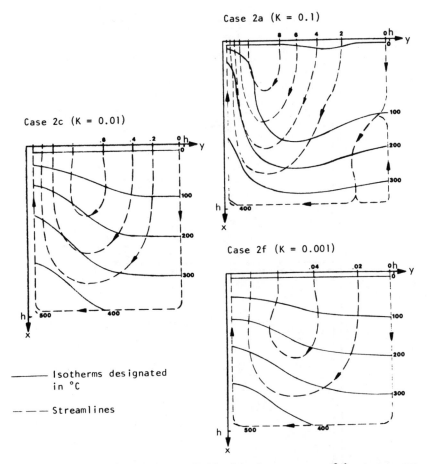

Figure 25-4. (a) Hydrothermal fields for three cases of homogeneous, isotropic permeability at $t = 15{,}000$ years.

time. The slopes for all convective cases are at all times greater than that for the conductive case, indicating that the average heat flux over the surface of convective models is always (for the 15,000-year period) greater than that over the surface of the conductive model. The heat conducted out in excess of that for the conductive model for cases 2b through 2f appears to be linearly proportional to permeability at all times, as shown in figure 25–7 for the single time $t = 15{,}000$ years.

Discussion

The emplacement of an intrusion of 200-m width every 15,000 years corresponds to a half-spreading rate of the sea floor of 1.33 cm/yr. This might

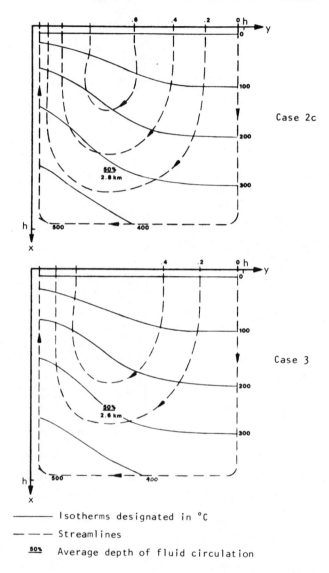

Figure 25-4. (b) Hydrothermal fields for cases of homogeneous (case 2c) and inhomogeneous, isotropic permeability (case 3), at $t = 15,000$ years.

roughly approximate spreading conditions at the Mid-Atlantic Ridge, where the average half-spreading rate is 1.1 cm/yr in conjunction with the eruption

Numerical Models

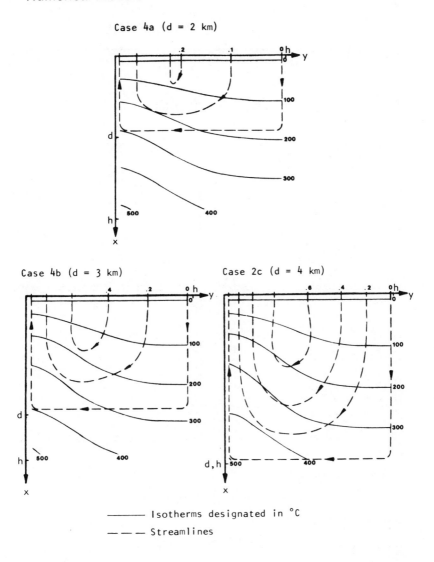

Figure 25–4. (c) Hydrothermal fields for three thicknesses of permeability layers at $t = 15,000$ years.

of pillow basalts at the ridge axis on 14,000-year cycles as proposed by Moore et al. (1974).

Rayleigh parameters for the models lie in the ranges

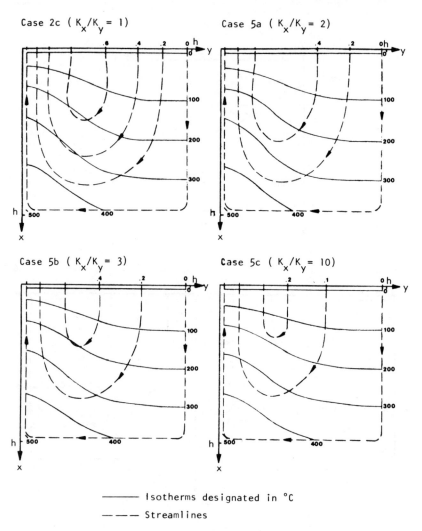

Figure 25-4. (d) Hydrothermal fields for four values of horizontal permeability at $t = 15,000$ years.

$$7.53 \leq R \leq 753$$

$$2.50 \leq R_q \leq 250$$

The critical Rayleigh number $R_{q_{cr}}$ for a porous medium with a homogeneous, isotropic permeability distribution, a constant heat flux across the lower

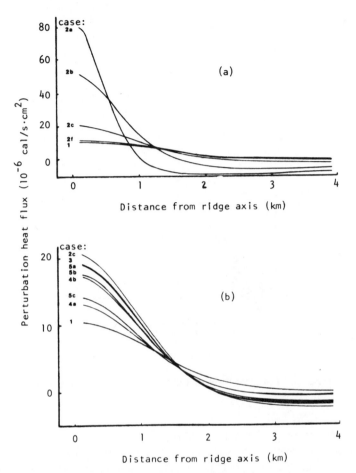

Figure 25-5. Perturbation heat flux through top boundary at $t = 15,000$ years.

boundary, and a permeable upper boundary is 17.7 (Nield, 1968). In case 2 (see table 25-1), therefore, convection ranges from subcritical, where it is driven solely by horizontal temperature gradients (cases 2e and 2f), to supercritical, finite-amplitude, where it is driven by both horizontal and vertical gradients (cases 2a through 2d). The somewhat anomalous appearances of the curves for cases 2e and 2f in figure 25-6 are probably related to the fact that R_q is less than $R_{q_{cr}}$ for these two cases.

Since the ocean bottom was assumed to be isothermal and perfect thermal contact was asssumed to exist between the circulating fluid and the rock, the models considered in this study did not allow the possibility of convective

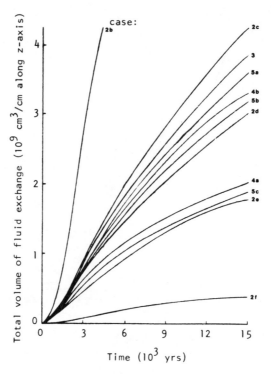

Figure 25–6. Total volume of fluid exchange as function of time.

heat transfer through the ocean bottom. For the range of permeabilities involved in the models, however, the maximum Darcy velocity that occurred was 5×10^{-6} cm/s (for $K_0 = 0.1D$). Assuming an intercrack spacing of 100 cm in the crust, Lister's (1974) theory gives a crack width of 0.045 cm. The actual fluid velocity in cracks of this width would be about 0.01 cm/s. For such a low velocity the assumption of perfect thermal contact is reasonable.

The case of a nonisothermal upper boundary has been treated by Ribando et al. (1976), and they suggest that its effect on heat transfer is small. Their models, however, do not include the intrusion zone. The assumption that the ocean bottom remains isothermal cannot be easily justified for the models of the present study. First of all, at time $t = 0$, extrusion of hot magma onto the ocean floor was neglected, and second, in view of the very high surface-temperature gradients that continue for some time after the emplacement of the intrusion (see figure 25–2), the ocean above should have been conductively heated so that convective heat transfer (by slow moving fluid in perfect thermal contact with the surrounding rock) would occur across the ocean bottom. Therefore, at early times the conductive heat fluxes of the convective models of this study are artificially high close to the intrusion.

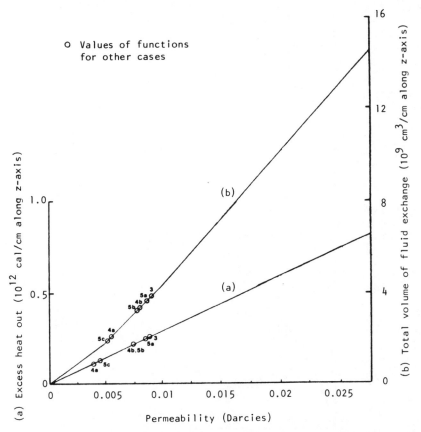

Figure 25-7. Heat conducted out in excess of that for conductive case (3.22×10^{12} cal/cm) and total volume of fluid exchange as functions of permeability for cases 2b through 2f at $t = 15,000$ years.

For no model was the heat perturbation introduced by the intrusion (5.64×10^{12} cal/cm) reduced by as much as 50 percent by the end of the 15,000-year cycle (see figure 25-8). Thus only a portion of the transient conditions building up to an equilibrium oscillatory state were modeled. Since in convective cases with homogeneous, isotropic permeability distributions the amount of heat conducted out in excess of that in the conductive case was directly proportional to permeability, curve (a) of figure 25-7 can be extrapolated to the value of the original heat perturbation. This extrapolation leads to the prediction that a convective model with $K_0 \geq 0.15D$ could conceivably reduce the heat perturbation introduced into the upper 4 km of the crust within the expanse of a single wavelength of hydrothermal flow before the end of the 15,000 year cycle. The model would give a mean (averaged over

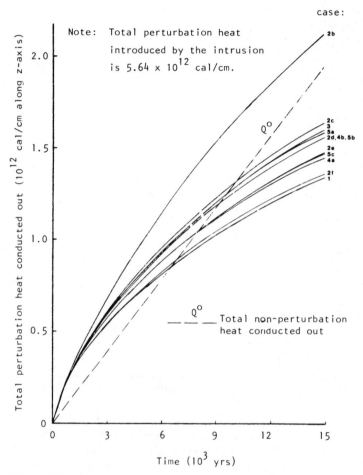

Figure 25-8. Total perturbation heat conducted through top boundary as function of time.

space and time) heat flux of 40 μcal/(cm$^2 \cdot$ s) from the ocean floor. For $K_0 <$ 0.15D modeling with periodic intrusions should be carried out until equilibrium cooling is attained, i.e., until the long-term rate of heat perturbation introduced equals the rate of its removal through the ocean floor. Such a calculation would involve lateral extension of the model away from the axis. The lower the permeability of the model, the greater would be the extension, and the broader the region of elevated heat conduction above the axis. In all cases, oscillations in the surface-heat conduction would occur at the axis with a period equal to 15,000 years. The amplitude of these oscillations would increase with K_0, and for $K_0 \geq 0.15D$, the amplitude would be zero for part of the cycle.

If such models of an oceanic spreading center are valid, measurements at a single time cannot, in general, be representative of conductive heat fluxes over the entire cooling cycle. Also, measurements only at distances from the ridge axis where sediments are sufficiently thick to accommodate instruments cannot be representative of conductive heat fluxes over the intrusion zone. Thus the redistribution of conductive heat flux by convective heat transfer within (but not from) a porous oceanic crust could account in part for the discrepancy between observed and predicted conductive heat fluxes. For those models, convective heat transfer from the oceanic crust should occur only directly over the ridge axis and only early in the cooling cycle. Discrete fracture models—i.e., hot springs (Lowell, 1975)—must be called upon to account for additional convective losses.

For cases of homogeneous, isotropic permeability, after 15,000 years the ratio of heat conducted out in excess of that for the conductive case to the total volume of fluid exchange—i.e., the ratio of the functions in figure 25-7—is approximately constant and equal to 70 cal/cm^3 of seawater. Thus an average of 70 cal of heat, in excess of that when no circulation occurs, is conducted out of the oceanic crust for every gram of water entering. Note that in these models, no heat is carried into the ocean by the circulating seawater. Rather, redistribution of heat within the oceanic crust has effected more efficient conductive heat transfer across the ocean floor.

For cases of other permeability distributions, enhanced conductive transfer through the ocean bottom is less efficient. Note in figure 25-7 that the positions of these cases on curve (a) are slightly to the left of their positions on curve (b). The ratio of each pair of values is therefore slightly less than 70 cal/cm^3. The least efficient cases are 4a and 4b (with restricted depths of circulation) with ratios 55 and 66 cal/cm^3, respectively. Case 3 (with linear decrease of permeability with depth) is less efficient than case 5a (with $K_x/K_y = 2$) probably because water does not move as freely into the higher-temperature regions (deeper and closer to the axis) as it does in case 5a (compare figures 25-4b and d).

A quasi-equivalent homogeneous, isotropic permeability may be determined from figure 25-7 for the cases of other permeability conditions. For example, case 3 has the same dependent variable values as a case of homogeneous, isotropic permeability with $K_0 = 0.009D$. Clearly, equivalent permeability must be understood in terms of flow patterns. Since the upper boundary of the model is permeable, more fluid flows near the surface than in other regions, and more fluid flows vertically than horizontally. Thus linear decrease of permeability with depth (case 3) and lower horizontal than vertical permeability (case 5) do not produce significant changes in cooling effects from cases of homogeneous, isotropic permeability. In porous-layer models of oceanic crust further from the ridge axis where low permeability sediments overlie the layer, more significant differences in cooling effects are to be expected. In such regions, vertical flow will not greatly predominate, and maximum flow will not be so concentrated near the surface of the porous layer.

A comparison of the flow patterns for cases 5a through 5c involving anisotropy in permeability (figure 25–4d), shows two noteworthy features: (1) the depth of average fluid circulation is independent of horizontal permeability, being constant at 2.8 km for all cases, and (2) the horizontal center of fluid circulation moves closer to the ridge axis with decrease in horizontal permeability. Those features are indicative of a decrease in aspect ratio (width/depth) of the convection cell with a decrease in the horizontal-to-vertical permeability ratio. Wooding (1976) has discussed the increase in aspect ratio with an increase in the horizontal-to-vertical permeability ratio for the geothermal areas in the Taupo Volcanic Zone. Such variations in aspect ratios should be considered in attempting to determine depth of circulation in the oceanic crust in accord with observed spatial periodicities in heat flow.

Acknowledgments

This work was supported by the oceanography section of the National Science Foundation under NSF Grant DES 74–00513–A01.

References

Anderson, R.N., Langseth, M.G., and Sclater, J.G. 1977. The mechanisms of heat transfer through the floor of the Indian Ocean. *J. Geophys. Res.* 82:3391–3409

Aumento, F., Loncarevic, B.O., and Ross, D.I. 1971. Hudson geotraverse: Geology of the Mid-Atlantic Ridge at 45°N. *Philosoph. Trans. Royal Soc. Lond.* A268:623–650.

Ballard, R.D. 1975. Project FAMOUS II: Dive into the great rift. *Nat. Geogr.* 147:604–615.

Bodvarsson, G., and Lowell, R.P. 1972. Ocean-floor heat flow and the circulation of interstitial waters. *J. Geophys. Res.* 77:4472–4475.

Corliss, J.B. 1971. The origin of metal-bearing submarine hydrothermal solutions. *J. Geophys. Res.* 76:8128–8138.

Corliss, J.B., Dymond, J., Gordon, L.I., Edmond, J.M., von Herzen, R.P., Ballard, R.D., Green, K., Williams, D., Bainbridge A., Crane, K.,

and van Andel, Tj. H. 1979. Submarine thermal springs on the Galapagos Rift. *Science* 203:1073–1083.
Fehn, U., Cathles, L.M., and Holland, H.D. 1977. Hydrothermal convection at mid-ocean ridges. *EOS* 58:514.
Lapwood, E.R. 1948. Convection of a fluid in a porous medium. *Proc. Camb. Philosoph. Soc.* 44:508–521.
Lister, C.R.B. 1972. On the thermal balance of a mid-ocean ridge. *Geophys. J. Royal Astron. Soc.* 26:515–535.
Lister, C.R.B. 1974. On the penetration of water into hot rock. *Geophys. J. Royal Astron. Soc.* 39:465–509.
Lowell, R.P. 1975. Circulation in fractures, hot springs and convective heat transport on mid-ocean crests. *Geophys. J. Royal Astron. Soc.* 40:351–365.
McKenzie, D.P. 1967. Some remarks on heat flow and gravity anomalies. *J. Geophys. Res.* 72:6261–6273.
Moore, J.G., Fleming, H.S. and Phillips, J.D. 1974. Preliminary model for extrusion and rifting at the axis of the Mid-Atlantic Ridge, 36°48' North. *Geology* 2:437–440.
Nield, D.A. 1968. Onset of thermohaline convection in a porous medium. *Water Resources Res.* 4:553–560.
Palmason, G. 1967. On heat flow in Iceland in relation to the Mid-Atlantic Ridge. In *Iceland and Mid-Ocean Ridges*. Reykjavik: Societas Scientiarum Islandica, pp. 111–127.
Parker, R.L., and Oldenburg, D.W. 1973. Thermal model of ocean ridges. *Nat. Phys. Sci.* 242:137–139.
Patterson, P.L., 1976. Numerical modeling of hydrothermal circulation at ocean ridges. M.S. thesis, Georgia Institute of Technology, Atlanta.
Ribando, R.J., Torrance, K.E., and Turcotte, D.L. 1976. Numerical models for hydrothermal circulation in the oceanic crust. *J. Geophys. Res.* 81:3007–3012.
Rona, P.A., McGregor, B.A., Betzer, P.R., Bolger, G.W., and Krause, D.C. 1975. Anomalous water temperatures over the Mid-Atlantic Ridge crest at 26°N latitude. *Deep-Sea Res.* 22:611–618.
Sclater, J.G., and Francheteau, J. 1970. The implications of terrestrial heat flow observations on current tectonic and geochemical models of the crust and upper mantle of the earth. *Geophys. J. Royal Astron. Soc.* 20:509–542.
Scott, M.R., Scott, R.B., Rona, P.A., Butler, L.W., and Nalwalk, A.J. 1974. Rapidly accumulating manganese deposit from the median valley of the Mid-Atlantic Ridge. *Geophys. Res. Lett.* 1:355–358.
Sleep, N.H. 1974. Segregation of magma from a mostly crystalline mush. *Geolog. Soc. Am. Bull.* 85:1225–1232.

Straus, J.M. 1974. Large amplitude convection in porous media. *J. Fluid Mech.* 64:51–63.

Talwani, M., Windisch, C.C., and Langseth, M.G., Jr. 1971. Reykjanes Ridge crest: A detailed geophysical study. *J. Geophys. Res.* 76:473–517.

Torrance, K.E. 1968. Comparison of finite-difference computations of natural convection. *J. Res. Nat. Bur. Stand.* 72B:281–301.

Von Herzen, R. Green, K.E., and Williams, D., 1977. Hydrothermal circulation at the Galapagos Spreading Center. *Geolog. Soc. Am. Abstr. Progr.* 9:1212–1213.

Williams, D.L., Von Herzen, R.P., Sclater, J.G., and Anderson, R.N. 1974. The Galapagos Spreading Center: Lithospheric cooling and hydrothermal circulation. *Geophys. J. Royal Astron. Soc.* 38:587–608.

Wolery, T.J., and Sleep, N.H., 1976. Hydrothermal circulation and geochemical flux at mid-ocean ridges. *J. Geol.* 84:249–275.

Wooding, R.A. 1976. Influence of anisotropy and variable viscosity upon convection in a heated saturated porous layer. Applied Mathematics Division, Technical Report No. 55, Department of Scientific and Industrial Research, Wellington, N.Z..

Index

Acoustical turbidity, 219–223, 225, 229, 231
Air/sea interface, 57, 58, 61, 73
Alkalinity in sediments, 8, 246, 309, 347
Aluminum in sediments, 369–375, 377
Amphipoda, metabolic activity, 161, 168–170
Anoxic sediments, 187–200, 203–216, 258, 261, 287, 297, 324, 330, 333, 346, 349
$^{40}Ar/^{36}Ar$ isotope ratios, 417, 418, 422–428, 434
Argon: basaltic concentrations, 184; in thermal springs, 417, 418, 422–429, 434; interstitial concentrations, 176–178, 182, 183; mantle, 418, 419, 426–428, 434; radiogenic, 428
Atmospheric gases and isotopes, 417–421, 426–428, 430, 434

Bacteria: methanogenic, 188, 191, 192, 210, 215; sulfate reducers, 188–192, 203, 210
Baltic Sea, 219–231, 339–353; sediment, 305, 309, 311–315, 341, 343, 346–350, 352
Barite, authigenic micronodules, 339–341, 343–345, 347, 349, 350, 352, 353; x-ray radiographs of, 341, 344
Barium: displacement by marine ions, 339–341, 347, 350–352; preciptations as $BaSO_4$, 347, 349, 352
Barotolerance, 150, 155, 158
Basalt, 369, 370, 372, 374, 376; interaction with seawater, 381, 382, 384, 396, 401, 407, 412; leaching, 369, 370, 373–376; major elements in, 396, 398–400, 403, 405–407, 412, 413; palagonitized, 397–399, 401, 405, 406; veins in, 381, 402–406, 409, 411; weathering, 381, 399, 401, 403, 406, 407, 412; x-ray mineralogy of, 396, 498, 403
Beaches: infaunal metabolic activity. See Macrofauna Models; Near shore sediments; Water movement
Benthic exchange, copper, 317, 326, 328, 333
Benthic regime: See Kinetics
Biochemical methane, 430, 432–434
Bioturbation, 62, 64, 65, 77, 83, 106, 107, 133, 161, 168–170, 265
Black Sea, ferromanganese nodules in, 275–283
Bottom photography, 382–386, 389–397, 411, 412
Bottom water, heated, 359, 360, 364, 367, 368, 472

Boundary layers (conductive or convective), 444–446, 450, 453, 455, 456, 461, 466
Box cores, 95–101, 103–105, 109, 111, 112, 196, 197, 287, 288
Breccia: basalt (tectonic), 386, 389–391, 395, 396, 402, 407, 413; hydrothermal, 388, 390, 395
Bubbles, methane, 195–197, 199, 200, 219, 226, 227, 229, 231
Burrowing organisms. See Macrofauna
Burrows (deep sea): abundance of, 96, 103; open, 96, 103; vertical, 96, 102, 103, 112

Calcite, 108–112, 341
California Borderland basins, uranium, thorium in, 180, 182
Calorimeter, 162–165
Calorimetry of sediments, 161–165, 170; semiconductor thermopiles in, 161, 162, 166–168
Carbon-14 dating, 110–112
Carbon dioxide: interstitial fluxes, 208, 209, 212, 213; interstitial, total, 204–208, 211; mantle, 432, 433
Carbon isotopes: exchange, 431–434; in spring waters, 419, 429–434; in sediments, 204, 205, 207, 211–213, 215; ratios, 204, 205, 211, 212
Carbon, mantle, 432
Carbon, organic: in sediments, 239, 240, 244, 246–248, 252, 288, 297, 299; interstitial flux, 208, 209
Carbonate dissolution (deep-sea), 95, 104, 106–110, 112
Cariaco Trench, 208, 213, 230, 231
Cation exchange, interstitial, 8, 339–341, 343, 347–351
Clams, metabolic activity of, 161, 168–170
Clarion Fracture Zone, 288, 359–368
Clay: brown, 258; layers, 359, 361, 364, 365; red, 240, 243, 246, 247, 249–251, 257, 258, 260, 262, 264, 272, 279–281, 287, 288, 296, 297; varved, 339–341, 343, 346, 347, 349, 350
Clipperton Fracture Zone, 288, 359, 361
Coastal sediments: Pacific, 258, 260; Pacific, terrigenous, 240, 242, 243, 245, 247, 249–252; water volume through, 48–50
Cobalt/manganese ratios in sediments, 282–283
Compaction, sediment, 305–307
Concentration gradients, interstitial (theoretical), 121–123, 131–137, 145

493

Continental-shelf sediments: infragravity waves, 59–61; internal waves, 59; oscillations of, 57, 58, 67, 69–73; pressure forces, 59, 61; surface wave action, 57, 58, 61, 73

Convection: as in porous media, 444–446, 449, 450; revised penetration theory, 441, 444, 450, 452, 453, 455, 461–467. See also Heat transfer

Copper: accretion on nodules, 299, 300; benthic exchange of, 317, 326, 328, 333; bottom-water concentrations, 294, 367, 368; in sediments, 305, 308–313; interstitial concentrations, 287, 289–296, 317, 318, 324–327, 331, 332; organic reactions, 329, 332, 333; interstitial remobilization, 317, 326, 328–333; interstitial removal, 317, 327–333; water-column concentration, 317, 322–324, 326–329, 331

Copper/manganese ratios in sediments, 282–283

Crab, 161, 166–168, 170

Cracking of oceanic rocks (thermal), 442–444, 450, 451, 453, 455–461; front, 442, 444, 461–463; fudge factors, 455, 463, 464, 466; scale of, 441–444, 446, 455, 461–463; speed of, 444, 446, 455, 461–463; temperature, 444, 448, 453–455, 461–463, 474

Crust: cooling models, 471, 474; cooling of, 441, 446, 455; helium from, 418, 420, 424, 425, 428, 434; insulation by sediment, 441, 452, 453, 456; permeable, thickness of, 446, 447, 449, 461. See also Hydrothermal circulation; Models; Permeability

Crustaceans, metabolic activity of, 161, 166–170

Currents: bottom (deep-sea), 105, 106; gravity, 32, 34–38, 41; pulsing, 32, 35–38, 51; "storms", 83

$\delta^{18}O$, 110–112

Darcy velocity, 474, 475, 477, 479, 480

Deep-Ocean Sampler (DOS), 150–153, 158, 159

Deep-Sea Drilling Project: 370–375

Deglaciation, 95, 96, 108–110, 112

Desulfotomaculum acetoxidans, 192

Desulfovibrio spp., 192

Diagenetic equation, 131, 179, 264

Diagenesis, 240, 241, 251, 252, 257, 275–277, 305, 306, 339, 350, 352

Diffusion: coefficient, eddy, 121, 122; coefficient, molecular (interstitial), 11, 115, 121–126, 131–137, 145, 179, 257, 264, 269, 271; interstitial impedance, 115, 130, 131, 134–137; molecular diffusive flux, 121–123, 134–137; near-bottom diffusive sublayer, 115, 116, 120, 122, 123, 126–132, 134–137. See also Benthic regime

Distance, near-bottom (dimensionless, Z^+), 119–123

Dolomite, 341

Dolomitization and interstitial alkalinity, 347

Donax variabilis, 161, 168, 170

Drag forces, 59–61

East Pacific Rise, 381, 389, 412, 413; deposit (EPRD), 369, 371–77; Galapagos Speading Center, 389, 407, 412, 472

Echograms. See Acoustical turbidity

Eh, 239, 241–247, 252, 290–293. See Oxidation potential

Electrical resistivity, soils and sediments, 10

Emerita talpoida, 161, 168, 170

Endogenic and exogenic methane, 432, 433

Erosion, 83, 86, 88, 89. See also Continental shelf sediments; Winnowing

Faults: in mid-ocean ridges. See Mid-ocean ridges; Tectonic;

Fe_{tot}: in ferromanganese nodules, 275, 278–281; in sediments, 239, 241, 244, 249, 250, 252, 257, 258, 261, 277

Ferromanganese concretions: and reactive Fe/Mn_{tot}, 275, 278–281; diagenetic formation of 275–277, 282; in the Black Sea, 275–283; in the Indian Ocean, 275, 277, 280–283; in the Pacific Ocean, 275, 277, 278, 280–283; trace elements in, 276, 277, 279–283

Ferromanganese deposits, 359, 361–364, 366–368

Ferromanganese nodules, 115, 127, 130, 131, 244, 245, 251, 252, 287–300

Fjords, 317–333

Florida Straits: current record, 81, 83, 84; erosion, 83, 86, 88, 89; sediment grain size, 78, 81, 82, 83; stereophotographs, 81, 85–91

Flux, molecular diffusive, 121–123, 134–137

Fluxes from sediments: calcium carbonate, 116, 131, 134–136; diagenetic, 309; manganese, 257, 258, 262, 269, 272; silica, 116, 131, 136. See also Copper

Fluxes, interstitial: carbon dioxide, 208, 209, 212, 213; carbon, organic, 208, 209; helium, 179, 180; iron, 257, 300; manganese, 257, 258, 262, 264; methane, 208–210, 212, 213, 229–230; sulfate, 208, 209

Formation factor and diffusion coefficients, 10, 11

Index

Fractures: in basaltic rocks, 381, 389, 396, 403, 405–407, 411–413; convection-related, 472, 474, 486; minerals, filling by, 451, 454, 455; thermal convection in, 441, 444, 448, 451, 461–467
"French Pressure Cell" effect, 158, 159
Freshwater, 191–193, 214, 215, 339–353, 417–434
Friction velocity (U*), 119, 123–127

Galapagos Spreading Center. *See* East Pacific Rise
Gases: carbon dioxide, 204–208, 211–213, 432, 433; juvenile, 417, 419; oxygen, 110–112, 314, 315. *See also* Anoxic sediments; Interstitial; Methane; Noble gases; Sulfur

Haustoriid amphipods, 161, 168–170
Heat flow, 441, 442, 444, 446–450, 451, 455, 465–467, 471, 472, 490
Heat transfer: conductive, 471–473, 479–481, 486–489; convective, 471–472, 479–481, 485, 486. *See also* Thermal transport
^{3}He/^{4}He isotope ratios, 417–428, 430–432, 434
Helium: basaltic concentrations of, 184; crustal, 418, 420, 424, 425, 428, 434; in springs, 417–431, 434; interstitial, 175–177, 180–183
Hot-thermistor techniques, 30–32
Hydrothermal: activity, 441, 446, 449, 450, 455; breccia, 388, 390, 395; circulation, 382, 384, 389, 396, 407, 412, 441, 442, 446–456, 461, 462, 464–467, 469, 471–473, 475, 478, 479, 489; crusts, 381, 382, 385, 388, 389, 403, 409, 413; deposits, 376, 377, 472; exchange, 381, 382, 396, 401, 403, 405, 406, 409, 411, 412; fields, 381, 383–385, 401, 407, 409, 432, 478; fluids (solutions), 369, 374, 375, 377, 381, 384, 388, 389, 395, 399, 409, 411, 412; hot columns, plumes, 448, 450, 452–456, 463–467; manganese, 381, 388–390, 392, 395, 399, 403, 406, 407, 409, 413; minerals, 381, 382, 384, 389, 395, 399, 407, 409, 413; models, 407, 409, 411–413; vents and conduits, 381, 382, 384, 389, 407, 412, 413, 472; water temperature, 444, 446, 448–452, 454–456, 464–466
Hydrothermal systems, 381, 407–413, 442, 449, 479

Iceland, 417, 418, 421–428, 434
Impedance: kinetic in pore water, 115, 128, 130, 131, 134–137; to mass transport, 115, 116, 130, 131, 134–137

Indian Ocean: ferromanganese nodules in, 275, 277, 280–283; sediments, 370, 372, 373
Internal transfer velocity, 134–137
Interstitial: electrode techniques, 10–13; ion-chlorinity ratios, 343, 346; production rates, noble gases, 175, 180–182; temperature effects, 6, 8
Iron: bottom-water concentrations, 294, 367, 368; diffusion or migration in sediments, 257; Fe$_{tot}$, 239, 241, 244, 249, 250, 252, 257, 258, 261, 275, 277–281; hydroxide deposits, 377; hydroxide precipitation, 276, 279; in sediments, 369–377; interstitial, in sediments, 257, 258, 260, 261, 289–294, 297, 300; oxide in basalts, 405, 406, 409, 410; "pyrite", in sediments, 239, 242, 243, 245, 249–251; "reactive", in sediments, 242, 245, 247, 249–251, 277, 278, 280; water-column residence time, 257, 269; weak-acid soluble in sediments, 257, 258, 260, 261, 277, 278, 280
Isotopes, *See* listings for individual elements

Japan Current, 252
Juvenile gases, 417, 419

Kamchatka, 417, 418, 420–425, 427–434
Kinetics: kinetic control, 130; pore water, impedance in, 115, 128, 130, 131, 134–137; surface reaction, 115, 128, 133–137
Krypton: basaltic concentrations, 184; interstitial concentrations, 176–178, 182, 183
Kurile Islands, 417, 418, 420–429, 434

Lavas (pillow basalts), 385, 389, 392–395, 397, 401, 406, 453
Lead-210, sediment-dating, 305, 312, 313
Lysocline, 108–110

Macrofauna: burrowing beach animals, 161, 168–170; metabolic rates of, 161, 165–170
Manganese: bottom-water concentrations, 294, 367, 368; in sediments, 369–372, 375–377; interstitial concentrations, 287, 289–297, 329–331; interstitial diffusivities, fluxes, 257, 264, 269, 271, 300; water-column residence time, 257, 269
Manganese dioxide (hydrated), 282
Manganese flux from sediments, 257, 258, 262, 269, 272
Manganese, geochemical balance, 257, 258
Manganese in sediments: diagenesis of, 257; diffusion or migration, 257, 258, 262, 264; in reducing sediments, 258, 261; interstitial, 257, 258, 260–264, 268, 270; Mn$_{tot}$, 239, 241, 245, 251, 252, 257, 262,

264, 269–271, 277, 278; Mn_{IV}, 242, 243, 252, 258, 264, 277, 278; oxidation and precipitation, 264, 265, 269, 271, 276; refractory, 260, 264, 265, 270; three-layer model of, 262–268; weak-acid soluble, 257, 258, 260–264, 270, 277, 278, 280

Manganese nodules. See ferromanganese nodules

Manganic oxides, 329–333, 357–377, 401–403

Mantle, elements of. See listings for individual elements of

Mass-transfer coefficient (β), 115, 124–126, 129, 130, 134–137

Meiofauna, effects of water movements on, 51–53

Metabolic measurements. See Macrofauna

Metals in seawater. See Trace elements

Meteoric waters, 419, 420

Methane: biochemical, 430, 432–434; endogenic, 432, 433; exogenic, 432; in cold springs, 430–433; in thermal springs, 417, 430, 432–434; mantel, 418–420, 430, 432–434; near-bottom, 198

Methane, interstatitial: bubble formation, 195–200, 219, 226, 227, 229, 231; concentration, 219, 225–228; consumption, 187–189, 193–195, 203, 204, 207–211, 214, 215, 230, 231; distribution, 187–189, 195, 198, 204–207, 213; fluxes, 208–210, 212, 213; freshwater, 209, 214, 215; models, 193–195, 209, 214, 215; oversaturation, 196, 198, 219, 226–229; oxidation, 204, 207–215; production, 187–195, 204, 207, 210–212, 215, 230, 231; transport, 187, 189, 195–197, 199, 200, 227, 231

Microbial respiration, 154, 155, 158. See also Sediments

Microbiological sampler (deep-ocean), 150–153, 158, 159

Micro-organisms, deep-sea, 154–159

Mid-ocean ridges, 381, 382, 397, 401, 407, 409, 411–413, 483; FAMOUS area, 388, 407; faults in, 381, 382, 384–389, 395, 403, 409, 411–413; heat flow, 441, 442, 446, 451, 455, 471, 472, 490; hydrothermal circulation, "passive", 441, 446–448, 450–453, 456, 457; rift valley, 383–386, 389, 405–407, 411–413; rock-seawater chemical exchange, 441, 442, 444, 446, 448, 449, 453, 455, 456, 467, 472; seawater residence time, 441, 444, 448–452, 454–456, 465, 466; seawater temperature in, 441, 448, 450–452, 454–456, 464–466; volcanic sediment on, 369, 370, 373, 376, 377

Mississippi delta, 58, 59, 61, 65, 73

Models: hydrothermal circulation, 441–469, 471–490; interstitial, 18, 19; manganese in sediments, 18, 19, 262–268; methane, interstitial, 193–195, 209, 214, 215; noble gases in sediments, 176–184; seafloor transport and diagenesis, 115–143; sediment mixed layer, 95–112

Molybdenum in sediments, 282, 283, 305, 312, 313

Momentum transfer, 117, 119–122

Near-bottom dynamics: sublayer, diffusive, 115, 116, 120, 122, 123, 126–132, 134–137; sublayer, viscous, 119–122, 127; turbulence, 115, 117–119, 122, 127

Neon: basaltic concentrations, 184; interstitial concentrations, 176–178, 182, 183

Nickel: accretion on nodules, 299, 300; bottom-water concentrations, 294; interstitial concentrations, 287, 289–296; interstitial fluxes, 299, 300

Nickel/manganese ratios in sediments, 282, 283

Noble gases. See listings for individual elements.

Nusselt Number, 445, 446, 448, 449, 461–467

Oozes: calcereous, 257, 258, 260, 287, 297, 299; siliceous, 257, 258, 260, 287, 288, 296, 297, 359, 366, 367

Oxidation potential: and sediment reactions, 330, 331; discontinuity, 52, 53; in Pacific sediments, 262, 272, 287, 288; in sediments, 13, 239, 241–247, 252; interstitial, 287, 288, 290–293, 297. See also Redox

Oxygen isotopes, 110–112

Pacific Ocean: See sediments, Pacific

pE, 308, 309, 330, 331. See also Oxidation potential

Percolation in near-shore sediments, 42, 43

Permeability (oceanic crust), 471, 474, 475, 479–482, 486, 487; effects on heat flow, 471, 474, 479–482, 487, 489, 490; effects on hydrothermal circulation, 472, 479–482, 486, 489, 490; models for, 472–474, 478–482, 489

Permeability (rock), 441, 444, 446–453, 455, 464–466

pH: in sediments, 11, 12, 246, 290–293, 309

Pillow basalts, or lavas, 385, 389, 392–395, 397, 401, 406, 453

Polarographic metal analysis, 319, 320

Polymesoda caroliniana, 168, 169

Porosity, 131–133, 144, 145, 179, 298, 306

Porous layer, circulation in, 471–473, 475,

Index

487–490. *See also* Hydrothermal circulation
Pressure: effects on micro-organisms, 158–159; forces; sensor, 57, 66. *See* Continental-shelf sediments
Pyrite iron in sediments, 239, 242, 243, 245, 249–251

Rayleigh: Number, 444–446, 448, 449, 451–453, 455, 461–467; parameters, 476, 484; stress, 118–120
Redox: characteristics, Pacific sediments, 239, 242–247; gradients, 330–332; interface (barrier), 239, 246, 250–253, 276; paleoecological indicator, 305, 311–314; redoxcline, 305, 307–310, 312. *See* Oxidation potential
Respiration: *See* Macrofaunal; Microbial respiration
Reynolds: Number, 467; stress, 118–120; turbulent mass flux, 121
Ridge axis, 471–473, 489, 490. *See also* Heat transfer
Ridge crest; seawater penetration of, 479. *See also* East Pacific Rise; Hydrothermal; Mid-Atlantic Ridge; Mid-ocean ridges; Ridge axis
Rift valleys, 383–386, 389, 405–407, 411–413
Rock; creep (against thermal stress), 453, 454, 463

Schmidt Number, 122–126
Sea-bed momentum transfer, 117, 119–122
Sea-floor: roughness, 127; spreading, 442, 446
Sediments: as insulation, 441, 452, 453, 456; "baked" crust, 359, 361–364; clay layers, 339, 341, 343, 346, 347, 349, 359, 361, 364, 365; compaction, 305–307; dating, 210 Pb, 305, 312, 313; deposition, depth of, 95, 96, 104–106, 109, 112; freshwater, permeation by seawater, 339–353; metal hydroxides, 382, 399, 401, 405, 413; oscillations, 57, 58, 67, 69–73; pelite fraction of, 240–249; ripples, Florida Straits, 77, 81, 83–91; stability of, 57, 58, 73; tsunamis, effect on, 106; volcanic debris, 240, 247, 249, 281, 361, 364, 365; x-ray radiographs of, 197, 200, 340, 341, 344.
Sediment-mixed-layer (deep-sea): homogenous layer, 95, 96, 102, 108; lumpy mixing, 95, 102; mixed-layer transition, 95–97, 102, 103, 108, 112
Sediments, Pacific: brown clay, 258; discolorations, 297, 298; hemipelagic, 240; organic carbon in, 288, 297, 299; oxidized (III), 243, 246–248, 250, 252, 287, 288; pelagic, 240, 242, 246, 250, 252; red clay, 257, 258, 260, 262, 272, 279–281, 287, 288, 296, 297; redox characteristics, 239, 242–247, 287, 288; reduced (I), 243, 246–252; terrigenous (coastal), 240, 243, 245, 247, 249–252; transitional (II), 243, 246–252
Shear strength, 58, 73, 103, 104, 108, 112
Silica, 6, 116, 131, 136
Springs, 417–434, 456, 472, 489
Stereophotography, 78, 80, 81, 85–91
Subtidal pumping, 42
Sulfate, interstitial, 188–191, 193, 195, 197, 204–208, 215, 224–228; fluxes, 208, 209
Sulfur, reduced: copper, precipitation as sulfide, 317, 327, 328, 330–333; forms in sediments, 242–249; hydrogen sulfide, 244, 246, 247, 249, 251, 306–310, 312, 314, 315; metal sulfides, 381, 405, 406, 409, 411–413; sulfide deposits, 377; sulfide derivatives, 244, 247–250, 310; sulfides in sediments, 310–312; sulfides, interstitial, 317, 423, 326, 327, 332, 333
Surface reaction kinetics, 115, 128, 133–137

Talus, basalt, 386, 389–391, 395
Terrigenous matter (TM), 369, 371–373, 375, 377
Terrigenous sediments, Pacific coastal, 240, 242, 243, 245, 247, 249–252
Thermal conductivity, 473
Thermal transport by circulation, 442, 444, 446–450, 455, 465–467
Thermistor techniques, hot, 30–32
Titanium in sediments, 369–374, 377
Trace elements: in basalts, 396, 398, 399; in sediments, 277, 280–282, 287–300, 305, 308–313, 359, 366, 367; interstitial concentrations, 287, 290–299; interstitial fluxes, 287, 298–300; mobilization mechanisms, 287, 289, 297–299; *See also* Copper; Iron; Manganese
Transgression, marine, 339–353
Turbidite, volcanic, 257, 258
Turbulence, near-bottom, 115, 117–119, 122, 127

Veins in basalts, 381, 402–406, 409–411. *See also* Fractures
Viscosity, kinematic, 115, 118, 119, 120, 122, 124, 125, 127
Viscous sublayer, 119–122, 127
Volcanic: ash in marine sediments, 240, 247, 249, 281; debris, 361, 364, 365; sediment, mid-ocean ridges, 369, 370, 373, 376, 377;
Voltammetry, anodic stripping, 318–320, 322

Water extraction from sediment, 3–6, 7, 8; by pressure filtration (squeezing), 8–10, 258, 288, 319, 340, 341; *in situ* sampler, 13–15 175, 184, 224

Water movement, near-shore sediments: annual volume filtered, 48–50; effects on meiofauna, 51–53; hot-thermistor techniques, 30–32; oscillating pattern, 35; percolation, 42, 43; redox potential discontinuity, 31, 52, 53; subtidal pumping, 42; surf pulsing, 35, 40, 42, 44, 51; swash pulsing, 32; water volume through coastal sediments, 48–50

Waves, 57, 58–61, 66, 71–73
Winnowing (deep-sea), 95, 105, 106, 112

X-ray mineralogy (basalts and sediments), 396, 398, 403
X-ray radiography. *See* Sediments

Zinc: accretion on nodules, 299, 300; in sediments, 305, 310–313; interstitial concentrations, 287, 289–296; interstitial fluxes, 299, 300
Zeolites in basalts, 398, 405, 406, 408, 410

List of Contributors

Ross O. Barnes
  Walla Walla College
  Marine Station
  174 Rosario Beach
  Anacortes, Washington 98221

H. Beiersdorf
  Federal Institute for Geosciences
    and Natural Resources
  P.O. Box 510153
  D-3 Hannover 51
  Federal Republic of Germany

Wolfgang H. Berger
  Scripps Institute of Oceanography
  University of California, San Diego
  LaJolla, California 92093

Barton Birdsall
  Mitchell Energy and Development
    Co.
  1 Shell Plaza
  Houston, Texas 77002

Kurt Boström
  Department of Economic Geology
  University of Luleå
  951 87 Luleå
  Sweden

Bernard P. Boudreau
  Department of Oceanography
  Texas A & M University
  College Station, Texas 77843

David C. Burrell
  Institute of Marine Science
  University of Alaska
  Fairbanks, Alaska 99701

J.M. Coleman
  Coastal Studies Institute
  Louisiana State University
  Baton Rouge, Louisiana 70803

Rita R. Colwell
  Department of Microbiology
  University of Maryland
  College Park, Maryland 120742

Kent A. Fanning (Editor)
  Department of Marine Science
  University of South Florida
  830 First Street South
  St. Petersburg, Florida 33701

L.E. Garrison
  United States Geological Survey
  Office of Marine Geology
  P.O. Box 6732
  Corpus Christi, Texas 78411

Norman L. Guinasso, Jr.
  Department of Oceanography
  Texas A & M University
  College Station, Texas 77843

H. Grundlach
  Federal Institute for Geosciences
    and Natural Resources
  P.O. Box 510153
  D-3 Hannover 51
  Federal Republic of Germany

Leonid Gutsalo
  Institute of Geology and
    Geochemistry of Fossil Fuels
  Academy of Sciences of Ukrainian
    SSR
  ul Nauchnaia 3a
  Lvov, 290043
  USSR

Rolf Hallberg
Geologiska Institutionen
Kungstensgatan 45 Box 6801
113 86 Stockholm
Sweden

Martin Hartmann
Geologisch-Paläontologisches
 Institut der Universität Kiel
Olshausenstrasse 40/60
D-2300 Kiel
Federal Republic of Germany

David T. Heggie
Graduate School of Oceanography
University of Rhode Island
Kingston, Rhode Island 02881

D. Heye
Federal Institute for Geosciences
 and Natural Resources
P.O. Box 510153
D-3 Hannover 51
Federal Republic of Germany

Peter Kruikov
Institute of Inorganic Chemistry
USSR Academy of Science
kv 1 ul Tereshkovia II
630072 Novosibirsh
USSR

Masashi Kusakabe
Department of Chemistry
Faculty of Fisheries
Hokkaido University
Hakodate 041
Japan

C.R.B. Lister
Department of Oceanography
WB-10
University of Washington
Seattle, Washington 98195

Christer Löfgren
Geological Survey of Sweden
75128 Uppsala
Sweden

Robert P. Lowell
School of Geophysical Sciences
Georgia Institute of Technology
Atlanta, Georgia 30332

Frank T. Manheim (Editor)
United States Geological Survey
Woods Hole Oceanographic
 Institution
Woods Hole, Massachusetts 02543

V. Marchig
Federal Institute for Geosciences
 and Natural Resources
P.O. Box 510153
D-3 Hannover 51
Federal Republic of Germany

Christopher S. Martens
Marine Sciences Program
University of North Carolina
Chapel Hill, North Carolina 27514

H. Meyer
Federal Institute for Geosciences
 and Natural Resources
P.O. Box 510153
D-3 Hannover 51
Federal Republic of Germany

P.J. Müller
Geologisch-Paläontologisches
 Institut der Universität Kiel
Olshausenstrasse 40/60
D-2300 Kiel
Federal Republic of Germany

Laszlo Nemeth
Nova University Oceanographic
 Laboratory
8000 North Ocean Drive
Dania, Florida 33004

List of Contributors

Jörg A. Ott
  Zoologisches Institut der
    Universität Wien
  Währinger Str. 17/VI
  A-1090 Wien
  Austria

Mario M. Pamatmat
  Tiburon Center for Environmental
    Studies
  San Francisco State University
  P.O. Box 855
  Tiburon, California 94920

Patricia L. Patterson
  School of Geophysical Sciences
  Georgia Institute of Technology
  Atlanta, Georgia 30332

Phillipe R. Peron
  Houston Oil and Mineral
    Corporation
  1700 Broadway
  Denver, Colorado 80290

William S. Reeburgh
  Institute of Marine Science
  University of Alaska
  Fairbanks, Alaska 99701

R.J. Riedl
  Zoologisches Institut der
    Universität Wien
  Währinger Str. 17/VI
  A-1090 Wien
  Austria

A.G. Rosanov
  P.P. Shirshov Institute of
    Oceanology
  USSR Academy of Science
  1 Letnyaya Ljublino
  109387 Moscow
  USSR

C. Schnier
  Society of the Application of
    Nuclear Ship Propulsion (GKSS)
  P.O. Box 160
  D-2054 Geesthacht
  Federal Republic of Germany

Robert B. Scott
  Geochemical and Geological
    Consultant
  223 West Main Street
  Marble, Colorado 81623

Erwin Suess
  School of Oceanography
  Oregon State University
  Corvallis, Oregon 97331

Joseph N. Suhayda
  Department of Civil Engineering
  Louisiana State University
  Baton Rouge, Louisiana 70803

Paul S. Tabor
  Department of Microbiology
  University of Maryland
  College Park, Maryland 20742

Darcy G. Temple
  Damson Oil
  260 North Belt
  Houston, Texas 77060

Shizuo Tsunogai
  Department of Chemistry
  Faculty of Fisheries
  Hokkaido University
  Hakodate 041
  Japan

I.I. Volkov
  P.P. Shirshov Institute of
    Oceanology
  USSR Academy of Science
  1 Letnyaya Ljublino
  Moscow 109387
  USSR

Thomas Whelan
   Coastal Studies Institute
   Louisiana State University
   Baton Rouge, Louisiana 70803

Michael J. Whiticar
   Geological Research and Services
   Petro-Canada

Box 2844
Calgary, Alberta, T2P 2M7
Canada

Mark Wimbush
   Graduate School of Oceanography
   University of Rhode Island
   Kingston, Rhode Island 02881

About the Editors

Kent A. Fanning is an associate professor of marine science at the University of South Florida. He has conducted research on sea-floor processes under sponsorship of the National Science Foundation, the Office of Naval Research, and the U.S. Department of Energy. His published research includes work on the chemistry of river plumes, deltaic sediments, geothermal springs, interstitial transport processes, and radionuclides in the sea.

Frank T. Manheim is a geochemist at the U.S. Geological Survey in Woods Hole, Massachusetts, where he has worked in conjunction with scientists at the Woods Hole Oceanographic Institute. Much of his research has concerned the chemistry of sediments. As one of the principal geochemists connected with the Deep-Sea Drilling Project in major ocean basins and with recent drilling projects on continental margins, he has contributed to the study of the fate of major marine cations and anions in marine sediments. Recently, his interests have also included the study of man's impact on the ocean.